Muriel Sibell Wolle

TIMBERLINE
TAILINGS

Tales of Colorado's
Ghost Towns and Mining Camps

Muriel Sibell Wolle

TIMBERLINE TAILINGS

Tales of Colorado's Ghost Towns and Mining Camps

SWALLOW PRESS
OHIO UNIVERSITY PRESS

ATHENS

First Swallow Press / Ohio University Press edition printed 1993
99 98 97 96 95 94 93   10 9 8 7 6 5 4 3 2 1

Swallow Press / Ohio University Press Books are
printed on acid-free paper ∞

Library of Congress Cataloging-in-Publication Data
Wolle, Muriel Sibell, 1898–1977
Timberline Tailings : tales of Colorado's ghost towns and
mining camps / Muriel Sibell Wolle.
1st Swallow Press ed.
p.   cm.
Sequel to: Stampede to Timberline.
Originally published: Chicago: Sage Books, 1977.
Includes index.
ISBN 0-8040-0739-X (cloth)–ISBN 0-8040-0963-5 (pbk.)
1. Ghost towns–Colorado. 2. Mines and mineral resources–
Colorado–History. 3. Colorado–History, Local. 4. Colorado–
Gold discoveries. I. Title.
F776.W67 1993
978.8–dc20   92–28343
                                                          CIP

To Francis, my husband,
in response to whose
enthusiastic urging and constant support
this book was written.

Contents

Illustrations

Maps

In addition, at the beginning of each chapter (except Chapter I), there is a map pertaining to that chapter.

Muriel Sibell Wolle

TIMBERLINE TAILINGS

Tales of Colorado's
Ghost Towns and Mining Camps

Rich Tailings

Timberline Tailings is made up of material—letters and interviews—of the hardy men and women who met the challenges of the gold and silver mining camps of Colorado with stamina, ingenuity, and a keen love of life.

In Chapter XX of *Stampede to Timberline* I asked for help in locating certain places that I'd been unable to find or reach and for information about them. To my delight a great number of people wrote me giving clear directions, informative data, and above all reminiscences of boom days in many of Colorado's mining camps. Some of these include stories told by their parents or other relatives and others are first-hand incidents experienced by the writers during the latter part of the nineteenth century and the early part of the twentieth. Most of these letters were written between 1949 and 1955, although some are of more recent date. They provide new material, correct inaccuracies in the text, and contain valuable personal recollections.

I felt that what they offered should be preserved. With this in mind, I contacted as many of these people as possible by letter, requesting permission to edit and print what they had written. None of those who replied refused. Naturally, a number of letters were returned undelivered, and some that came back were sent by relatives who indicated the persons addressed had died. Even so, over eighty percent of my inquiries brought affirmative answers. My editing of the letters was confined to some deletions and a few bracketed explanations; otherwise I have reproduced them verbatim, retaining the writer's spellings, etc.

Because many of the people reading this book are already familiar with *Stampede to Timberline,* the material here is arranged in somewhat the same geographical format as the chapters and locations of the earlier volume, although I have resequenced slightly for better progression.

As to the title *Timberline Tailings,* the term "tailings" indicates not merely piles and dumps of waste rock but potential treasure. Most mine dumps have been worked and reworked two or three times, and, since reduction processes have constantly improved, ore that once was considered worthless has been salvaged from many a dump and has yielded good returns to the company or individual lessee who sifted the gritty piles of discarded ore and dirt.

In much the same way, new materials I have received in letter form or through personal interview pertaining to the life, wealth, boom, and demise of many of Colorado's camps, have proved treasures that warrant preservation since they add new facets to those already gathered and used in the book *Stampede to Timberline.*

Josephine H. (Mrs. Frank D.) Peirce of Leicester, Massachusetts, was one of the first to write me (Dec. 12, 1949); she had interviewed a man who was born in Leadville and had also lived in Kokomo and Robinson:

His father managed several mines for Thomas F. Walsh including the Camp Bird. His name was Lawrence J. Cronin and his home was at Rouses Point, New York. He still owned property in Kokomo Gulch and in 1934 went back to look things over. He stated that on the way from Denver to Leadville the mountains were dotted with old tunnels, cabins and prospect holes—monuments to the broken hopes of men.

He said: "I knew Leadville when its inhabitants numbered 5,000, when ore wagons drawn by four and six horses rumbled through the streets from daylight to dark, carrying the wealth of many an Eastern Potentate from the mountains surrounding the city. At Kokomo I found 40 people in what used to be a prosperous community containing mines, smelters, stores, and hotels with a fine degree of social and fraternal life and about 1500 souls.

"Robinson is a ghost town now, the buildings listed to the four winds of heaven, speaking mutely of the men who had passed and the days of long ago."

The Camp Bird property above Ouray. Road to Sneffels above buildings in center and on right.
Courtesy Josephine H. Peirce

Everyone wrote freely and with enthusiasm and often included diverse memories or bits of information as they came to mind—all interesting and valuable in their way. For example, Clifford W. Kingsley of Los Angeles wrote (Dec. 2, 1950):

A lady who now lives in San Francisco was born in Ouray over 66 years ago. Christened Wilhelmina Meyer. When Chief Ouray invited the settlers to a meeting of Friendship, her folks went, and Chief Ouray admired the long straight baby so much he named her "Volantah" which name means "tree." She adopted the name.

Another such letter came from Lucille E. Foltz of Fort Collins, Colorado (July 30, 1962):

I have my faith now restored in Colo. Historians because you told (on TV) the Morrissy and other Irishman's stories, just as my mother told them to me and she was the daughter of a Cornish "mine Captain" of Leadville in

the 1880's. He, Harry Slockett, was of the group of Cornish and Welsh deep rock miners who migrated from England to America. . . .

My father being a son of a freighter, drove teams of mules, one pair belonging originally to a cousin of Jesse James, in or around "St. Joe." My dad's name was Ben Foltz, the son of Joe Foltz, who freighted goods on the Western Slope. My father climbed the hill to the "Sunnyside" mine with Tom Walsh, when they both worked there when father was a boy. Dad and Mom are both gone now.

I belong to the Pioneer Ass'n. here and have written a few articles about our Pioneers here, having interviewed them for Mr. Hafen of the Denver Museum [LeRoy R. Hafen of the Colorado State Museum in Denver].

You might be interested to know that the old retaining wall behind the Shirley-Savoy [Denver] on 17th and Lincoln was the wall to the Capitol Hill home "front yard" of my Cornish grandparents, the Slockets, at the time Mrs. Augusta Tabor lived across from them on the other corner, I believe. "Mom" told me these things but my memory naturally wasn't as good as hers as she knew these people and I'd never seen them.

Although my correspondents were strangers whom I had never seen, yet the style of each had a flavor that told me something about and created a mental image of the writer. Some letters were fluent, others terse. Some described such tragedies as fires and snowslides that had damaged the communities where they lived and gave vivid details of the incidents witnessed. The men wrote mostly of mines and business, of roads and trails, of ore mills and freighting; the women recalled schooldays, the isolation of remote camps, living conditions that to us seem appalling, and the cold terror in every woman's heart at the mine whistle's shrill blast announcing disaster—fire, cave-in, or underground flooding.

A number of letters contained such a delightful sense of humor that I should have liked to have known the authors. But the characteristic dominating all this correspondence is the interest shown by the writers and their pleasure at having an opportunity to share highlights in their own lives with someone who is as eager to hear about the past as they are eager to tell it.

I am indebted to a number of persons who generously sent old photographs with their letters and gave me permission to use them; these are credited in the captions. All other photographs, uncredited, were taken by me. The drawings and paintings reproduced here are, of course, as in *Stampede to Timberline,* mine.

Gertrude Ridgeway of Riverside, California (Feb. 10, 1952) sent two photographs taken in 1903 of the property where her first cousin, Robert Higday, worked; the "he" in her letter refers to her cousin:

> We received your book Stampede to Timberline and are enjoying it. Your descriptions of the country are very true and satisfying to us. You see, we know Colorado, the western and southern part. I lived while going to school for ten years at Grand Junction and he was in western Colorado for about thirty years.
>
> He was in the mines around Ouray for about five years, up to 1906. He was at the Revenue-Virginius in 1902-1903, most of the rest of the time at the Atlas. So he is glad to be able to send you a copy of a photo of the Revenue-Virginius that was taken in 1903 when he and his step-father worked there.

Revenue-Virginius mine and mill, 1903. *Courtesy Gertrude Ridgeway.*

The Governor mine had closed down when he was there, its boom over.

The electricity there was the first used in any mines in the West, the reason he knows this is because he is a steam engineer and electrician. It has been his work.

Miss Ridgeway's last statement may be contested by Telluride, for Lucien L. Nunn of that city built a generator in a log cabin at Ames and another at the Gold King mine near Telluride and in 1891 perfected the first electrical power by means of alternating current that was used in the West for mining purposes.

Ouray and the country around it offer so much in scenery as well as history that it is not surprising that others besides Miss Ridgeway wrote about the area. Surrounded as it is on three sides by high mountains, the roads and trails up several canyons lead to mines from which millions of dollars have been extracted. The Camp Bird, five miles up Canyon Creek from Ouray, is probably the best known, but a short distance farther up the canyon at Sneffels are the Atlas and Revenue-Virginius properties and what is left of their great mills. An old shelf road up the edge of the mountain to above timberline leads to the Mountain Top, Humboldt, and Virginius mines. Even in a jeep it provides a thrilling ride.

Road down to Sneffels from the Virginius mine, 1964.

In 1955 R. H. Pearson of Denver wrote his impressions of it:

Last fall, with friends, we made a jeep trip to the upper Camp Bird and around to the Mountain Top Mine, and the old Atlas M. & M. I have been on shelf roads before but that one excelled anything I had seen.

In 1964 my husband and I took a jeep trip from Ouray with other adventurers up beyond the Ruby Trust and the Governor mines to the Virginius mine at the top of the mountain. Shortly after we returned home, we heard from one of the couples on the trip, Dr. and Mrs. Audre L. Stong of Pasadena, California (Sept. 25, 1964):

For posterity it's too bad you did not know about the hole in the mountain just across the ravine from the Revenue Mine, where my father put all his money from his mortgaged jewelry store to no avail. My grandfather Lewis went over the Camp Bird property looking for minerals, not including gold, which he did not find, but Walsh and his group found the gold a short time later.

Some day you should include the stories of the American Nettie and Batchelor high on the mountains to the northeast side of Ouray. I understand these mines are not yet accessible to a jeep. But the story of old man Armstrong, whose son or grandson still runs the Armstrong garage in Ouray is a fascinating one. My father took the trip to Yellowstone Park in a pullman train, paid for by Charlie Armstrong. For years after this Charlie was still looking for a grubstake in Ouray to prospect for more gold, which as late as 1940 he never found.

While pursuing my ghost town hobby, I talked with all sorts of people and from them learned valuable facts or heard amusing stories.

At the Colville's home in Del Norte I visited with Grandma Colville. She spoke of the early days and of coming to Colorado in a wagon train and of how thankful everyone was to arrive safely, for the train ahead of them and the one behind them were all massacred by Indians.

I met Charley Jordan in the Fairplay Hotel. He had lived in Ouray when a boy and had worked at the Camp Bird mine. At that time whiskey was delivered in gallon kegs, which could be rolled under a bed or be hidden anywhere. He and the other boys would ride their burros up to the Revenue and Atlas mills and hunt for empty kegs, which they collected and stuffed in big sacks. These would be thrown across the backs of the burros, and since the boys usually found a good many, they'd have to walk back to town leading their animals. In Ouray they would sell the kegs for twenty-five cents apiece. The purchaser would then refill them and sell them again.

Sam Tooker, the thin, rangy cowboy who guided me to Liberty in the San Luis Valley, had a dry sense of humor; and as he drove me across the sand dunes toward Short Creek, he spoke of other people he had taken on trips into the mountains. He mentioned a group of easterners he had guided along the crest of the Sangre de Cristo Range on a pack trip. "It was too much for one man," he commented laconically. "I had fifteen head of horses and ten head of tenderfeet." Another time, on a similar trip, one woman got scared and refused to ride down a steep grade. Even her son couldn't persuade her to try it. So Sam and her son suggested she dismount and walk down the steep snowbank. "Of course," said Sam, "as soon as she started to walk, she lost her balance and slid down to the bottom. We knew she would, but we had to get her down."

While in Platoro I was eager to learn about its early days and was told to look up Mr. Hawkins. I knocked at the door of his house, but there was no response; so I walked to the back and found a man working in his garden. "Are you Mr. Hawkins?" I asked.

"The same," he replied.

"Do you know if a Chicago engineer named Axell is staying in Platoro?" I inquired.

"He left last week," he replied. "Are you the lady he's been corresponding with? He was going to try to see you on his way east. Come in and sit down."

After some conversation, he continued, "You should see Otto Blake. He knows Platoro. His cabin is here. I think he'll be up today, so you'd better wait and see him." Since that didn't fit my time schedule, he went to the phone and called the man in Monte Vista.

"Are you in Montie?" I heard him ask. "Are you coming up today? You aren't? You're going to Moosie [Alamosa] this afternoon? There's a lady here wants to talk to you. She's writing a book." He handed me the receiver, and I talked to Mr. Blake and found he wouldn't be home until late evening and was leaving for Pueblo the next morning. So I thanked Mr. Hawkins for making the phone call and tried to reimburse him, but he refused to accept anything. "I have to pay eighteen dollars a month whether I use it or not," he told me. "I won't take your money."

It had also been suggested that I look up Walter Carter while in Platoro, and I found him at home. He was quite deaf but very helpful and told me many things that I needed to know. He had arrived in Platoro in 1911 and from his "old Uncle Jim" had heard many stories, one of which centered around a saloon that stood below Platoro where the Summitville fork joined the main road. The place was both saloon and dancehall. According to Uncle Jim, it saw an awful lot of tough times. His uncle had mentioned a dude, an Englishman, who had just come to Summitville and went down to this place wearing a derby, a necktie, and patent leather shoes. The miners in the bar told him to dance, but he refused, saying he didn't know how. They said they'd make him dance. One shot a cap-and-ball pistol at his feet. He jumped up. Every time they hit the floor with a pistol shot, he jumped. In this way they kept him jumping until he was exhausted. Then they put him up on the bar and bought him drinks. They said he was a regular fellow and one of them. He was delighted that they had accepted him, for he had thought he was a goner.

Many roads that my hobby demands I travel are unimproved and hard on tires, so I check the tires frequently. On one trip, after exploring some rough back country, I asked a garage man to do this. "I haven't had a tire gauge for years," he replied. "When we want to check our tires, we just run over to Gunnison"—a thirty mile drive!

Many Scandinavians worked in mines, and Telluride had an unusually large percentage of Swedes and Finns. Once when I was there I looked up a friend and former student of mine, Eino Pekkarine, who introduced my driver and me to his aunt Ino, his father's sister, who had come from Finland to live with them five years earlier. Her muslin dress was ageless, her stockings sagged, but whenever we saw her she was hurrying about in high-heeled black slippers, talking excitedly in Finnish and smiling happily. We had dinner with them in the apartment above the dry goods store they ran, and afterward she showed us her treasures—the teacups, pictures, and other trinkets she had brought with her from Finland.

Next she took us for a walk so we might see some of the sights of the town. Our first stop was at the Finnish Hall, a long, narrow frame building with seats around the walls, a stage at the far end of the room, and a potbellied stove near the door. The stage still held some of the scenery used in an entertainment, and crepe paper festoons hung from the rafters. Aunt Ino skipped up on the stage and seating herself at the piano sang snatches of Finnish songs. She showed us everything in the Hall, talking volubly, with Eino interpreting for our benefit. She spoke of the "Name Day" parties that were given for each person on his birthday. Eino described Basket Socials and Toe Socials. At the latter the girls exchanged shoes and thrust their feet out under a curtain, making the men bet and guess which partner they were getting for the evening.

After we had expressed our thanks and said goodnight, Aunt Ino clutched my arm and said, "Won't you have a glass of wine in the hotel bar?" Finally we left them and went upstairs to fall wearily into bed.

Before we start sampling the unexpected richness of the tailings revealed in the spontaneous letters to follow, I should like to contribute two bits from Cripple Creek obtained through the Rev. Ray Grieb, who in 1960 was priest of St. Andrew's Episcopal Church there.

He and his wife Alice invited us to dinner in their extraordinary apartment above the bank on Bennett Ave. Because of Cripple Creek's steep hills, I was not surprised to find that the apartment entrance on the alley behind the bank building was really at the third floor level. The front door, therefore, opened directly into a hall, on either side of which were bedrooms and a bath. Farther to the front a short flight of stairs led down to the second floor into a big square dining room with dark woodwork, a beamed ceiling, and leaded glass windows close to the ceiling on either side of a massive fireplace. Still further forward was a forty-foot-long living room, also with a fireplace, and with lighting supplied only by stained glass skylights. In front of this room was a small study and to its left a long, narrow bedroom and a bathroom of similar proportions. Just beyond these was the front room with big windows looking out over Cripple Creek and down on Bennett Ave. All in all most unusual living quarters.

It had belonged to the A. E. Carltons, when they had lived above the bank which he owned. Every room was equipped with an alarm bell, and there was one on each side of their bed. The study, which was over the teller's cage, used to contain a large, glass panel in the floor, so that anyone above could look down and see into the bank. On one or two occasions, when the bank had been tipped off and warned of a probable robbery, police with guns sat upstairs and peered down through the glass. Mrs. Carlton, who died in 1958, had redone the apartment, and it still contained her furniture, dishes, books, and paintings. It was a perfect period piece of the late 1890's.

St. Andrew's Church is not far away. It is old and architecturally attractive but difficult to heat. Our friend, the priest, repeated a story told him by one of his older parishioners. Some years before, a new priest was holding his first Sunday service in the church one cold winter day. The big stove near the center of the building was glowing hot. The service had been going well, and the priest was about to begin his sermon. Suddenly, to his amazement, everyone stood up and walked to the other side of the room. "They're all about to leave," he thought, but instead they simply changed seats. They had developed this pattern so as to warm themselves on both sides, and they did it all at once in order to make as little confusion as possible.

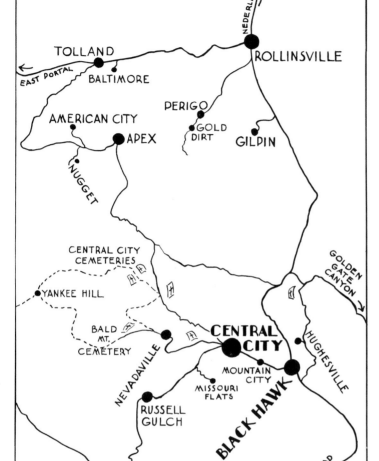

CHAPTER II

The Central City Area and Gilpin County

Black Hawk, Mountain City, Nevadaville, Russell Gulch, and a few smaller camps dotted the rich mineral-laden hills above North Clear Creek at the start of the gold rush of 1859-1860. In later years, after quantities of gold had been washed from the streambeds and torn from the earth, the area was often called the Little Kingdom of Gilpin.

So much has been written about these places and so few of the pioneer families still live in them that I have received few letters describing their colorful days when heavily laden ore wagons rattled down the steep roads to Black Hawk, where the ore mills were located along North Clear Creek, and when the Teller House in Central City was known as one of the best hotels between the Missouri River and the Pacific Coast.

I wish I had known Clarence Reckmeyer before I wrote *Stampede to Timberline,* for his knowledge of Gilpin County and of the Central City area was both authentic and informative. It was 1952, however, before the first of several useful letters from this affable old man reached me and two years more before I met him. That many years of his life had been spent in mining camps and that his interest in mining history was genuine and sustained is further shown by the diversity of places that he knew about. Take for example these excerpts from his first letter (Jan. 13, 1952):

> Clarence B. Richardson, my "Sour Dough" 1898 Klondike friend has just presented me with a copy of "Stampede to Timberline." Mr. Richardson was selling two doughnuts and a cup of coffee for a dollar on Chilcoot Pass when I went into the Klondike in 1898. . . . I have visited and camped at most

of the places that you tell about in your book since about 1908.

I suppose that you know that the town of Kokomo burned entirely to the ground—I recollect that it was in 1882—and that what the people now call Kokomo is Recen on the county records. When Kokomo burned what was left of it moved to Recen, but the Post Office Department refuses to change the name. Henry A. Recen, who now lives in Recen (Kokomo), relates that his father and uncle founded Recen—he has many old newspapers of the early days.

Kokomo

In 1930 I erected a copper marker in Granite at the site of the spot where the old log Lake County Court House stood, in whose court room County Judge Elias F. Dyer was killed, during the Lake County Feud (on July 3, 1875). Some years afterward an adjoining building was burned and the marker was destroyed. In the same year I erected another marker at the site of old Hamilton, which was removed when they placer mined the country.

Granite

P.S. If you ever get up this way I will be glad to tell you all that I have been able to find out about Missouri City—"The only town in the county that would ever amount to anything"—and all of the other old historical spots in the county—better let me know in advance WHEN you will be up here, so that I will be home. I live just above the Black Hawk Post Office—right across the street from Otto Blake's Service Station. Better not come while the roads are so icy—they are dangerous now.

You will find Missouri City on a map in "Colorado's Little Kingdom" by Captain Donald C. Kemp.

Had Mrs. Marjorie Thompson Lee of Denver known Mr. Reckmeyer, he might have been able to answer her inquiry about her grandfather. Perhaps some reader of her letter (March 14, 1956) can still do so:

Black Hawk

I am wondering if in going over records of Black Hawk you ever came across anything of my grandfather, John Alden Smith. I have several clippings from different papers concerning him. He came to Colorado in 1864 from Kennebec County, Maine. He first located in Central City where he was editor of the Register. In 1872 he was appointed to the office of territorial geologist. He was also associated with the history of Idaho Springs. He was one of the owners of the Champion Dirt mine.

One clipping states that he was finely educated and was considered one of the best authorities on geology and mineralogy in the state. He was superintendent of the American Mine in Boulder County. He was associated with Gen. Frank Hall.

All this as I am so anxious to know if there is anyone living who could show me my mother's (daughter of J. A. Smith) home in Black Hawk if it is still standing. I will appreciate it if you have any additional information on my grandfather.

Enclosed was a business card:

J. Alden Smith
Mining Engineer
Late State Geologist.   Boulder, Colorado.

As Clarence Reckmeyer's letters continued to arrive, I decided to visit him on my next trip to Central City. Accordingly, I stopped at his house in Black Hawk one May morning, but there was no response to my knocking. Three months later I tried again, and to my delight the door opened and there stood a smiling, elderly man. When I introduced myself, he said, "Come in. I thought you'd never come to see me." "I did come by early in May, one morning as I was starting back to Boulder," I replied. His face lighted up as he said, "Was that you? If I'd known it, I'd have gotten up and let you in."

Later that fall I wrote him for information about Mountain City, the small settlement that grew up around the Gregory Diggings. Reckmeyer sent me the material I needed (Feb. 2, 1954), together with maps and diagrams he had drawn of the city's streets and businesses:

Mountain City

I ran across some dope on Mountain City that I thought you might wish to have. I took this from County Records several years ago. Mountain City seems to have been a city by *common consent*. As far as I know there never was any plat of it. It would also appear that different people had different names for the streets. I don't know where the sawmill or the express office stood, but, of course, the Masonic Temple gives us a starting point—for some of the lots.

Here is a list of businesses . . . (Can't locate the exact location)

Lane & Dodge's Saloon	Gregory Butcher Shop
Express Bldg.	Mountain City Boarding House
Piper & Co., Storehouse	Smalley & Bryant's Carpenter Shop
Steam Sawmill	B. H. Perren's Blacksmith Shop
Smith & Blood, Meat Market	

John L. Merrick, Restaurant—one door west of Kohler & Patton's Store
Smith & Rariden, Attorneys at Law, Mountain City, Jeff.
Allen, Slaughter & Co., Dealers in Gold Mining Claims and Real Estate, Office in Dr. Burdsell's office.

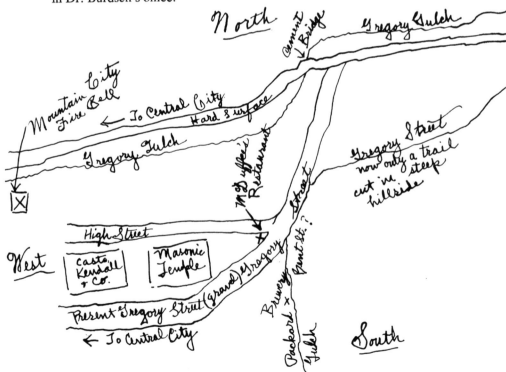

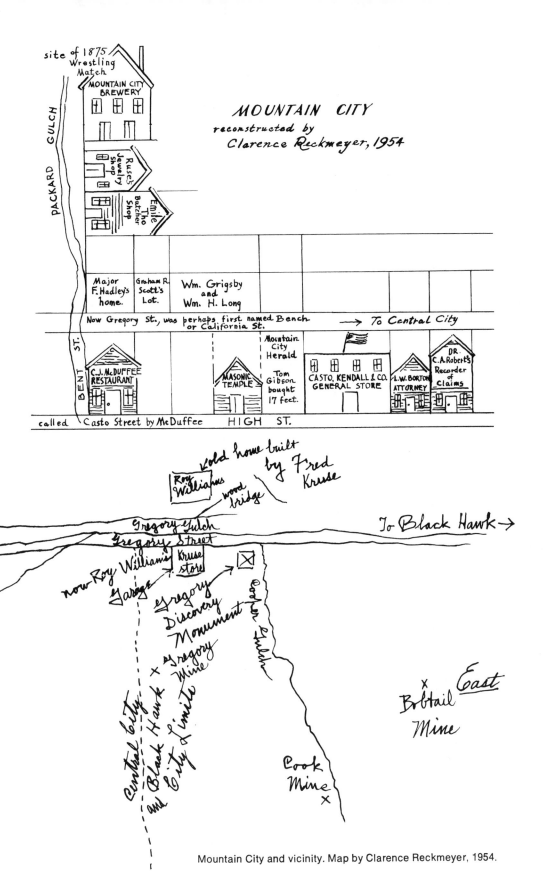

Mountain City and vicinity. Map by Clarence Reckmeyer, 1954.

While the Post Office was located down near Lawrence Street in later years, I IMAGINE that it was on one of those lots between the Masonic Temple and McDuffee's Restaurant in early days of 1860. I never found any record of claim to the two lots between Masonic Temple and McDuffee's. U may have figured some of these out, but I thought this might help U.

My reason for locating the office of the Mountain City Herald next to Casto & Kendall's store is because Casto, Kendall & Co. deeded him the 17 feet on the same day that the first edition of the paper came out, 8-9-59.

Am going to put two stakes showing where C. J. McDuffee's Restaurant stood in 1859—third 40 feet east of Masonic Temple. Howard Wherry, who lives on the old Brewery site, says the old house with the lace trimming *was old* when he came there in 1895. . . . It is right back and uphill from the Masonic monument. Wherry and wife both say they found many bones where Emil Tho's butcher shop stood in 1859. Am wondering if Frank Ricketson & Co. will consider a proposition to rebuild Mountain City as it was in 1859.

P.S. The 1875 wrestling match, over which a man was killed in Black Hawk, was about where the Brewery was.

After studying the maps Mr. Reckmeyer sent me, I wrote him asking if there had been a street north of the present highway that leads up to Central City but not so high on the hillside as the Casey road. He replied:

I don't know as I understand JUST what U mean by the question. If U are asking if there was a road on the NORTH side of the creek (right hand side looking up stream), I can't answer, but *perhaps* in the earliest days the miners meandered through the undergrowth wherever they could get through. PERHAPS they did have a road on the NORTH side of the Creek, where the hard surface highway now is, sooner or later, but I am only theorizing.

The only road that the location notices seem to locate FOR SURE seems to be High Street (which I think C. J. McDuffee, the Restaurant Man, calls Casto Street)—which runs in FRONT of the Masonic monument and the old stone Buell Mill ruin, where Casto, Kendall & Co.'s store stood. I suppose that the Bent Street that C. J. McDuffee refers to means the crooked street that went up Packard Gulch—you know McDuffee's ad about his restaurant says it was Corner Casto and Bent Streets. I suppose that there was also a street BEHIND the Masonic Temple—which is now Gregory Street. I will draw a plat showing HOW Gregory Street ran Behind the Masonic Monument—there is still a sort of trail where Gregory Street left the present street and kept to the hillside, descending to HARD SURFACED Lawrence Street just above Gregory Discovery Monument—right across the street from where Roy Williams now lives, which was the home of Fred Kruse, who built the Gilpin County Tram. If U happen to be up here I can point places out to U. I don't know whether any part of Mountain City was ever on the north side of Gregory Gulch, but perhaps it started to die out when Central City grew up.

Apex and Nugget

Reckmeyer also wrote about Apex and Nugget, two camps north of Central City that are reached from the upper end of Eureka Street where the road branches at the cemeteries. The right fork leads, after about eight miles, into Apex. According to Reckmeyer (Jan. 2, 1954):

Apex, 1951. Miners' Hall.

As Apex and Nugget were started *about* 1895—after many of the histories were written, I don't know much history of them, except what the old settlers tell me. Otto M. Blake who runs the filling station across the street from me says buildings lined both sides of the Apex business street and that there were many cottages there. I happen to have a list of the business men there. You will notice the name of Wilson & *Fritz.* Lumber and Hardware dealers. Mrs. Luella Fritz, who now runs the hardware store in Black Hawk, is the daughter of Mr. Fritz of the Apex hardware. . . . The principal mine there was the Evergreen and there are those who claim that there is rich mineral there—which remains to be proven. The Evergreen Mine had an immense smelter on the lefthand side of the road as you drive UP (West Side), about a half mile below Apex business district—the old foundation is still there. Otto Blake, who with his father ran the Black Hawk Livery Stable in about 1895 to 1900, says two six horse stages ran up there for a couple of years.

Concerning Nugget he wrote (Jan. 13, 1952):

On page 38 of your book you say that you were unable to locate "a place called Nugget, west of American City." In August, 1946, the National Archives wrote me as follows:—

"A Post Office was established at Nugget, Gilpin County, on November 21, 1895, with Gustav Meyer as postmaster. The Post Office was discontinued March 15, 1901, when the mail was sent to Apex."

However, I had never been able to locate Nugget until last fall when our County Judge, William S. Barrick, who lived at Apex around 1900 when his father lost $120,000 on Apex mines, went out there with me and pointed out the location. You will remember the Mackey Mine a little west down the hill from American City junction. As you get to the bottom of that hill a side road turns down Elk Creek. Nugget was less than a half mile west of where Elk Creek road turns off. It is on the south side of the county highway. Only three old foundations mark the site. Gustav Meyer ran a store there and he was also interested in a mine about a half mile north of there, where an old log cabin is still standing. It is said that Meyer afterwards died in the Pueblo Insane Asylum.

Mammoth on Mammoth Gulch is shown as a settlement on the 1866 map in "Colorado's Little Kingdom." It is not the map in the back of the book. It is said that a number of fellows who did not want to enlist in the Civil War established Mammoth.

Reckmeyer continued:

Kingston

Kingston was only a settlement near the London Mine—about a mile west of the point three miles west of Apex, where the highway makes an abrupt right angle turn toward Tolland.

A letter written by Numa F. Vidal of Sharon, Pennsylvania, gives a more recent account of Apex (July 27, 1951):

Apex cont'd.

Have just finished reading Stampede to Timberline. It gave me many hours of pleasure—I could almost smell the smells of the tunnels and deserted shaft houses.

Back around 1910-1912 I spent several summer vacations on Pine Creek about a mile below Apex and out from Black Hawk. My uncle Fred Thompson had a drift mine, a stamping mill at its mouth and a small dam and pipe line all under the magnificent title of the Pioneer Mining Milling Power And Tunnel Co. I think the money came from Boston. My impression is that he never uncovered anything very rich and many years ago he moved back to Denver. I understand he is working on some kind of a mining history. Perhaps you have run across him.

Our summers in Pine Creek were delightful. Even then the operations around Apex and American City were falling in on themselves. But the wild strawberries and huckleberries were out of this world.

My mother was born in Silver Cliff, and since I have never visited the place, or for that matter ever been sure where it was, it was of unusual interest to me to read your complete history of it.

Hughesville

Not far from Black Hawk as the crow flies but rather difficult to reach was Hughesville, where silver mines were discovered in 1870 or slightly before that date. Knowing that I wanted to visit the site, Clarence Reckmeyer sent me a map that he had drawn complete with directions and a little information about the place. Armed with this material, I did locate it and found only prospect holes, part of a cabin, and a few foundations. According to his notes:

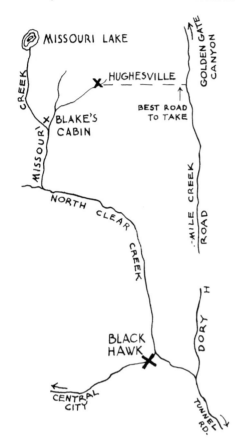

"At one time two hundred men were working there. It had no Postoffice or store, but there was once a small school house there. There is nothing there now but a couple of old mine shafts and one old tumble down log shack.

"It is about 2½ miles from Black Hawk to Hughesville. You can drive 2 miles up North Clear Creek to the mouth of Mis ouri Gulch to Blake's cabin—from there u would have to walk the rest of the way. It is best to go up Four-mile Creek as shown. Turn off Dory Hill road at ranch—it was once the Haubruch Brewery. Paddy Hughes lived in Central City in 1942. He died six years ago (1948)."

Hughesville.
Copy of map by
Clarence Reckmeyer,
1952.

Two other small places mentioned by Reckmeyer were Gold Dirt and Wide-awake. A short distance south of Rollinsville on the Peak to Peak highway a dirt road turns to the west up Gambell Gulch, and if taken far enough up a steep grade it passes through Perigo, a mining camp whose old buildings have all disappeared. One mile farther was Gold Dirt and beyond that was Wideawake. Of the latter Reckmeyer wrote:

Gold Dirt and Wideawake

Mrs. Cotter's father was at Wideawake in '59 or '60. (Mrs. Cotter lives in Rollinsville). The camp got its name after several meetings were held to decide upon one. Still there was no decision. The men were then told that one more meeting would be called "and you'd better be wideawake."
"Why don't we call it Wideawake?" someone shouted. And that was it.

Three no-longer-existent camps, two of them on the Tunnel Road, are described in another letter from Reckmeyer (Jan. 21, 1952):

Chinatown, Cottonwood, and Roscoe

The only camps that ever existed below Black Hawk on the Tunnel Road were a place known as Chinatown and another known as Cottonwood.

"Chinatown" was sort of a temporary place of board shacks where Lin Sou ruled a number of Chinese who were placer mining North Clear Creek along about 1885. It was about two or three miles below Black Hawk. Lin Sou had a son, who was running a restaurant in Denver some years ago, and as far as I can hear, his grandson now lives in Denver and runs a restaurant. Nothing now marks the site of "Chinatown."

Cottonwood was at the mouth of Cottonwood Creek, which enters North Clear Creek from the left as you face downstream, about five miles below Black Hawk. One time when I was going through the Gilpin County records I ran across a plat of Cottonwood, which some ambitious man had held, but about the only excuse for a town there happened along about forty years ago, when a couple of men brought some tellurium ore over from Boulder County and salted a mine at Cottonwood. This caused considerable excitement and an excuse for a town which existed there for about two or three months. The old Colorado Central narrow gauge railroad, which reached Black Hawk December 17, 1872, listed Cottonwood station as being two miles above the Forks of North and South, or main Clear Creek. When I came here to live in 1942, there was a small three room house at the mouth of Cottonwood Creek, but that was torn down about five years ago. Howard Wherry, who lives in Central City, owned the site of a mine and the three room house at the mouth of Cottonwood—the next time I see him I will ask him what he knows about Cottonwood. It may be that the place you refer to was Roscoe, which is listed in the old Colorado Central Railroad time table as being in Clear Creek Canyon, 2¼ miles below The Forks. It was a small placer mining camp that existed for about three years. It is said that they had a good placer there, but that the spring floods usually washed out their workings. I have been down there a number of times on the *new road* up Clear Creek Canyon, which will be opened for traffic this coming spring. When Roscoe was there, the only way to get there was to walk from the Forks or ride the train, as there was no wagon road there. I think it was about 1900 that Roscoe was in its prime. A company was organized to build a ditch from near Floyd Hill to Roscoe for placer mining purposes, but the company went broke before the ditch was completed. The remains of a portion of the ditch are VERY CLEARLY visible up on the mountainside south of the middle tunnel between the Forks and Floyd Hill—I will draw a sketch of the remains of the Roscoe Ditch, the Forks, Roscoe, Cottonwood and Chinatown and enclose the same.

Yes, when you come up here in warm weather, I can point out all of these places to you better than I can describe them.

Missouri City was a *real* town, but Missouri Flats seems to have been a sort of mythical location wherever a number of Missourians camped. Some say it was near the lone grave—others say it was close to Russell Gulch—the other day I asked Earl Quiller, who has run a grocery in Central City for over forty years, and Earl says it was up near James Peak.

By the way, I can point out to you the spot where George M. Pullman had a cabin in 1863, about a mile north of Russell Gulch.

Missouri City   Another small camp above Central City but south of it was Missouri City. I had seen the child's grave just off the road at the top of the hill above Central City on the way to Russell Gulch and had been told that the spot where it stood was

Missouri Flats. Another of Clarence Reckmeyer's letters (Nov. 10, 1953) helps clarify this point:

> I suppose that you know that the same company that built the Colorado Central Railroad secured a franchise to build a Toll Road up Clear Creek to Central City, including a branch of the Road from the mouth of Russell Gulch up Russell and Lake Gulches to Missouri City, but by the time the railroad was built to Black Hawk all of the gold had been taken out of the Gulches at Missouri City and the Post Office was abandoned. The First Assistant Post-master General wrote me as follows on Nov. 26, 1945:
>
> "The records of this Bureau show that Missouri City, Colorado, post office was established in Arapahoe County, Kansas Territory, March 24, 1860, with Mark Moore as Postmaster. It was changed to Colorado Territory and located in Gilpin County."
>
> Fossett's 1880 Colorado, on page 587, gives Missouri City a population of 597 in 1860, when Central had a population of 598.
>
> Bill Russell of Central City has a paper pamphlet published in 1862 which gives "Incorporation Act and Laws of the Consolidated Ditch Company," together with the Secretary's Report for 1861. It brought water from Mill Creek and Fall River, meandering twelve miles around the hills to the lake at the head of Lake Gulch (near Missouri City), to Russell Gulch and Nevadaville. Its headquarters were at Missouri City and A. H. Owens, who built the submarine on Missouri Lake, and Green Russell, after whom Russell Gulch was named, were among the Board of Directors. . . .
>
> People in the old days conflicted Missouri City with Missouri Flats, just as they do today. It should be remembered "Flats" was only a name by common consent. Missouri Flats, as far as I am able to learn, was located near the Saratoga Mine near Russell Gulch. Missouri City is the ghostliest town in this vicinity.
>
> P.S. I met "Poker Alice" in Omaha at the 75th Territorial Anniversary in 1920.

The last letter I received from this remarkable white-haired man whose memory remained so sharp and interest in the past so keen was in 1954. Not long afterward I read of his death. It was winter, and he had driven up to Central City from his home in Black Hawk to buy some groceries. On the return trip he lost control of his car on the icy road; it plunged into the gulch, and he died from injuries. It seems fitting at this place to quote the conclusion of one of Reckmeyer's letters:

> I am not particularly interested in being given credit for the information that I have given you—as the more ignorance we profess, the more we are able to learn. I do it for the enjoyment of it. When I was studying Nebraska history some 20 years ago, Dr. Sheldon, then Superintendent of the Nebraska Historical Society, said to me one day, "Reckmeyer, you know more about Nebraska history than I do, we will have to give you the degree of Doctor." "No, Addison, you won't," I said, "I don't want any barnacles on my name. Old history is a good lubricant for an active mind, but it does not deal with the fundamental principles of life."—and then I dropped quietly out of sight. Let me be only one of the common herd.

Central City,
Black Hawk
cont'd.,
Nevadaville,
and
Russell Gulch

Two of the most gratifying and informative letters I received about the Central City area were written by Myrna Davis (Mrs. Howard G.) Beehler of Denver, for they picture the towns when they were alive:

Nov. 3, 1950

I am writing in reply to your letter to my mother, Mrs. E. W. Davis. She is 80, and unfortunately does not remember much about her early days in Nevadaville and Central City. She was born in Nevadaville in 1870 and lived there and in Central City until 1917. Her father, W. J. Lewis, came to Nevadaville about 1867, met and married my mother there, and was a mining and mill man in Gilpin County for many years. He was a member of the second legislature after Colorado became a state. For many years he ran the Hidden Treasure mill on Clear Creek just above Black Hawk. My father was a young Welshman who came to Russell Gulch in the late 1880's and was interested in mines and later was cashier of the old Rocky Mountain bank in Central City. It was in the lower corner of the Teller House building. I was born in Central City in 1900 and lived there until 1917, so remember it well when it was still a busy mining town and saw it gradually go down. . . .

In Donald Kemp's book "Colorado's Little Kingdom," Plate 25 is of Nevadaville in 1899 and clearly shows the Episcopal and Methodist churches, the schoolhouse and the mines on the hill behind. They are the Hubert and Slaughter House mines. Plate 44 is a very good picture of my grandfather's Hidden Treasure Mill.

Feb. 27, 1951

I don't know how many mines were working in the period from 1910 to 1915, but quite a few were in the early part of that time, and gradually they shut down, especially after World War I began in Europe, till by 1917, when we left, there were very few, if any, still running.

The train came to Central City to the old station around eleven each morning and six in the evening. After supper (the evening meal was called supper in those days) everyone went to the Post Office for the mail. On Saturday night particularly Main Street was crowded; mostly men, some very well started on their evening's drinking.

There were picture shows in the old Opera House on Wednesday, Saturday and Sunday nights. One show a night. "The Iron Claw" a thriller-diller serial, and "The Perils of Pauline" with Pearl White in a different hair-raising exploit each week were two I remember, probably because they were so especially awful. Always there was someone playing a piano accompaniment to the picture. Sometimes a traveling company would come for a week with a different play each night, or a vaudeville troupe, all quite second-rate, but surely no worse than many of our movies today. This, of course, was after the time when really good companies came to Central City as they definitely had done much earlier.

We had some music teachers in the schools who put on plays and operettas and these were given in the Opera House. I was in several of these, and it was fun dressing in the old dressing rooms and performing on the old stage. The whole place was very dirty and run-down, but looked pretty good with the lights on and bright costumes when the plays went on. We were all afraid to go through under the stage from the orchestra pit to back stage, as

Central City, 1930. Store facades on Main Street.

Central City, 1953. Eureka Street,
showing Washington Hall (left), built in 1860's, now the City Hall;
lawyer's office; Williams Stable; Teller House (on right).

Nevadaville, 1928. Christ Episcopal Church on Main Street.

Nevadaville, 1928. Mines.

Nevadaville, 1931. Main Street; City Hall with bell (on right).

Nevadaville, 1931. School interior.

there were so many white rats under there. The story was that a traveling vaudeville act had some trained white rats which got away and they never could catch them. They got into the flume which still runs under all those buildings. However they got there, they were there. There were home talent minstrel shows and, by the way, it was surprising how many good voices there were in that little place. Among the miners were many Cornish and Welsh and they loved to sing. The Elks Flag Day programs and school commencements were held there too. The house was always full for these special occasions and the picture shows particularly on Sat. nights had quite a good attendance. I have a booklet put out in 1932 which states that the Opera House had not been used since 1910, but that is not correct as I was in plays there in 1914, '15, '16.

There were many other things to do also. The high school had basketball teams and played against nearby towns. They played Idaho Springs, Golden, Arvada and Louisville. These games were in either the old Turner Hall or Armory Hall. The High School in Central City was Gilpin Co. High School, so the young people from Russell Gulch, Nevadaville, Black Hawk all came there.

Just below the old station, where Spring St. and Nevada St. met at the end of Main St., there was a baseball field with a covered grandstand where there were baseball games on Sunday afternoons in the summer. There was a tennis court across from the Court House on Eureka St. which is all washed away now. We roller-skated Saturdays in Turner Hall.

The Elks and Knights of Pythias had dances in their lodge rooms and there were dances also in Turner Hall and Armory Hall. Every fall the Catholic church had a week-long bazaar and carnival in Armory Hall and it was crowded every night. It was called the Catholic Fair.

In the summer there were two big picnics, the Elks Picnic and the Grocers' and Butchers'; on Clear Creek above Black Hawk. These were definitely community affairs. All the stores and offices in town closed those days and pretty well everyone in Gilpin Co. went. The big barbecue pit was fired up the day before, and all night a whole beef would be turned on a spit over the fire for the big feast. There was free meat and coffee for everyone, races for young and old, and a baseball game in the afternoon. Several years ago I went by the old picnic ground and stopped to see if I could still find the pit. I could see where the big barbecue pit had been and the pit for the fire under the huge copper coffee vat. And, of course, like all mining towns there were the fourth of July and Labor Day celebrations in town with the hook and ladder and fire cart races, drilling contests, etc. Like rodeos nowadays.

There were four churches, all active, with ministers, Sunday schools, young people's societies, socials, guilds or Ladies Aids. The women played whist or 500 and had very, very elaborate card parties (if they belonged to a church which didn't frown on card playing).

Black Hawk was alive too, as the mills to process the ore from the mines around Russell Gulch, Nevadaville and Central City were there. The mill you asked about above Black Hawk on Clear Creek was the Hidden Treasure Mill and my grandfather, W. J. Lewis, ran it for years.

. . . Russell Gulch was a village about like Nevadaville. Quite a colony of Welsh people congregated in Russell Gulch. My mother was born in Ne-

Russell Gulch, 1951.

vadaville but my father was born in Wales and came first to Russell Gulch
with an uncle. Later the rest of his family came over and all settled there.

Water was always a problem in Russell Gulch. The deep mine shafts
drained the hillsides and gulches. I remember visiting an aunt there once when
I was 12 or 13 and I couldn't get used to having to wash in a cupful of water.
Water for household use was hauled from Central City in two-wheeled carts
and sold by the barrel, so they were very careful not to waste any. There was
a good brick schoolhouse, at least two churches that I remember, several stores
and, of course, as usual in all mining towns, at least as many saloons as
stores. . . .

I look back on happy years there.

Not infrequently, visitors to the old towns found some buildings still furnished
or at least strewn with relics from the past. Wallace X. Rawles of Los Angeles
reported (July 18, 1946):

In the autumn of 1928 the then Mrs. Rawles and I spent two or three
days prowling around Central City, Black Hawk and Nevadaville. . . .

Nevadaville of course was an old mining town. In one old house we sat
down at a large German-made grand piano. A touch of the keys produced a
remarkably distinct *in key* tone—and flushed a pack rat larger than a squirrel.
Memory of the one living human in Nevadaville, a caretaker of the Masonic
Temple I believe, was such he could not tell us much except that "that there
pianner come 'round the Horn. . . . The woman next door shot her husband
with a shotgun when he come back from Cripple Creek with an imitation
blonde." . . .

CHAPTER III

Boulder County Camps

Most of the Boulder County camps were early settlements that sprang up in the 1860's, wherever prospectors found sufficient placer gold in streambeds or in groups of mountainside lodes rich enough to promise steady production.

It may seem odd that fewer letters were written to me about the Boulder County camps, with which I became familiar in the late 1920's and early 1930's, than about any other mining area of the state. Perhaps it is partly because when I visited them, it was possible in many cases to talk to some of their older residents, who eagerly discussed the "old days" which they or their parents had known so well. As I think back to these early visits, I recall sights and bits of information that gave me a "feel" for the mining camps and the life within them.

Ward   Early one March, when winter winds were still strong off the high peaks and snowdrifts were banked deep against the north side of buildings or in the ravines where sunlight never reached, I spent a weekend in Ward with a friend, and we stayed in the old C. & N. Hotel. It was closed to guests for the winter, but its owner, Mrs. Thompson, who lived in it, agreed to let us stay if we did our own cooking and didn't expect a heated bedroom. It was bitter cold, and we wore our coats in the kitchen where we cooked our sketchy meals. Our beds were sleeping bags laid on the floor of an upstairs room directly above the one that contained the living-room stove. Its flue entered our room through a round hole in the floor and escaped through a similar hole cut through the ceiling. It did not provide any warmth, but

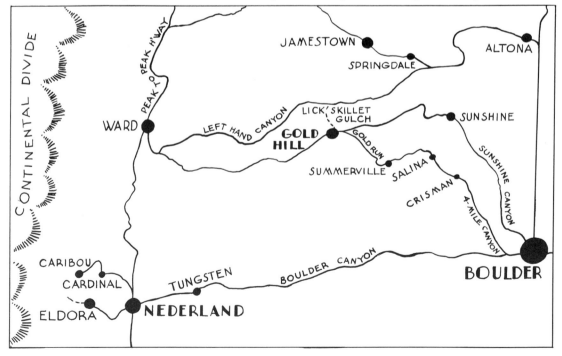

at night it glowed red from the banked fire below us. In addition to its ruddy color, whenever a particularly strong blast of wind hit the building, the stovepipe shook. Only exhaustion permitted me to fall asleep with the menacing pipe so close to us.

The first time I was inside Casey's old square-hewn log hotel at Gold Hill, it belonged to the Holiday House Association of Chicago and was known as Blue Bird Lodge. To it each summer came young business and professional women from Chicago, that they might spend their vacations in the mountains. While sketching portions of the interior of the sturdy old building I was shown a small, round card-table, at which, I was told, Eugene Field had composed the poem "Casey's Table D'Hote." Though no longer frequented by the Blue Birds, the exterior of the hotel is unchanged. The newer log building to the west was built as an annex to the Lodge, but in recent years it has been a popular restaurant—the Gold Hill Inn. *Gold Hill*

In the 1930's a one-story, weathered wooden building stood on the southwest corner at the junction of the Four-Mile Canyon road and the Sunshine Canyon road. I was shown the interior, a long, narrow room with a stage at one end and a kitchen behind that, and was told it had been the Congregational Church of Gold Hill. I wandered through the cemetery, which was below the town and above Gold Run, to read the inscriptions on the old stones and the names of the faraway places from which so many of those buried had come.

The old road up Boulder Canyon, the main artery to and from most of the county's mining camps, was narrow; it crossed and recrossed the creek many times between Boulder City, as it was first called, and Nederland. As mining increased in importance and ore was freighted to mills and smelters, traffic also increased. To avoid collisions with vehicles traveling in opposite directions, turnouts were provided, and for a further safeguard and warning, a set of bells became part of the harness on the lead horses of the heavily loaded ore wagons that rolled down the steep, twisting grades. Years after the canyon road was rebuilt and paved, a set of these harness bells was presented to the Little Theatre of the University of Colorado; partly to preserve local tradition and partly for practicality, they were rung *Boulder Canyon*

Ward, 1895. *Photo by J. B. Sturtevant, Boulder. Courtesy Denver Public Library, Western History Department.*

Ward, 1930. Center of town; schoolhouse (upper left).

Ward, 1941. Community (Congregational) Church.

before each performance and at the end of intermissions to alert the audience to go to their seats.

The Colorado and Northwestern Railroad, better known as The Switzerland Trail, ran up Boulder and Four-Mile Canyons as far as Sunset, where it split into two branches, one terminating at Ward and the other at Eldora. The road was built primarily to serve the mines near the Divide and to freight their rich ores down to the valley where they could be shipped to smelters and processing mills. Upon its completion it also became a popular attraction for picnickers and excursionists, who enjoyed summer trips into the high country, especially when the wild flowers were at their best. The Switzerland Trail is gone, and its right of way can be seen only in fragments along the sides of Boulder and Four-Mile Creeks. But one relic remains in Eldora. A friend of my husband's spent the summer of 1913 working for the railroad, and he told us of his daily duties. Each day he would clean out and dump the ashes from the firebox of the locomotive after its run into the mountains from Boulder. Next morning he had to rise at 3 a.m. to start the fire under the boiler so that a full head of steam would be up in time for the train to start its run down the canyon to Boulder. About fifteen years ago my husband and I drove to Eldora and walked past the location of the old station. After some poking about in the weeds, we found unmistakable remains of the coal and ashes where many fireboxes had been emptied.

Eldora

One of the earliest camps in Boulder County was Caribou. While elk hunting in 1860, Sam Conger stumbled upon some silver float on Caribou Hill above Nederland. Eight years later, when he saw a car of similar ore on a Union Pacific freight train in Laramie, Wyoming, he picked up a piece to examine it. When ordered to drop it, he told the irate trainman that he knew where there was a whole mountaintop of ore just like it.

Caribou

"Is that so?" was the man's reply. "You're looking at high grade silver from the Comstock Lode in Nevada."

Back in Colorado, Conger revisited the spot where he had found the float and uncovered a vein which he called the Conger. In the spring of 1869, he returned to do further prospecting and uncovered another outcrop which he called the Caribou. He sent the ore to Professor Hill's smelting works at Black Hawk, where it assayed high in silver. Convinced that he had made a good strike, he took five partners—William Martin, George Lytle, Hugh McCammon, John H. Pickel, and Samuel Mishler—and together they began developing what proved to be the richest silver mine in the district. Naturally, the good news leaked out, and by June 1870, a stampede was on. More lodes were located, including the Poorman and the Seven-Thirty. About the latter, William H. Burger of Evanston, Illinois, wrote me (Aug. 30, 1950):

> We left Caribou when I was one year old. Of the stories told by my mother about the only thing that has been retained is that father made a strike —the "Seven Thirty" mine—and he bought mother a beautiful gold watch and chain. Mother passed this on to me about 1925.

A Seattle lawyer, Hugh E. Pickel, wrote (Jan. 29, 1955) in hopes that I could provide him with additional data about two of Sam Conger's partners:

Re: Discovery of Caribou Mine 1869—Boulder County
    My great-uncle, John H. Pickel, along with his kinsman, Hugh C. Mc-

Cammon, were among the discoverers of the Caribou back in 1869. . . . I am particularly interested in finding source material on:

(1) John H. Pickel who in addition to above was Vice-Pres. of Rocky Mt. Telegraph Co. 1874-1876 along with Sen. Teller, Pres.; and even ran for State Treasurer in 1880 on the Greenback Party Ticket (sounds like quite a switch!)

(2) Hugh C. McCammon, director of 1st Natl. Bank of Central City, legislator (1874), a founder of the Univ. of Colo. at Boulder & rancher of both Boulder and near Blackhawk.

Mrs. Justini Fritze Irwin of Des Moines tells of her grandmother's occupations both in Caribou and Nederland:

Since I spent nearly all my childhood summers in Nederland, I love that town and its surrounding country. . . .

My first love is Caribou. For it was there that my Grandmother, Bertha McKinney White, taught school in the old schoolhouse that when I last saw it was almost a pile of rotten lumber. Later, she opened a women's ready-to-wear store in Nederland and built the little three-room cabin, which after we enlarged it we named "Just-In."

I have spent many hours prowling through old homes in Caribou reading the dates on newspapers hung on walls as wallpaper and insulation against the cold wind, and the epitaphs on the Tombstones in both Nederland and Caribou. . . .

Marie E. Kittell of Longmont has stimulated my imagination . . . with her tales of my Grandmother as she remembers her. My Grandfather, Dr. W. J. White, of Longmont, has helped some, since he practiced medicine in the area for 52 years before his retirement. But these people have not filled up the holes in my files.

Altona   Altona, only a mile from the mouth of Left Hand Canyon, was a trading center which served both the prospectors who searched the canyon for gold and the quarrymen who worked the red sandstone deposits along the face of the front range of the Rocky Mountains. The small townsite centered around the ranch of Peter Haldi, a pioneer, who was also blacksmith and postmaster. In its heyday the settlement had a schoolhouse, a grocery store, a bank, and a tollgate. When stagecoaches made regular runs between Boulder and Jamestown, a change station for the horses was also located at Altona.

Like so many people at the beginning of the twentieth century, the father of the family mentioned in the following letter left the east coast where all the kinfolk lived and went West for his health. He settled in Altona. His son, Leon G. Baldwin of Fulton, New York, wrote (Oct. 1, 1961):

For my 70th birthday I received a copy of your book "Stampede to Timberline." I was born in Greenpoint, Brooklyn, 18 India St., 1891 and in 1898 we moved to Altona, Colo. My father had been with Brooks Bros. of N. Y. and went West for his health. In your article . . . you mention Peter Haldi whom I knew very well and we purchased the Old Tole Gate from a Mrs. Camp of Altona. My father was Irving Baldwin, Sr., there was my dear mother and a brother Irving, Jr. We moved down to Hygiene where we had

Gold Hill, 1940's.

Jamestown, 1940.

a store and P. O. and left Colorado in 1905 for Poughkeepsie, N. Y. My father's family all came from Dutchess Co., N. Y.

Jamestown   Some miles above Altona is Jamestown, another of the county's older camps. It is situated on Jim Creek, three miles from the Left Hand Canyon road. George Zweck, the first settler, built his cabin there in 1860. He prospected for gold while his cattle grazed on what is now the townsite. By 1860, after a few galena mines were located, a camp was established, but it grew slowly until 1875 when a rich gold deposit was uncovered.

According to one acocunt, it was Frank Smith and Indian Jack who found the surface float which carried a high percentage of gold and located a claim which they called the Golden Age. Before they realized what they had found, they sold it for $1,500. A letter from Mrs. J. B. Richardson of Granby (June 10, 1950) credits Indian Jack with the discovery:

My husband's grandmother boarded Indian Jack who found the Golden Age at Jimtown. Indian Jack sold the mine while drunk for fifty dollars and a barrel of whisky. When he became sober, he was very angry, walked out of Jimtown, came back with gold equally good, but enough different from the Golden Age ore that everyone knew it came from a different mine. Indian Jack swore he would never tell a white person where the new mine was located. So far as we know he never did. He came nearer telling my father-in-law, then a young lad, when he told him it lay on the ridge between the Golden Age and Red Dirt Hill.

The only other letter I received that made any mention of Jamestown came from L. E. Butler of Loveland, who provides vital statistics about Ray Pulliam, a native Coloradan and member of a prominent family, who died in 1951:

Dear Author of "Stampede to Timberline": . . . am sending a clipping taken from *The Reporter Herald* of Loveland, Colo.

I came to Boulder in 1916, during the Tungsten Boom, and drove on stage runs and worked around some of the mines, and was familiar with quite a number of places mentioned in your book.

[Enclosed newspaper clipping]   Nov. 5, 1951. Loveland, Colo.

RAY PULLIAM DIES SUNDAY OF LEUKEMIA

Ray Pulliam, 68, 531 East Tenth Street, died at the Presbyterian hospital Sunday evening, after a lingering illness of leukemia.

He was a native of Colorado, having been born in Jamestown, Colo. in 1883, during the mining boom, and was the first child born in that village.

Later he moved to Berthoud, where he farmed and contracted. In 1924 he moved to Loveland, where he was employed by the Great Western railroad. He was a member of the Loveland Eagles. He was married to Miss Louie Keener, in 1903.

He is survived by his wife; a daughter, Miss Mabel Pulliam, who teaches at Lakewood; a nephew, Elmer Courtney, Denver, whom he reared as his own child; and a brother, E. A. Pulliam, Oroville, Calif.

Funeral services will be held Wednesday at 2 p.m. at the Hammond drawing room, with the Rev. A. E. Plummer, pastor of the Methodist church, officiating. Interment will be in the family plot near Berthoud.

CHAPTER IV

Georgetown and Its Neighbors

The stretch of Highway I-70 west from the foot of Floyd Hill, through Idaho Springs to Georgetown, is flanked on both sides by gold-mining properties, easily recognized by their sizeable russet dumps and by the ruined shafthouses, ore chutes, mill foundations, and scraps of rusting machinery strewn about. It is easy, however, to be driving too fast on the highway to pay much attention to these vestiges of an active mining area.

Nevertheless, if you take the Fall River exit just beyond Idaho Springs it will *Alice* lead you to an unpaved road that climbs to the stream's source in St. Mary's Lake and, en route, passes the side road to the old camp of Alice. It was deserted when I visited it in the 1940's, and recently realtors divided the townsite into lots for summer homes. Of Alice and its mining days, Elsie Eaves writes (July 18, 1945):

> I have read your account of your Ghost Hunting in the July University of Colorado Alumnus. Alice was a childhood headquarters for me. My mother, Katherine Eaves, Brush, Colorado, was the first school teacher of Alice but, as I remember her account of it, the Alice mine closed before she got paid and the remuneration was slight finanically but should be rich in her excellent memory....
>
> My grandfather John Elliott had the cabin with the first plate glass window, which later belonged to the incomparable Pete Sweeney. Alice is a wonderful place of rugged individuals.

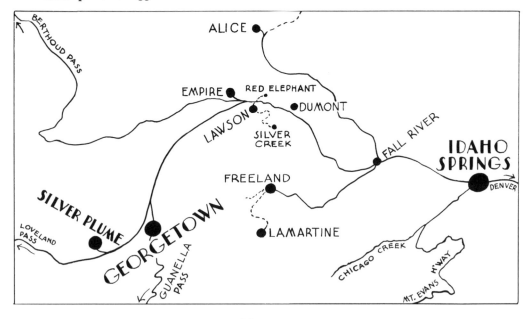

You don't list "Sweet Home," the crumbling cabins over the hill from Alice on the way to Chinn's Lake and the absence of Yankee Hill.

Interested in your hobby.

(Chinn's Lake was also called Silver Lake and was closer to the summit of the mountain than was Alice. Yankee Hill was across the summit on the Central City side.)

Freeland and Lamartine

Directly across Clear Creek Canyon from Fall River is the mouth of Trail Creek, up which a steep, narrow ledge road climbs four miles to Freeland and continues to Lamartine. In the early 1940's Freeland had a number of well kept frame houses, a small population, and at least two mills. Since then most of the buildings have disappeared. Lamartine is more difficult to find. On an earlier trip I had reached the mouth of Trail Creek by a different route. From Idaho Springs I took the old road on the south side of Clear Creek, west past the Stanley mines and the Alma-Lincoln property to the mouth of Trail Creek, so as to visit with Mrs. Jean D. Lindsay whose house, built against the mountainside, was half hidden by vines and shrubs. My knock was answered by a handsome woman with deep-set eyes and a low-pitched musical voice. When we were seated in her small parlor with its piano and comfortable chairs, she told me that she was born in a cabin near Idaho Springs at Spanish Bar but that she had spent most of her life in the house where we were conversing and that her three children had been born there. On a table was a copy of *Stampede to Timberline;* seeing me glance at it, she smiled and told me that she read some of it every night and that it brought back many memories. From between its pages she removed two old photographs, one of Mr. McLean, an early Idaho Springs photographer; the other of Mr. Chavanne, of the Lamartine mine, who at one time was also the keeper of the tollgate that stood at the entrance of Trail Creek. "At that time," she told me, "there were big barns, a mill, and weigh-scales at the tollgate, right near this house."

"Do you remember seeing a big house on the left side of the road just after you'd passed the Stanley mines buildings at the west edge of Idaho Springs?" asked Mrs. Lindsay. "It was used by the superintendent and the higher officials. You said you remember the two mills close to the highway at Fall River, across the creek from here. The road beside Clear Creek had tollgates and many bridges. Do you remember seeing a house several miles up the Fall River road with a colonnaded railing on the second-story porch? It was brought to St. Joseph, Missouri, by train and from there by oxcart. But you were asking about Lamartine.

"You could reach it either by climbing up Ute Creek which branches off Chicago Creek out of Idaho Springs or by going up Trail Creek. The camp was built on the side of a steep canyon with the buildings terraced one above the other. When the mine was working, there was both a schoolhouse and a store up there. The ore taken out of Lamartine was packed down Trail Creek. Not far from the town was French Camp where log cutters lived. All mines have to have logs for their mill machinery, and shafts were run by steam and logs provided fuel for the boilers."

Directions as to how to get to Lamartine were included in two letters sent to me in 1946 by Mary Lou Cox, a member of the Colorado Mountain Club, who had reached the abandoned town seven times and had approached it by the Trail Creek road. Detailed information was necessary, for as she said, "The area is honeycombed with mining roads and trails."

Freeland on Trail Creek, 1941.

Follow the road up the canyon; at 4.3 mi., take left fork; follow main-traveled road uphill; at 5.3 mi. turn left (this is the third of three similar "roads" leading off to the left within a few rods of each other). At 5.4 mi. park on a mine dump on left side of road. Continue walking on road; in about five minutes you will come to an intersection. (You may need more explicit directions at this point as to which of several indistinct trails to follow). If you take the correct one, in a mile or two you will come to Lamartine in a gully to your left, just where Mt. Evans bursts into view. Go down the middle of the gully through the main part of Lamartine. If the road disappears follow the path along the telephone line. At the far side of the powerline clearing you will pick up an old road. (Take time out to find this road; if you miss it you will be in a mess.)

Lamartine is rather hidden from view as you walk along the main road. The mine shaft is at the saddle above the gully in which Lamartine is located. I would guess there are about two dozen houses still standing and foundations of dozens more. There was no mill. Some of the houses were papered with newspapers and we enjoyed reading the ads especially. One headline read: "The Invasion of Cuba Has Been Put Off For A Week," dated May 8, 1898.

Lamartine, 1947. Remains of homes; view down Ute Creek.

I still don't know the name of the town west of Lamartine. It apparently was almost as big as Lamartine, but it isn't named on the map. It's on an exposed slope instead of in a gully, so there are only a half dozen or so buildings still standing. Could this nameless settlement be French Camp?

L. Hanchett of Salt Lake City wrote a long letter (Aug. 6, 1949) correcting statements made in *Stampede to Timberline* and recalling incidents from his past when he lived in Empire and knew intimately the territory between Idaho Springs and Georgetown.

In 1887 my father was placed in charge of the Lamartine Mine and in 1888 met with a severe accident, which resulted in my injection into mining work, as I acted as his assistant. Upon his death in 1894, Mr. Himrod made me manager of the mine, and during the next ten years until I removed to Utah, the mine was the principal one in the county, and in the sale of its ores, etc., I came in contact with many of the leading business men of the state.
. . . Page 111 (of your book) states that the Lamartine was sold; the Lamartine has never been sold, and is owned today by the grandchildren of Peter Himrod—upon the death of their father Fred E. Himrod. The heirs formed a corporation, Himrod Mines Co., and placed the title in that company and retained 100% of the stock.

Dumont   Dumont was another small town a short distance west of Fall River, dating back to 1859. Its Mill City House, built in 1868 and still preserved, was a stage stop on the old toll road. Hanchett's letter continues:

Page 111 of your book mentions the Mill City House and aroused memories of the man who operated it for many, many years. When Mill City was a booming camp, the good women of the place organized a Union Church, and when mountain raspberries ripened in August, they held their First Church Festival, each good woman contributing one or more cakes to the feast—at close of the banquet there were four cakes left and someone proposed that they be auctioned off—a young mule skinner accustomed to use of much profanity was asked to be the auctioneer. His name was Chinn—when he came to the last cake, he asked, "How much am I bid for this beautiful frosted cake made by one of Mill City's prettiest women?" No response, and he repeated his entreaty, when from a voice near the doorway came a bid of 25 cents.

Chinn was stunned and said, "What's that? You mean to say you only bid twenty-five cents for this beautiful cake? Why, you dirty son-of-a-bitch, you ought to be ashamed of yourself!"

His tersely appropriate remarks caused someone to say, we ought to make him a Deacon of this Chuch—and thus he came to be known as "Deacon" Chinn for the remainder of his long life.

When the town was renamed Dumont, he declined to change the sign on the west gable of the hotel "Mill City House" and this old lettering is still visible on the west end of the building.

From Dumont a group of large mine dumps near the top of the mountain on the south side of Clear Creek canyon were pointed out to me on different occasions as from the Seven Sisters mine at Silver Creek. Margaret Shaffer of Idaho Springs had mentioned Silver Creek as a mile and a half south of the town of Lawson, and added that in 1890 it still had one saloon, one store, and two mines that were good producers of silver and lead ores. She told me that the road was in poor shape and that it was a four-mile hike to the townsite. Fred M. Mazzulla, a Denver lawyer, showed me an old photograph of the place with the Reynolds mine and mill, and below its dumps a few cabins and the schoolhouse.

Silver Creek

Silver Creek. Joe Reynolds mine; schoolhouse behind smokestack (just right of center). *Courtesy Fred M. Mazzulla.*

Red Elephant

At first Silver Creek was known as Chinn City, getting its name no doubt from the "Deacon" in Dumont.

When I finally reached Silver Creek in 1964 and scrambled on top of one of the Seven Sisters dumps, I looked across the Clear Creek valley to a similar group of dumps—those of the Red Elephant mine. It was a mining man from Idaho Springs who told me about Red Elephant. It was a small camp established in the 1880's one mile north of Lawson, and its life span was brief. By the late 1940's few of its buildings remained except a mill, with a trestle over the highway from which dangled a sign depicting an elephant.

Dailey

My Idaho Springs friend, Margaret Shaffer, wrote to me about the Dailey District where the government was carrying on a molybdenum experiment. She described the location as a "scattered mining district seven to eight miles west of and further up a branch of Clear Creek from Empire." "There is no village," she added, "only cabins at different mines." This experiment that she mentioned was the Urad mine, which was further developed by the Climax Molybdenum Company. On November 30, 1974, the *Boulder Daily Camera* reported:

> The Urad molybdenum mine of AMAX Corporation was closed permanently Friday with its ore reserves exhausted. The company said the mine's ore supply was known to be limited to six or seven years. The mine was opened in 1967 at a cost of $30 million.
>
> The 130 employees of the mine will be transferred to AMAX's Henderson mine in the state. The mine is scheduled to begin production in mid-1976 with production estimated at 50 million pounds a year. [Full production subsequently rescheduled for 1980.]

Empire

The most complete description of Empire, one of the oldest mining centers in the Clear Creek valley, and other towns nearby is found in Mr. Hanchett's letter:

> First I should qualify as a competent witness by reciting incidents of my life in Colorado. My father as Manager of a Mining Company owning prospects in Georgetown District came to Colorado in 1872 and moved his family from New York in Jan'y 1873, leaving them in Denver temporarily, until he could procure shelter for them near Georgetown—On April 28, 1873, my mother with my sister and myself took the narrow gauge train from Denver on the Colorado Central to Floyd Hill station, then the end of the line, where we transferred to stages for the remaining part of the trip; the stage rested at the Beebee House in Idaho Springs for noonday meal for its passengers, and while there, a light buggy from Central City arrived carrying two very dusty passengers, dressed in long dusters; one of them was President U. S. Grant, and the villagers soon assembled and formed into a reception line, in which I joined, and I had the great honor of shaking hands with the President (a thrilling experience for a five year old boy, on his first day in Colorado mountains). Soon after my father removed his family to Lower Empire, where my boyhood was spent. In 1873 my father established a general store in Empire which he conducted for ten years.
>
> And now to refer to items in your book:
> Pages 116, 132 mention John Coley; in the early days miners used ordinary black powder for blasting; and used to tamp in the holes usually with wooden sticks—Coley one day used an iron bar which struck a spark and

caused an explosion that destroyed the sight of both his eyes; he lived in Empire until he died, and I knew him very well; tragedy seemed to follow the Coley's, for his son Joseph was killed by a premature blast in Lamartine shaft in 1894.

Page 141—Harry Cairns—Paul Lindstrom. Cairns was employed by Lindstrom in his Brewery, and often came to father's store; he was quite a philosopher, and often told me "If you want to make a man believe you are lying to him, just tell him the plain truth." Your story about his concealment of Bishop confirms his statement. Cairns while not of the underworld, had many friends among the toughest element and often did them a good turn.

Page 106, in re George Jackson's discovery of gold Jan. 4th 1859, says he dug $9 worth of gold—in fact he dug 19 ounces of gold dust worth over $300. While resident in Idaho Springs I made a search of the facts concerning the discovery and obtained accurate information from his daughter and other pioneers, and succeeded in locating the spot from which the first gold was discovered; I then bought a tract 100' square from J. G. Roberts who owned the patented placer ground, and deeded same to Town of Idaho Springs to be used for location and erection of a monument commemorating the discovery.

The Jackson monument marking the spot is located about quarter mile above its junction with Clear Creek, on Chicago Creek, and is perhaps 100' west of the highway to Echo Lake. On your next visit to Clear Creek County see this spot, for here began the greatest mining growth of Colorado in January 1859.

You ask for old locations you may have missed that were historically important. On your next visit to Clear Creek County see the slag dump and foundations of the Swansea Smelter, which was the first smelter built in the County and which was built before Sen. N. P. Hill built his smelter at Black Hawk—it is located on the creek about 1000' from the junction of US 40 and the road branching to Georgetown at Empire Junction. The Swansea Smelter was built by Richard Pearce, an experienced smelter operator from Swansea, Wales, and owned by a British Company who had sent Pearce to Colorado to build a smelter (See Hall's History Vol. IV Page # 539 Richard Pearce). There is only left today part of the slag dump and some of the foundations, but as a small boy I visited the smelter many times.

Pearce was an expert metallurgist and owned patents covering a method of separating gold from silver and copper, which Prof. Hill very much coveted, so when the Boston & Colorado Co. moved from Black Hawk to Denver to its Argo Plant, Hill made a deal with Pearce to become his Company's metallurgist, and to close the Swansea Plant.

I believe the Swansea was the first smelter in Colorado to treat complex ores. When I was a boy the place was known as Swansea, and when the railroad extended to Georgetown in 1876, the name of Swansea was dropped and the place designated as Empire Station.

* * *

Read Page 317 - Vol. III Hall's History, telling that Edgar Freeman and Henry Cly Cowles were the first white men to walk into the valley at Empire —I knew both these pioneers very well—when they extended their exploration to a point near the foot of Berthoud Pass, Cowles discovered a skeleton of an

Indian who had been scalped, so he kept the skull and years afterward kept it on his desk, as a smoking tobacco container—using the scalped portion as the entrance.

He delighted in having some tenderfoot decline a pipe-filling from the gruesome container.

* * *

On your next visit to Empire go to the cemetery and locate the stone erected over the grave of the first man buried there, on which is inscribed "Murdered over a dog fight at North Empire"

* * *

Empire was settled when Civil War was rampant in the east and the mountains were named—Lincoln (after Abraham Lincoln), Douglas (after Stephen A. Douglas), Covode (after John R. Covode, Representative from Missouri), Democrat, Republican, and Columbia.

Living in Empire were many of the old pioneers whom I personally knew.

Your recital names men with whom I have had business dealings, and some who were only acquaintances, and reading their names stirred up many, many memories—I list below. First of course Gen. U. S. Grant, followed by names of men with whom I had business relations:

David H. Moffat	Fred Walsen	Capt. Yankee
Spencer Penrose	A. E. Carleton	Sen. H. M. Teller
Chas. N. McNeil	Jno. M. Dumont	Louis Dupuy
John Quincy Adams Rollins	Anton Eilers	Sen. N. P. Hill
Gov. Charles S. Thomas	Gen. W. A. Hammill	Rev. T. A. Uzzell
Capt. L. D. C. Gaskill	Peter Himrod	Peter McFarlane
W. S. Comer	Harry Cairns	Paul Lindstrom
Hugh R. Steele	George Dory	E. H. N. Patterson

Of men I knew personally but had no business with are:

Sen. H. A. W. Tabor—his wife "Baby" Doe and their daughter Silver $.

Gov. Alva Adams	Gov. Jas. H. Peabody	John Campion
Congressman J. B. Belford	Gov. Routt	Thomas F. Walsh
"Pap" Wyman	"Soapy Smith"	David Frakes Day

"Soapy" Smith had an invalid wife to whom he was devoted, and in the summer months boarded her at an Idaho Springs hotel, and would come up from Denver Saturday afternoon train, and spend Sunday wheeling his wife around town in a wheel chair.

"Soapy" was very much a gentleman and a nice person to know if you avoided meeting him professionally. . . .

Very cordially yours,

L. Hanchett

Georgetown   By chance I was in Georgetown in August 1950, just in time to attend the Old Home Day dinner at the Grubstake Inn with Mrs. James Grafton Rogers. Almost everyone present had lived in Georgetown and eagerly reminisced after dinner about life in the city when it was called the Silver Queen of the valley. I did not know the speakers, but their recollections were so colorful that I jotted them down.

One man remembered watching jack-trains leaving each morning for the mines—the jacks all milling around and some lying down in the street. He became a jack-whacker.

A Mr. Eckeley recalled going up into the Argentine District to get the loads off the jacks. "They'd get back to town three to four hours before the men," he added.

Another spoke up: "Some of the mines were pretty inaccessible, so the companies had to build roads up to them for the pack trains and later for the ore wagons. You can see such a road up Saxon Mountain, starting from the lower end of town and climbing steadily the length of the valley to its mouth and then up to the mine."

"We sure needed a smelter in the early days," said one old-timer. "Between $25,000 and $40,000 was taken out of little stakes around Georgetown. When the Chamberlain Dillingham Ore Company had interests here, they induced the miners to bring in even small amounts of ore—one hundred tons at a time. Take the ore from the Wide West, for instance. They'd weigh it up, give the man a check for $800 a ton, and pay him the difference when they'd run it. The Dives-Pelican shipped $2,000,000 during the time of one foreman."

"What was the size of the Georgetown population in the seventies and eighties?" asked one of the guests. "About 1,600," was the answer. "Women didn't vote then. Today it's 402. In 1940 Silver Plume's population was 190."

"Speaking of Silver Plume," said a new voice, "you all know the Buckleys. Mrs. Buckley was widowed and left with thirteen children—nine sons. She was a Roman Catholic. She formed the Buckley Boys Company and all the money they earned was pooled. How she could swear! I'm one of the sons and this is my wife."

Georgetown, 1940. Star Hook & Ladder Co., now the Town Hall.

"Do any of you remember the Halloween joke we pulled on a Frenchman who had an ore wagon?" asked an elderly man with a twinkle in his eye. "The boys decided to steal his wagon and drive it three or four miles out of town, down the road, and then leave it there, so he'd have to go after it the next morning. We just got it to the place where we were going to abandon it, when the Frenchman sat up in the bed of the wagon where he'd been lying and said: 'Now, you can drive it back again.'"

"That reminds me," said another. "Remember the cold night when the jail burned and one of the boys brought lots of coffee to the fire fighters? He brought it in a tea kettle and he had a basket of cups—his wife's best Sevres china—to pour it into."

The Georgetown postmaster then spoke up. "I remember in the old postoffice an old man came down to town from up on the hill to get his mail. The postmaster couldn't remember his name, so he asked him: 'What's your name?' 'Oh,' said the man, 'you'll see it on the letter.'"

"When I was twelve years old I picked silver ore off the dump of the Pelican-Dives mines and averaged from a dollar-and-a-half to three dollars a day," reminisced another man. "Twenty-five cents (two bits) was the smallest change in use here. It was good for a cigar or a drink of whisky. I remember when road companies traveled by stagecoach visiting all the mining camps, usually hitting Denver, then Central City, Idaho Springs, and Georgetown, in that order. The cheapest seat was one dollar."

Another speaker was Lady Blythe Marvin, known as the "Original Mechanical Doll of Stage and Screen." She was born in 1890 and was sponsored in her career by Mary Elitch. She made her initial public appearance in 1897 at Elitch's Gardens in Denver, where fifty years later she also gave her final public performance, this one under the sponsorship of the Pioneer Men and Women of Colorado. In her early years, Blythe had traveled with her mother, and her performances on the Orpheum Circuit had taken her all over the world. Ripley had featured her, and she had performed before two presidents; on one occasion, when she was exhibited in a store window, a woman, convinced that she was a mannequin, tried to buy her.

Speaking of performances," said one of the men, "I knew the man who built the old McClellan Opera House which stood on Taos Street on the vacant lot next to the Hotel de Paris, in 1869. It burned down in 1892. I went to McClellan's funeral in Denver—it was a long trip in those days. While I was gone someone entered my house and took my cookstove."

Silver Plume   Silver Plume like Georgetown has kept much of its architectural charm with the majority of its buildings dating back to the seventies and eighties. The many big dumps on the mountainsides above the town are proof of the richness of the mines, and with the extension of the railroad beyond the Georgetown Loop through Silver Plume and on to Graymount the shipping of ore posed no problem.

Through an interview with Mrs. Alice Griffin Buckley, one of a pioneer family at the Plume, I learned that she was the first white child born in Nevadaville and that her family moved to Silver Plume in 1870, just when the camp was being laid out. Trees were cut to build the town; most of the cabins were high on the mountainsides; and not until the valley became cleared and drained was it practical to build homes and business houses close to the creek level.

Owen Feenan, step-father of Mrs. Buckley, was the real discoverer of the Pelican mine. While prospecting in 1868, he discovered a rich deposit of silver but told no one about it. In the meantime he worked in another mine at $7.00 a day, walking over from Central, where he lived, on Monday mornings and walking back at the end of the week. Sometime later, while very ill and fearing he would die, he told two friends, and they opened the Pelican mine in the spring of 1871, without setting aside any shares for him.

The Pelican House was run by Mrs. Buckley's mother. The beds were never empty. As one shift left to go to work another shift would be coming in ready to tumble into the empty beds.

Silver was $1.29 an ounce and small fortunes were made every day. The men were paid on Saturday nights. On these evenings the front room of the Pelican House was always crowded with excited children, for the men took whatever small change they had and tossed it into the room just to watch the kids scramble for it. Some youngsters gathered as much as ten dollars in an evening.

Silver Plume, 1941. False fronts.

The variety of design achieved by those who fashioned the wooden false-front facades on buildings delights me, and the store front in Silver Plume, shown on page 5 of *Stampede to Timberline,* is one of my favorites. Those who are familiar with the building will enjoy this letter from Arthur C. Davis (April 8, 1960):

This store, cottage and other buildings happen to be the properties belonging to my wife's mother, Mrs. Martha Jenkins Mundy, who had a little business selling candy, souvenirs, post cards, etc., also serving meals particularly to Tourists on Sundays. The cottage—to the right—is the birthplace of my wife Florence.

Her parents had operated a General store across the street in what is now an Antique shop. Owing to Mrs. Mundy's health they sold out and returned to England (Cornwall). Her father became ill, passed on and is buried there. He was a Mason. The mother and daughter returned here, and Florence continued her education in this country finishing in Greeley in preparation for her teaching profession. Mrs. Mundy was really a very remarkable person meeting the many hardships, to give her daughter an education, acquire her property and stock her little store, in spite of her hard-of-hearing handicap.

My thought, if there were to be another Printing, you might find it interesting to identify the owner of the False Front and buildings as shown on page 5.

G. V. Benson, secretary of the Gunnison Pioneer Society, in a letter written in 1950 questions the accuracy of a statement in *Stampede to Timberline:*

> You state in your article on Silver Plume (p. 130, 2nd paragraph) that "stone for the Denver Capitol was cut from the granite quarry just beyond the edge of the little mountain city." Probably we are talking about two different Capitols, but I have always been under the impression that the Colorado State Capitol was built from stone quarried at Aberdeen in this [Gunnison] county. In the "Twenty Years Ago" column of the Denver Post under the date of Dec. 30, 1929 there was the following item. "All granite needed in construction of Denver's new city and county building was offered as a gift by J. W. Wyatt, quarry operator at Silver Plume."
>
> I am sure you can put me straight on whether or not the granite for the Colorado State Capitol came from Gunnison County. I would appreciate the information.

My statement in *Stampede to Timberline* was incorrect. The granite from the Silver Plume quarry was used in constructing the Denver City and County Building, while the stone for the state Capitol came from the Aberdeen quarry in Gunnison County.

Mrs. K. P. Laumann, who wrote to me from Colby, Kansas (Aug. 13, 1950), gives her vivid memories of Silver Plume in the early 1900's:

Dear Mrs. Wolle,

We left Salt Lake City in the Spring of 1917. Seven of us in a Model T Ford—I was almost ten—and the big sister. Two brothers next, and the babies were twin girls not quite two years old.

Wyoming was as wet and squishy that spring as Kansas is right now. No auto camps nor tourist cabins. The hotels were nearly always full because salesmen and others were waiting for the rain to stop. This past summer we drove over much of the same route and covered in a few hours the distance that took a complete day in 1917. We sort of stayed with two other cars—to help each other out of mud—another Ford from Oregon, and a Maxwell from Dakota.

One night at Bosler we all had to stay in an upstairs dance hall where the dance had been held the night before. The next morning one of the men went out for a pail of fresh milk, the women cooked oatmeal, fourteen of us feasted. Even now I'm truly fond of oatmeal—it was so wonderful that morning.

There had been a night when we couldn't reach a town and had to sit in the car all night while it drizzled and sleeted and we were hungry.

At Hannah, Wyo., a kindly mother with a large family and a small house gave my family a whole room, to make beds on the floor, because the hotel was full.

It took us eleven days to make the trip from Salt Lake to Silver Plume.

Our furniture had gone by freight and was sitting, crated, in the front yard. Dad had been staying in Silver Plume several months before he came for us. He did a very thoughtful thing. The last night of the trip we stayed in Golden. Thus, we were fresh and rested for our trip into the real mountains. I had day-dreamed all the way and it was even better than I dreamed.

We lived in the "company" house at the east end of town from 1917-1922, and the big Mendota mill was just across the creek from us. When it shut down for some reason in the night we'd all wake up. When the mill was torn down it was rebuilt closer to the mine—at the west end of town. The scales and the scale house in which they weighed the wagons of ore was just outside and to the left of our front fence. After the mill was moved, my sisters used the scale house for their "play house" and doll furniture.

I didn't realize that the ore wagon is almost extinct until I saw one and read so in my last "Colorado." Nor that my sons have never seen one. All day long the wagons drove to the mill, full of ore—up the ramp of the mill after weighing their load at the scale house right by our front yard. The level flooring at the top of the ramp would accommodate two ore wagons at once. The men who drove, unhitched their team and brought it to an empty wagon waiting between our house and the creek. There were usually two men to unload each wagon into the crusher room. If they were not rushed they would sometimes let us "kids" help unload. We were not allowed in the crusher room. When the wagon was empty, one man would hold and guide it by the tongue, back it down the ramp and turn it into the lot beside our house. We had a log to keep it from going back too far. That was always exciting to watch. The interior of the mill had a clean, wet, earthy smell that I like. I liked to watch the big, flat tables that had riffles on them, and a thin layer of water that kept gently shaking the concentrates of ore. Whys and wherefores I know not. I was fascinated by the sampling room—it had a hardwood floor. The piles of concentrates were put into that room before they were shoveled into freight cars on the other side of the mill. Mostly it was lead and silver. They have the warm gray look of pewter when they are concentrates. One time the men in the mill fished out a big roll of reed from the "sump." Evidently some tourist at the pavilion grove up in the west end of town had had it drift downstream. The men gave it to me and I made a basket.

The Schoolhouse was far toward the west end of town. That summer the town's men had been making a walk for the school children, of stone and gravel and elevated a little above the regular road. I heard a girl say, "When it's finished, the men will get a pasty supper." I didn't understand but I just pondered it. It sounded to me like she said "a pass to supper." That was within my knowledge for I once got a "pass" to roller skate in the hall at noon, which was a special privilege in Salt Lake. Do you know a pasty? A good juicy meat pie with potatoes and onions and made oblong. A good size is one that fits into a man's dinner pail. They are good and they belong to the "Cousin Jacks."

I don't know why they called these people "Cousin Jacks" and "Cousin Jennies." They came from Cornwall, England. In speaking they omit the "h" at the beginning of words and often add it to words that begin with a vowel. Since I've done more Bible reading, I am startled to come upon word groupings in the King James version that are identical to phrases I have heard only in the speech of these Cornish folk. The most outstanding ones are in James 4:13 and James 5:1, "go to now"—emphasis on "to."

Just recently I've been reading about Billy Bray. He was Cornish. Whoever recorded his speeches has caught them as I recall them in the folk I knew in Silver Plume. "un" for "it," "e" for "you."

Concerning a three-cornered cupboard that Billy Bray tried to buy at auction for a pulpit:

"The very thing, the very thing. I can cut a slit down the back of un, and strengthen the middle of un, and put a board up in front of un, and clap a pair o' steers behind un, and then the preacher can preach out of un pretty." (Someone outbid him).

"Well, Father do know best."

"I'll be gone down an' tell Father about un." (He followed the purchaser's wagon out of curiosity. It would not go through the doorway).

"Here's a mess," said the purchaser angrily. "I've given seven shillin' for un, an' shall have to skat un up for firewood."

Now was Billy's chance. "I'll give 'e six shillin' for un if you'll carry un down to my little chapel." (Gladly it was done).

"Bless the Lord!" cried Billy, " 'tis just like Him. He knew I couldn't carry un myself, so He got this man to carry un for me."

Quoting thusly makes it clearer than I could any other way. After I was grown, an old lady stopped me on the street to inquire where someone lived; after I told her, she asked me my identity and added, "Be 'e young, or be 'e married?" That tickled me so, because I considered myself both.

I've been in churches of all denominations in Kansas and Nebraska the past two years, but it's hard to find congregational singing as lusty and gladsome as I remember among the Cornish in the little frame church in Silver Plume.

And Christmas! The church tree must touch the ceiling—all the family gifts were brought to the foot of it. During the evening while the program was given, men in the congregation built up for the climax of Santa's arrival exclaiming, "Here 'e comes!" A pause, while everyone turned toward the back door—"No, 'e ain't!" It was wonderful.

Some more things I remember about these people—the most dreaded omen or superstition was "tommy knockers" in the mines. Supposed to precede and warn of a cave-in.

The women showed the utmost deference and carefulness for the comfort of their men, waiting on them in countless ways. I remember one woman who saddled the donkey every morning, led it to the porch and held it until her husband was in the saddle, then handed him his well-filled lunch pail as he set off to the mine.

Those women were just about *all* super cake and pie cooks. Something very special is saffron cake and saffron buns. All their homes were spotless, without clutter, and full of good cooking smells.

We had neighbors who wintered in Silver Plume and spent their summers *high* up, at Argentine. They went over the top of the hill with a donkey train of 6 or 8 animals loaded with their duffle and themselves. My mother took their picture as they set out one spring. I don't know what has become of those pictures.

When we first came to Silver Plume to live, the people who lived in the square house in the right foreground of your sketch on page 131 kept a herd of donkeys in their basement. I was so amazed the first time I heard one bray.

The steep hill behind the two-story building in your sketch had been mowed clear of houses time after time by snowslides, but people kept going on back to live on it. There was a bad one in, I think 1921—walls of broken houses with clothing still hung on hooks, came to rest against that two-story garage and lodge hall. Sheet music from a home I'd been in many times was out in front of our house, but the drifts of snow around the front of our house had muffled out the sound of the snowslide that stopped so near us. . . .

Sunday afternoons in summer, people strolled in their Sunday best down the "Georgetown road." When my family went we took a rope because we had a new baby sister, born in Silver Plume, and pushing the big heavy buggy downhill was fine. But coming home, Mom or Dad pushed, and five brothers and sisters pulled on the rope tied to the front axle.

That baby sister was born during the flu epidemic of 1918. Flu had invaded the remote mines where men lived in boarding houses—as well as the towns—and doctors had long, tedious journeys. It was November and very wintery. Our doctor used a motorcycle. We lived just below the "Georgetown" road's beginning. There was a morning when my mother heard that motorcycle start for Georgetown and realized that Dr. Templeton was headed for the Urad mine, where a full war-time crew of miners needed him. But it was her day, and she knew the doctor would be needed; so she phoned to the drug store in Georgetown, hoping he *might* stop there for medicines. He did. And came back. So little Carol was born. (She married a mining engineer; has been living in Guatamala for 7 years; her baby was born a year ago in Guatemala City, but now they live in a mountain camp that can only be reached by jeep.)

Guess this is plenty.

Sincerely,

Katherine P. Laumann

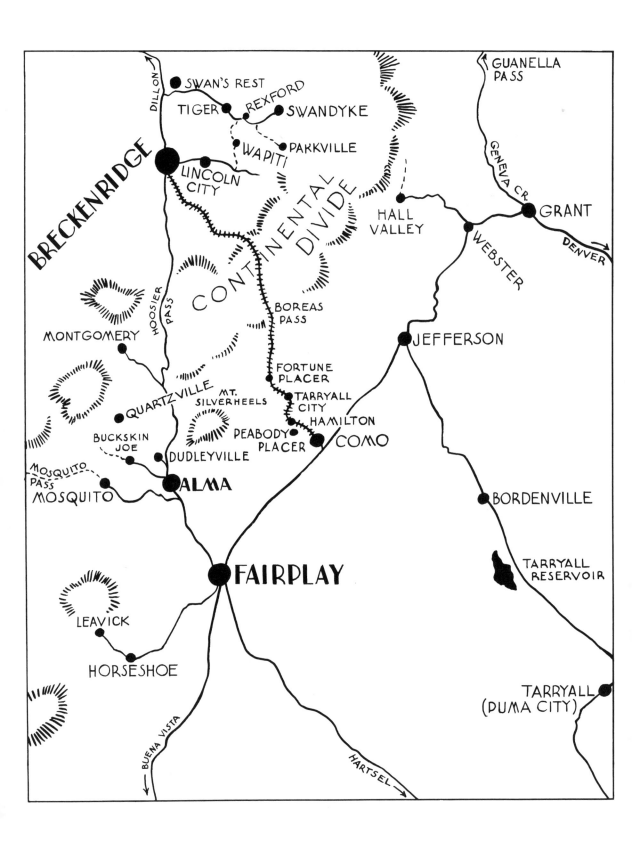

CHAPTER V

Dredge Country

Fairplay lies in the heart of South Park with dredge dumps in its front yard and Mt. Silverheels behind it. It was one of the first towns to emerge in the high mountain valley, and despite booms and depressions it has served ever since 1864 as a supply center for the mining camps that surrounded it. Later on it also became the trading center for the several dredging operations in the vicinity as well as for the ranches which fill the huge 10,000-foot-high basin.

One of my first recollections of South Park was the view from the summit of Kenosha Pass, where I watched a train inch carefully down the grade to the valley floor and steam southwest through Jefferson toward Como and Fairplay. I had discovered the roadbed of the narrow-gauge at Bailey and had seen how it skirted the North Fork of the South Platte as far as Grant and then began its circuitous climb to the Pass. A short distance west of Grant I'd also looked for the small railroad town of Webster at the mouth of Hall Valley, but the highway had obliterated the townsite. Needless to say I was glad to receive a letter from Harold T. Smutz (April 7, 1951); it described the Webster-Hall Valley area as he had known it.

Webster and Hall Valley

. . . For a time after World War I, and after my graduation from engineering school, I was connected with metal mining in Colorado; one of the mines was the Whale which you mention on page 102 (Stampede to Timberline). It was reopened in the latter days of the War and a 50-ton flotation mill was constructed. The road through the lower part of Hall's Valley—as we always termed it—was not too bad; the grade really began at the Shayler boys' cabin, and you can imagine the difficulty of getting ball mills, tables and other mine and mill equipment up to the Whale which is above timberline. It could only be operated during the summer months because of the heavy snows; the road had to be cleared of slide rock each spring before the teams and wagons could get up there. It was an expensive operation.

In those days there were some old kilns at Webster, where we used to stable our horses, most of which we kept down there so they could come up light in the morning and go down in the afternoon with the ore wagons loaded with lead and silver concentrates. There was a telephone at Joe White's in Webster, but we had to go to Cassell's, below Grant, for our mail. The old narrow gauge is gone now, but I can still see the freights, with three engines, laboring over Kenosha Pass. Once, when it washed out, I rode over the Divide into Montezuma for snuff and horseshoes, and such sundries as I could carry on horseback, following the trail you mention. We were down to making bread of Cream of Wheat, but the men seemed to miss their snuff and tobacco more than they did food, although I made another trip and bought a sheep from a herder. My horse certainly didn't relish the idea of having the carcass in front

49

of the saddle, but we finally made it. We were happy when the trains started rolling again.

The Shayler boys, two of them, worked for us for day's wages during the season so they could hole in during the winter and work their own claim which was even higher on the Divide than the Whale; everything they took into it had to be packed in. As far as I know, they never made a dime out of it.

I have also visited many of the other places which you mention, and it is like renewing old friendships to read about them.

Fairplay, 1953. Everett Bair, Park County historian (right), and Alex Warner at courthouse; stone jail (left background).

Fairplay   On my many trips to Fairplay no one was more helpful than Everett Bair, historian of Park County and for many years its deputy treasurer. He was a writer of short stories and articles on the West, as well as the author of a novel, *West of Pikes Peak*. He knew details of local history that were new to me, such as the location of the two smelters at Dudley, the older—a wooden structure—at the east end of the town, and the concrete foundations of the newer one further west. He spoke of another smelter at the mouth of Sacramento Gulch, built on the meadow where Indian tribes had gathered each year throughout the 1830's and 1840's to trade. He also mentioned how Democrat Mountain got its name: during the mining era, a black prospector, who was a good cook, had a cabin on the mountainside close to the road, and teamsters would stop at his place to get food. As they didn't want to call him "nigger," they spoke of him as the "Democrat," and soon the mountain was known by that name.

More than once I spent an evening in Everett Bair's home listening to his vivid portrayal of historical events. On one occasion, having heard that I was in town, he even came over to the Fairplay Hotel after dinner on the chance that I might be in the lounge and have questions to ask him.

Prune's grave in Fairplay, next to the Hand Hotel, is unique, for, in addition to the monument which memorializes a burro that worked in so many of the mines of the district, it contains also the remains of Rupert Sherwood, the animal's master. Sherwood, who died in the Fairplay hospital, requested that he be buried in the same grave with Prunes, and his wish was carried out.

Fairplay, 1942. South Park Hotel.

In an entertaining letter (Jan. 30, 1954), Russell K. Havighorst, Colonel of Artillery, USAR, writes of Sherwood and of Baby Doe Tabor, both of whom he knew:

> . . . It is interesting to note that R. M. Sherwood seems to have carried the name Rupert during at least the last half of his life. He was originally Reuben Sherwood, a wiry tireless quiet individual, who could take care of whatever situation might arise. He used to ride a fine buckskin horse groomed to perfection, and rode him so well that Bill Anderson, who edited and printed the Park County Bulletin in Alma, entered a personal item that Prince Rupert Sherwood had come down from his claim in north Mosquito to buy some equipment at Moynihan's store. Many residents figured that Rube wouldn't like it but it seemed to tickle his vanity a bit and he signed himself Rupert for what now appears to have been the rest of his life.
>
> He was a dead shot with his 38-55 singleshot rifle, which with its big octagon barrel weighed about as much as a 105 howitzer. He was also an expert with a divining rod or "Dowser," as it was known in those days. He tried to teach me to use it and was quite disgusted when I could get no response whatever. In his hands I could distinctly feel the pull on his wrists, but

there it ended. I may say that in spite of an engineering education and its practice for many years I have never seen a scientific explanation for the dowser nor can I formulate a lucid one of my own. Be that as it may I have seen it work—not once but many times. I cannot question its effectiveness.

My father knew H. A. W. Tabor well in Leadville and after. Tabor did not add to his stature nor reputation when he took the little actress Baby Doe as his wife. However she always deported herself modestly in Denver and other cities and was generous and considerate. In 1934 I made one of my rare visits to Leadville and learned that she was living at the Matchless in destitution. I asked some questions and was told that it was not healthy to visit her as she might cut down on callers with a shotgun. I took the risk—and the precaution of calling to the tiny cabin before I arrived. Mrs. Tabor came to the door and upon hearing that I had known Mr. Tabor invited me in most graciously. During the ten minute chat I sought for ways of helping her out but was informed that there was always plenty "of everything," a statement not borne out by appearances, although the housekeeping was immaculate and the general impression one of the dignity of life.

Horseshoe and Leavick

In 1941 I reached Horseshoe and Leavick, small deserted camps in the mountains just southwest of Fairplay. At that time both contained buildings. Today the Four-Mile Creek road, a short distance beyond the South Platte river crossing, leads to the sites of both of them and on beyond to the Hilltop mine. Carl F. Mathews of Colorado Springs (Jan. 21, 1951) provided me with excellent information about both towns and pinpointed where East Leadville had been:

Leavick and Horseshoe were two separate settlements, East Leadville being a station on the narrow-gauge line to the Leavick mill and about 2½ miles east or down stream. It is listed in Crofutt's "Grip-Sack Guide to Colorado," which is one of the best authorities on the towns of the state in 1881, the year it was published. Men who came over the range from Leadville in 1879 naturally named it East Leadville. Later it was called Horseshoe. There were lots of cabins all around and a smelter at the end of the meadow.

The camp was founded by Col. J. H. B. McFerran (grandfather of Mrs. Dorothy P. Shaw of this city, who passed away just recently). Crofutt says (1881) "It is a new town, population 300 (in the summer months 800?). It has 2 stores, 2 hotels, 2 saw mills, 1 smelter, also another smelter in course of construction. Teaming, lumbering and mining are the principal occupations of the people. The minerals of the vicinity are very rich, some running as high as 1,000 ounces of silver. The ores are carbonates, galena and sulphurets, also some iron and galena mines. Principal producing mines are the Peerless, Badger Boy, Last Chance, Dauntless, Crusader, Mudsill and Sacramento. The stores are credited with doing $5,000 a month business each. Hotels were the Leadville and Palmer."

On April 15, 1880, the "Fairplay Flume" said: "A new hotel of large dimensions is to be built soon by John Grazier." On April 22, 1880, one Hanlon is named as proprietor of the East Leadville Hotel. On June 24, 1880, the "Flume" had the following item: "The demand for supplies is so great that a store is demanded and Messrs. Sykes and Hanlan will supply this want. In fact a large quantity of goods was sent up last week by J. W. Sykes of this

Leavick, 1942. Sheep Mt. in background.

place (Fairplay) and were opened in a new building provided by George A. and M. R. Hanlan. The Hanlan Bros. also own and operate the East Leadville House, which is doing a large business; from 20 to 50 are accommodated nightly."

The Colorado Business Directory of 1881 lists the following business houses: Henry W. Allen, General merchandise; Crusader Mining Company; De Garmo & Palmer, Hotel; East Leadville Mining & Smelting Company; Hanlan Bros., Hotel-General merchandise; F. D. Hill, Hotel; C. M. Van Cleve, Assayer; C. Wagner, saloon.

I have the day book and single-entry ledger of some storekeeper in the town for 1882 and the entries are most interesting. (Too bad he did not place his name in the books).

I well remember the night of May 30, 1927, which was spent alone in the deserted cabins of East Leadville. Pitched my tent by the side of the old Model T and camped on the ground; about midnight a terrific wind came up (or rather down) from the high peaks on the north and began to whip the tent like a whip-cracker. I crawled out and gathered up all the old iron lying around and pinned down the tent with it. Awoke in the morning to find about six inches of snow on the ground and my canteen on the running board of the car was frozen solid. Got a fire going and thawed things out, then cooked breakfast and hiked on up to Leavick; back to the car in the afternoon and pulled out to find hardly a trace of snow in South Park after reaching Fairplay. At that time there was a two-story store and post office building, as well as several others and the rails were still in place on the narrow-gauge (this was not a branch of the Colo. & Southern, but had been built by the Hill Top people to handle ore, etc.). A Denver couple bought the buildings a couple of years later and started a summer camp for "dudes," but did not last long; last year [1950] there was nothing there but a few foundations.

Leavick was a perfect ghost town when I saw it in the early 1940's: many frame buildings, at least two mills, and a tram with buckets which connected the Hilltop property some distance beyond and above the townsite with one of the mills. In the 1960's I again drove to Leavick only to find a government marker: SITE OF LEAVICK. 1882. The buildings were gone.

Frances Smith, who lived in Boulder in one of the quonsets provided by the University of Colorado for married students during World War II, wrote (Sept. 28, 1952):

> . . . Some time ago my father conceived a plan to reopen the Hill Top, consolidate all the adjoining property under one management and overcome prohibitive shipping costs by drilling a tunnel through the mountain to Leadville. It was too big a plan for one man, and he no longer controls it; but the tunnel is daily pushing deeper into the mountain and the great iced chambers you spoke of are accessible again. In the dim light of a carbide lamp, they are immense and awe-inspiring. . . . My uncle is president of the company and I could probably get permission for you to see them if you are interested.

Park City,
Mosquito,
and
Mahoneyville

From Fairplay west to the Continental Divide on Hoosier Pass, State Highway 9 provides access to what were once a number of flourishing mining camps. Just before reaching the outskirts of Alma, a road cuts off to the northwest through what was Park City and then through Mosquito, on its way over 13,188-foot Mosquito Pass.

Russell K. Havighorst, whose letter (Jan. 20, 1954) I excerpted earlier, also comments on the naming of the mining camp of Mosquito:

> . . . I lived many years in the mining camps of the West (Colorado particularly). . . . You have been remarkably successful in winnowing the material for accuracy. . . . Many of those old sour-doughs lose the partition between their memory and imagination early in life and some of the tales told of events which I myself witnessed show the specter of old Ananias himself throughout the whole recital.
>
> For example your account [*Stampede to Timberline,* p. 92] of the christening of the Mosquito Mining District is entirely factual, though perhaps it does not go far enough. The book in question was a large leather bound

ledger in which the resolutions were of course hand written. That pertaining to the naming of the district had been written out beforehand, leaving a blank for the insertion of the name. It was right in this blank that the unfortunate mosquito was crushed and when the name was later inserted the outline of said skeeter was traced also. Doubtless this book was long ago lost but I was shown it and its entry about the turn of the century. I never forgot it.

I was fortunate to have an interview in 1958 with Havighorst, whose father was one of the group that set up the Mosquito Mining District.

"I knew Rupe Sherwood, the man who owned Prunes, when he lived near Park City," he began. "He was a small, wiry, indefatigable man. He had a clay bank pony and rode like part of the horse itself. He hiked a lot too, and thought nothing of running over to Leadville in a day and back.

"I remember once he was in front of my dad's store in Park City. He'd killed a deer walking on the far side of Mosquito Creek, at a range of four hundred yards. He used a single-shot rifle with a charge of black powder—it was a good deer gun. No one thought he could get the deer, but he killed the buck with one shot. We kids went over and dragged it in and skinned and dressed it.

"Rupe lived in a house alone above Park City, by the road. I saw him there in 1936. He went up to the North London mine—he had a gravel spot where he worked when he was down on his luck. He put up a cradle and washed out three or four hundred dollars worth in a season. He'd go in the first of June and quit placering in early September. Three or four hundred dollars was enough to keep a man for a year in those days.

"Beyond the town of Mosquito the road forked, with a branch following both North and South Mosquito Creeks. About one hundred and twenty yards east of the fork was Mahoneyville, founded in 1872 but not incorporated by E. X. Mahoney. I remember him when I was a kid. The 'town' was just his own place—a couple of woodsheds and tool houses. He worked in a nearby mine.

"There was a railroad from the junction to the North London mine. The right-of-way was along the bottom of the cliffs of Loveland Mountain. Everyone expected the country would bloom with mines. The Black Flag was in operation at that time. It had an eighteen-stamp mill and it made a million dollars. Then the ore pinched out. There was a wooden flume from the mine to the creek and a tram down to the mill. The Hock Hocking was another good mine not far from Park City.

"The old stone mill, founded by L. S. Smith and John Conrad—Smith's name was on the cornerstone—was a stamp mill, but no stamps were installed. The men went broke building the mill. Another pair of adventurers brought in a boiler, with a cast iron base for the slide-valve engine. That was all they could do before their money ran out. The next group set the thing up, mounted the boiler, and didn't invest a dime—and quit. From then on the stone mill just stood there. In 1900 the roof caved in and the boiler was buried under it. I was eight years old then, and we kids played in the ruins. The next group who bought the mill brought in a boiler with an eight-horse team. The mill was intended to do work that the Black Flag mill continued to do for years. The last property that dad had was the Langtry mines up on the cliffs of Loveland Mountain."

Mrs. J. B. Richardson of Granby shared another experience from the past (June 10, 1950):

While doing test work for a plant at old Mosquito I had the pleasure of showing two old ladies around. They had lived in Park City in the early 1870's or so. They were old and frail and in true sister style contradicted each other as to place, but they told me the old Park County records were once at Park City, they had been there when the town had been named, and said one of the big stage stops on Mosquito Pass road to Leadville was there, and there was a huge hotel at one time and quite a town.

There was nothing left that they knew except one log building, sided over, which had once been a school.

Alma   All the original diggings along the South Platte River northwest from Fairplay to the site of Montgomery at the foot of Hoosier Pass were placer camps. At Alma, five miles from Fairplay, the river had been worked and reworked by placering, hydraulicking, and dredging for the gold that lay hidden in its sandy bars and banks.

The original main street, still lined with some of its older houses, stands on a higher level than the present highway through the town. Today Alma's population is small, but at the turn of the century it had at least one newspaper, as described by Russell K. Havighorst (April 3, 1954):

> The real purpose of this letter is to suggest to you that any remaining copies of the Rocky Mountain Hee Haw might furnish material for future research. The only reservation I make is whether or not a single copy remains. I was quite young and I do not recall the purpose of the rather non-descript magazine, which was about 5″x7″ and never carried more than ten pages. The organizer—name unknown—came to Alma in about 1906, gathered much material and printed several authentic stories of the Shelby, Hocking and other mines. I saved them for years but finally lost every one of them.
>
> The verse on the cover page of the first issue should have survived, for it was dedicated to the Colorado burro. It is my loss that I have forgotten it or could not save that one issue; for it could well fill in as the theme song for all prospectors since the beginning of time. Its last verse ends with the line—"that song of patient hope, harken to his hee-haw mountain song." The public Library in Denver knows nothing of it so it is probably forever gone.

Buckskin Joe   The last time I saw Buckskin Joe the cemetery was its best preserved area. As I drove up the hill to its shady grove, I met a carful of people who had left flowers on one of the graves. Even in 1949, according to W. G. Mussey, Jr., "there were just two cabins left, one on one side of the valley, which is supposed to have been Tabor's grocery store, and the other on the other side of the creek was an old dance hall." The Tabor store is no longer there, but the dance hall-saloon was still standing. Speaking of Tabor, another informant wrote: "Over the crest of Hoosier Pass on the Blue River side, about two curves down, was H. A. W. Tabor's log cabin." I was never able to pinpoint it.

Montgomery   At the end of the valley, in amongst the trees at the base of Mt. Lincoln and spilling down onto the marshy flat below the present highway over Hoosier Pass, stood Montgomery, a camp of the sixties which on one occasion polled the most votes that were cast in the whole mining district.

When I investigated what was left of the camp in the 1940's, and scrambled up a dirt road to a high valley behind the present Magnolia mill (which is visible from the highway), I discovered cabin frames, foundations, and hollows marking

Buckskin Joe, 1966. Dance hall.

Montgomery, 1942. Mayflower mill at head of valley.

where buildings had once stood. From the mill I looked down to the valley toward Fairplay and saw three or four skeleton cabins and rough tracks on the ground which marked former streets. At least one of the buildings had been preserved.

Martin Rist, professor at the Iliff School of Theology, Denver, wrote in 1966:

> In 1867 Rev. John L. Dyer purchased a log cabin "hotel" in Montgomery. It was moved log by log to Fairplay to be used for a Methodist Church and parsonage.
>
> Recently our Rocky Mountain Historical Society had it moved to a new site, where it is preserved as a historical "shrine."
>
> Is it possible to identify this Montgomery Hotel? When was it built? Who was its owner? What was its name?
>
> On p. 82 of *Stampede* you wrote: "at one time Montgomery had three streets, a two-story hotel, and the largest theatre in the region."
>
> Possible, but not definitely, this is the hotel Dyer purchased and had moved to Fairplay, in 1867. Do you know anything more about this hotel?

I do not. Maybe some reader can supply further data. After its first boom Montgomery was quiet until 1881 when silver was discovered on Mt. Lincoln and Mt. Bross and mining activity resumed. After the demonitization of silver in 1893 most properties closed down and little work has been done since.

The valley is changing. From Sewell Thomas, a mining man from Denver, I learned (May 8, 1957):

> . . . Five years ago (1952), as consulting engineer for the Colorado Springs Blue River Diversion Project, I located a storage reservoir over the site of Montgomery. This dam is now built and Montgomery is under 80 feet of water. I wonder, did you investigate the little graveyard there?

Mr. Thomas added:

> My father, Governor and U. S. Senator, had his early days at the practice of law in Leadville, from 1870 to 1884. He was attorney for one side or the other, usually opposing Charles J. Hughes, in all the important mine litigation all over Colorado, in the Coeur d'Alenes; Butte, Montana; in Nevada and many other places. He was attorney for Tabor, for the Guggenheims, Dave May, Levi Leiter and many others. He handled the Camp Bird Mine sale for Tom Walsh, who ran a large Boarding house for miners at Leadville before he stumbled on the Camp Bird. His stories of the early days were many; they were tragic and amusing and everything in between these extremes.
>
> I think you have caught the spirit of the "old timer" in his exaggeration of the mines and their values, size of veins, etc. Mark Twain describes a mine as a "hole in the ground owned by a liar," and no definition was ever truer.

Dennis F. Smith of Denver who has tramped over many of the South Park mining areas wrote (Aug. 19, 1970):

I am a native and have visited various old town sites such as Quartzville in the Alma district. Many of the old miners I knew are gone now.
I remember walking up the Dolly Varden cable line many times from the
mill and smelter. One gentleman, Chester Coan, now living in Colorado
Springs, was one of the last to pack ore on a burro down from Mt. Bross
in the 1930's. I believe they were working at the Moose mine.

Today Real Estate Developers are subdividing the valley. Already
Dudleyville has been torn down. Alma Park Estate Realtors are selling
five-acre sites and the [lower] Montgomery townsite is being cut up into lots for
summer homes.

Como

About ten miles northeast of Fairplay on U.S. Highway 285 beyond Red
Hill Pass is Como, a once important coal and railroad town, which prospered
during the 1880's and 1890's when it was also a division point for the narrow-
gauge trains that crossed Boreas Pass to Breckenridge. The roundhouse was
located there, and fuel for the engines came from a nearby coal bank owned
by the railroad. Even as late as the 1930's a good many people lived at Como,
but today only a fraction of its original buildings remain, and those that are
occupied serve chiefly as summer homes.

In 1962 Mrs. Luella McGonegal of Fort Morgan wrote of Como as she
remembered it:

I was born in Como, Park County, 1885. My father and mother
were married there. My Dad was a conductor on the South Park narrow
gauge railroad. It ran from Denver to Leadville, Gunnison and Alma,

Como, 1941. Railroad hotel and station at right.

and Como was the division point. They had big railroad shops there—employed several hundred men besides all the men who worked on the trains. I had an uncle who was Passenger Conductor for 32 years from Denver to Leadville. Have you ever taken that trip up Platte Canyon over Kenosha Pass, then you drop down into South Park where they raise all that fine wire grass hay. . . . We made a trip to Como two years ago and I went to the school house and sat on the steps where I was taught my ABC's, and shed a few tears.

We lived in Victor all during that awful strike. My dad was yard master there. I remember well the blowing up of the depot. They wanted to get the train but they were too slow. My son was born in Victor, 1905.

I love the mountains and would like to live again in Victor but I am old now and alone. I just have a limited income, I get the pension, and the price of everything you buy is so high. We just about have to count our pennies.

Tarryall   Clinton W. Kanaga of Shawnee Mission, Kansas (Oct. 30, 1962), kindly sent me a copy of a letter written in 1860 from a young man spending his first winter in a diggings near Tarryall on Deadwood Creek. Kanaga's letter reads:

For years I have been interested in early Kansas history but have never known much about that part of Kansas Territory which became Colorado. Such material is scarce. I have but seven envelopes and one letter from there in that period. However that letter is interesting so am sending you a copy of it to add to your Colorado Americana.

Dead Wood-Kansas Terry-Dec. 7, 1860

Dear Mother & all

I rec'd a letter day before yesterday from Father, Sam'l & Harvey. I have not time to write to each of them just now and I fear if I should write to *one* of them the others might be *jealous* or something else, so I seat myself to trouble *you* with a few foolish ideas, etc. etc.

I presume you are pretty acquainted *about my whereabouts, circumstances,* etc. for I think I have written half a dozen times all about them—still Father writes to *be more particular.* Well now you just tell him—these *few facts*—there are four of us in company—we have a cabin 12 by 14 feet—are 75 miles from Denver City—75 from Pikes Peak, 4 from Hamilton and 3 from Tarryall where at the present time our nearest neighbor lives. We intend to live through the winter if we can make our living which we are doing now, and a little more—but not getting rich very fast. Guess there is little danger of my freezing, starving, or getting "snowed in"—there is a good supply of provisions in town, etc. etc.

By the same mail I rec'd a letter from Frank Morgan which gave me a very definite history of affairs in W_____ particularly of the young folks, the marriages *past* and in *prospect* and I conclude there is no chance for *me there*—so I shall remain away awhile—perhaps some of them will become widows sometime.

I think of you every *Thursday*—think perhaps it is Thanksgiving Day—should like to be with you—I would *almost* promise to be home *next* Thanksgiving but I have told about going so often that I shall say no more about it

until I get ready to start—but I *really think* one winter in the Rocky Mountains will do me—snow 18 inches deep and more coming most of the time.
I should judge from the present appearance that we shall have but one snow
storm that will last about 7 or 8 months. We have had until within a few days
much finer weather than we anticipated. If we had a few of your *potatoes*—
and *apples*—a little of your *butter, cheese, cider* and *sugar* etc. I think we
could *appreciate* their *merits*. Since I left Kansas City I believe I have eaten
nothing that the *"Softer Sex"* had *paddled*.

There are but few persons in the mountains compared with their numbers
during the summer. We look for them back in the spring—still there are
enough remaining to support a *store*, P. O. etc. There is a Debating Society
at Tarryall—and in *Spring Gulch* in a log cabin covered with dirt a *real
Melodeon. Ain't* we civilized? There are about half a dozen women down to
town. Dancing parties are quite common— . . . .

I suppose I can write but little but what you have heard before—there is
but little interesting transpiring in the mountains—occasionally we hear of a
shooting affray—or of one's freezing a little etc.—(I see in the newspapers
accounts of suffering in some parts of Kansas). We hear nothing about it here
but what comes from America— . . . .

Make them write. Give my love to all and accept this from
Yours truly
H. F. Dix

Dec. 13th
Since writing the above we have heard of two *trains* for Hamilton, and
one of them has 200 bbls of flour so that provisions are likely to be very
plenty. It is worth 40 dollars per bbl now. (Our *snow storm* has *gone over* and
we are all right again). I go to town tomorrow for mail matter and will mail
this. Still in good spirits.

Henry

Boreas Pass

The rugged trail over Boreas Pass, west of Como, was a much-traveled thor-
oughfare across the Divide from 1859, when placer gold was discovered in the bed
of the Blue River, until the completion of the railroad in 1882. Without this route
the inhabitants of Breckenridge and other mining camps that were emerging on
the Western Slope would have been completely isolated. Even maintaining mail
service over Boreas Pass was extremely difficult, especially in winter when snow all
but closed down the road, year after year. In 1938 the last train ran down the grade.
This last run is referred to in a letter from Fae (Mrs. Ben) Kuhlmann (Aug. 13,
1961):

> Mrs. Geo. Champion at Como showed us George's collection of relics of
> the old R. R.—he ran the *last* train—the day we called he was gone, on a
> jeep trip with Caroline Bancroft, trying to find the old town of Dyer. Father
> Dyer's church is now at Fairplay.

It is only nine miles from Como to the summit of Boreas Pass. Since 1953,
when an army battalion and the U.S. Forest Service got together and made an
automobile road out of the narrow-gauge right-of-way, it has been a scenic shortcut
to Breckenridge. Upon leaving Como the Road crosses dredge dumps which nearly
fill the Tarryall valley and then swings west beside them for a short distance to a

sign marked "Peabody Switch." At this point the road makes a big curve and begins its ascent to the pass.

Peabody Placer   The name "Peabody" struck me, for in studying an old map of placer camps on the Tarryall, I'd seen the words "Peabody Diggings." Then in 1950, when at the gathering of pioneers in Georgetown, I had heard speak Lady Blythe Marvin, the "Original Mechanical Doll of Stage and Screen." (See previous chapter, under heading Georgetown.) I made no connection between these two items until eight years later when I received a note from the lady with an enclosed postcard picturing the Doll. I learned that her grandfather, Leland Peabody, was one of the first to work placer claims on the Middle Fork of Tarryall up Deadwood Creek, six miles northwest of Como. His claims were extensive and ran on both sides of the creek for about two miles. Her letter:

May 23, 1958

My Dear One,
    Just a few lines to send you my Doll picture. And I Hope that you will accept it. I read parts of your book *Stampede to Timberline*. I am so heart broken that some of the Peabody's were not around at Peabody Placer out of Como. Grandpa Leland Peabody was the first to have chinamen work in Placer in '59, he had three hundred at one time. But nothing in your book. But perhaps all were gone when you were gathering Data. . . .
    I am Very Truly Your Doll.

Lady Blythe M.

Boreas Pass   I have a letter (Nov. 13, 1954) from James A. Anderson of Fort Collins who
cont'd.   spent years railroading over Boreas Pass and whose father worked as blasting foreman on the South Park line:

Thanks very much for your letter. You wanted a bit of information about the Pass from the time the railroad operated over it.
    The Pass on top of the Divide is located 11 miles from Como and 12 miles from Breckenridge. The first station, 3 miles out of Como, was *Peabody,* a gold washing project—hydraulic—very good at that time, first clean up May 1913. A gold brick weighing 9 pounds was shipped from there by train—heavily guarded. Next station was *Halfway*—only a siding, then 4 miles west was a watertank—only a stop called *Selkirk.* Four miles from there to top of the Pass. On those four miles were four snowsheds, the last one on the top of the hill ½ mile long—Depot, Siding and Repair Shop for R.R. in the same structure. The only outside buildings were a *Section* house and small cabin for the R. R. Inspector.
    On the western slope about the only stop was *Baker's Tank.* Further down on Nigger Hill was *Mayo*—a gold camp. The trouble with this gold was that it was flour gold imbedded in Blue clay, very hard to get out. The next station was *Puzzle,* then *Gold Pan* and last *Breckenridge.* West from Breckenridge was the Tonopah Dredging Diggings, very rich. The first clean up they got 14 lbs. of gold.
    On top of the Pass were two mines, one on Little Baldy, owned by three Swede Brothers. It was above timberline a couple of hundred feet—11,950 (elev.) Straight south of the Pass is a mining shaft with all the buildings on the left. In my days I was afraid to explore it, it was very deep.

I had charge of the Rotary snow plow from Como for two years which meant keepin' those little R.R. branches in running order. I know nearly every foot of the country up Platte Canyon. My father helped to build the So. Park Railroad, 1879-1882—that is why I hate to see the old landmarks torn to pieces—there should be a law prohibiting it.

There is very much can be said about the country up there. I know every foot of it in Park and Summit Counties and many people that lived there. In 1910 Boreas Pass was closed until 1912 when the C.&S. R.R., then the owners, received a Court order to open the road from Fairplay. At that time I was Engineer Extra on passing No. 70-71 over what was then called the High Line, far as I know there never was any other Road than the R.R. The passing crew had orders if they found anyone walking to pick them up and bring them to safety. I was up there until 1916.

Far as Leavick is concerned—Dist. 14 miles from Fairplay—the first stop was *Tie Landing,* operated as a tie camp by Sam Cohen whose office was in the Miners Exchange building in Denver. Millions of ties were shipped from there. Next was *Old Town*—a mining camp at one time. There are still a lot of tunnels and old mines further back in the hills—mostly lead. Then we come to *Mudsill,* a short distance east from Leavick—a number of prospector's cabins left—mostly from Tabor's times.

Well Mrs. Wolle, this puts you on top of timberline again on 2 places. During my times up there I could have written a book to cover a lot of Happenings in those hills. One thing I will say is that in Park County our laws were a sheriff and the People, at that time mostly the people, and I believe that everyone was more behaving then than they are at present time.

See what you can find out and do with an old camp (Coal Town) called *King,* owned by Union Pacific R.R., 3 miles S.E. of Como. . . . I will leave it to you to study.

No one was more helpful nor sent me more complete information about the problems connected with railroading over the Continental Divide than did W. C. Rupley, chief train dispatcher for the Colorado & Southern at the beginning of the twentieth century.

June 14, 1946

. . . My knowledge of the ghost mining towns in Colorado is . . . from the angle of my connection with the Colorado and Southern railroad. I think the earlier name of that South Park division was THE UNION PACIFIC DENVER AND GULF (U.P.D.&G.).

In my former letters I mentioned the fact that the old South Park line ran from Denver to Como, then a line branched off to the left to Gunnison, and on to Baldwin coal mine. That is really the glamour line to me. The other line ran from Como through several mining towns to Leadville. This latter branch was a more important branch from the standpoint of business and revenue than the Gunnison line. Leaving Como on the Leadville line, the first railroad point was "Half-way" but there was nothing there but a side track for trains to pass, and there was never any mining activity near there that I am aware of. Next came Boreas on top of the Continental Divide. I doubt if anyone lived there except the section crew. It was a terrible area for snow in the winter. The operator in the depot didn't shovel paths or cuts from the

Boreas Pass, 1966. Railroad section house at top.

door to nearby points, he tunneled through when it got bad. There was a real advantage to that, even though the initial labor was much greater, for after he had his tunnels established it could snow all it wanted to but the tunnel remained undisturbed, while cuts and paths would "blow shut" in—well, sometimes a few minutes. The cold and blizzards up there were severe. On many occasions, when four engines were coupled together equipped with snow plows, 2 headed east on the east end the two on the west end headed west, would roar from Como to Boreas, when they actually tore by Boreas, the great volume of snow they would throw out each way would knock the depot door in and literally almost fill the room with snow. That would be a mess for there was no place to put the snow. Leaving Boreas westbound, about nine or ten miles below we came to Breckenridge, which relatively was some town. I never did know anything about its real size but RELATIVELY it commanded quite a bit of respect; possibly, to some extent at least, because of the connotation of its name. (A few prominent men have had that name).

Around Breckenridge was some quite pretentious placer mining. Even at that time they seemed to have quite an extensive plant and the "head" of water which they had harnessed and appropriated was the talk of the whole immediate country. I do not know whatever came of it as a mining project. Incidentally there was some work being done between Boreas and Breckenridge, but it never attracted much attention in my time. It was all 4% grade from Como to Breckenridge, which is a very heavy R.R. grade. The South Park division had sixty (60) miles of 4% grade, a great deal more than that 3% grade and much 2½, 2, etc. Breckenridge west the grade eased down through Braddocks, maybe 4 or 5 miles west of Breckenridge, Braddocks being only a passing track for trains. Next came Dickey, and from Dickey it was up hill westward toward Leadville and also eastward toward Como. From

Dickey a branch went to Dillon back in the hills about 2½ miles as I remember. A rather sequestered place as I mentally pictured it, although I never saw the town. I do not know why the railroad went there, it may have a "history" antedating my sojourn.

Dillon does however bring to mind an incident that I do not forget. One of the very famous artists, I wish I could remember which one, came to Dillon one summer to hide out and to paint. It may have been James Montgomery Flagg or maybe Frederic Remington, or some one else. He probably was accompanied by his wife and family, if any, but I do not know. In getting his house-hold goods in there (They came by railroad) they were pretty badly damaged. This artist wrote a long, several page letter (a claim for damages) to the general office of the railroad company, and in his delineation of the damage he set aside liberal areas along both margins of each page, alternately, and drew pictures of the various pieces of furniture showing the nature and extent of the damage to each. The whole thing was a classic and it took a lot of will power on my part to restrain myself from "losing" that letter when it passed through my hands. His comments along with each picture made one of the most convincing and compelling claim letters I have ever seen. I do not know what ever came of his claim, but he should have been well paid if only for his superb effort and portrayal. Some of the furniture, as he pictured it; for example a bed-stead showed a face in the wooden bed end on which was portrayed agonizing pain due to the rough treatment it had suffered. A chair racked with pain and minus two legs, was trying to hobble along on the other two, and one of the legs broken in two and the two ends lapped over and tied together with a string, crouched in a suppliant posture with an outstretched hand holding a tin cup in which he hoped the railroad would put something to mend his broken members. It would have to be seen to be appreciated.

Going on from Dickey toward Leadville, there is some very choice scenery for a few miles. There is a side track 3 or 4 miles west of Dickey but I have forgotten its name. There is a beautiful cascade along there also. The water cascades over bright hued rocks, it seems to me of a greenish-yellow tint. On up the hill is Kokomo (accent on the first o). By the way the population of Breckenridge, I just learned, is 375. Kokomo is about 25 or 40 people. One winter the mother-in-law of the section foreman died there in her home. The snow was very, very deep and it was practically impossible to get the body out or to find a way to bury it, so they made a rough coffin and buried her in the snow until Spring. The landscape sort of leveled off around Kokomo. On toward Leadville, from the train, off to the right one got a fine view of the "Mount of the Holy Cross."

Five or six miles west of Kokomo was Climax, which as its name suggests was the top of the hill for the railroad. To the left from the train was the Saw-tooth range of mountains—impressive, and back in some of those draws is where General Fremont camped one Winter and played hide and seek with the Indians. I will not guarantee my history, however, but the statement is generally true I think. We had a telegraph operator at Climax. I was for a long time incredulous when, in getting the afternoon weather report, say in the Spring and Autumn, he would give me 45 degrees above zero a half hour before sundown and then half an hour after sundown he would give me

15 below—a drop of 60 degrees in less than an hour. I made it a point to check on it and found it to be fully verified.

About 1½ or 2 miles on down hill on the R.R. toward Leadville was a place called WORTMANS. Named after George C. Wortman, rather a strange character who lived there and did some prospecting. He had some ability as an inventor. Wortmans had a side track, or rather a spur, for it was connected up with the main track at only one end, and I think it would hold only 2 or 3 cars, it was that short. Wortman seemed to try to develop mineral there. During the war I understand it became a very important mineral producing point. There was mined there some "rare" ore [molybdenum] that had a peculiar value to the war effort. I mean the recent World War. I do not know what that ore is, but I happened on to an article about the activity. Eleven miles on beyond Climax is Leadville from which came almost all the year-round tonnage from that line for the railroad. Those mines in Leadville were then in good production and the total tonnage was divided up and allocated so much to each of the three railroads, the South Park, the Colorado Midland and the D.&R.G.

I was never in Leadville but once and I slept at the Vendome Hotel, a famous old hostelry then and more especially earlier, but it would be a sorry specimen now, for a place to stay. Gambling and other characteristic frontier activities were wide open. Some of the gutters in Leadville along the curbs were walled up and water, clear as crystal, rushed through. The ditches, walled on both side, were probably 36 to 40 inches wide and 8 or 10 feet deep and quite a volume of water was running in them. It seemed an ideal way to dispose of waste water, litter of all kinds.

Many of the way-bills for that ore carried the name of a shipper that I well remember, a name one might expect to emanate from that setting. All there was to the name officially and on the way-bills and on their letter head etc. was R.A.M. A poor old miner, so the story goes, had struggled for years to develop the property and he dressed in rags and lived miserably, but finally it became a big producer and out of all that came the name—R.A.M. which is eminently appropriate. The property may have been worked out years ago.

For your Pictorial Record I am afraid that what I am relating has questionable value, unless a little of it might be used for filler. It would give me pleasure to write what might add to your work or your production, but I am not at all optimistic about the value of what I can offer.

Sincerely,

W. C. Rupley

Breckenridge

It is hard to describe the mining camp of Breckenridge, for today few of the old landmarks have survived the surge of building that has produced the present modern ski resort. Even some of the mounds of washed stones, spewed in curving piles by the stacker of the gold dredges, have in one area been leveled to make a landing field for small planes.

I first saw the town in the late 1930's when the main street was lined with false-fronted stores. The old firehouse housed hand-drawn equipment, and the Silverthorne, Occidental, and Brown (now the Ore Bucket) Hotels were among the larger buildings. A dredge floated in its pond behind the main street. And, had I

Breckenridge, 1938. Dredge dumps in Blue River; Tenmile Range (now ski area) in background.

arrived two years sooner, I could have heard the whistle of the narrow-gauge train as it snaked around Nigger Hill and watched its engine chuff into the station across the Blue River. On these first visits I talked with men and women to whom Breckenridge had been home for many years, and I tried to see it through their eyes.

My first interview was with Mrs. Duane Miner, who mentioned in passing that the Silverthorne Hotel had been started by her grandmother and that she herself had lived in the town since 1878. She pointed out the Tenmile Range to the west, where the ski area is now, and said it began at Frisco. The peaks are named One, Two, Three, etc. from north to south, with the highest peak near the foot of Hoosier Pass called Quandary. Did I know, she asked, that a strip of land from Hoosier Pass to Kremmling was not legally a part of the United States until 1936? The Louisiana Purchase had unintentionally failed to include it. She talked of the dredges that began to gouge up the bed of the Blue in 1895 and of how they needed ponds in which to operate. As I was about to leave she brought scrapbooks, old newspaper articles, and a paper that she had written about Breckenridge and urged me to take them to the Brown Hotel that I might read them that evening.

"But I'm leaving town very early tomorrow morning. How can I return them to you?" I asked. "Just leave them on the porch. They'll be safe," she said. "I'm glad I can help you."

Another source of information was Clifford W. Kingsley, whose letters were full of stories from the several camps in which he had worked. On one occasion he went from Bonanza to Breckenridge in winter:

Nov. 13, 1950

Breckenridge I went to in winter, a most delightful looking town in the still air of winter with snow on the massive range and a Vermontish look of the town. The three weeks we were snowed in, the narrow gauge was doing its best to get through. They had a Broad Gauge plow on a Narrow Gauge track, three or four engines, and would back up a mile and an eighth or so, and slam into the snow. Then have to jack up an engine or so—as the blow would act as crack-the-whip business.

As you've heard the mule-skinners were a hardy bunch who loved their teams much. On the harnesses would be numerous celluloid rings. There were six horses or eight on the ore wagons, which consisted of two wagons. The leaders had a bell on, and then in Breckenridge, where the thermometer reached 44 degrees below, you'd hear the tinkle of the bell and also the steel rings of the wheels in the snow, two or three miles away.

There was in Breckenridge, as in Leadville, a set of Near-do-Wells. They rode to work in carts. We poor slobs walked, and in order to keep the frost from entering our shoes and freezing our feet, we wore perhaps two pairs of socks and those wool lined sort of moccasins, and then rubber two three buckle overshoes.

When I was in Breckenridge, Bob Foote owned the Denver Hotel and made his money off Farncomb Hill. The Swan dredge was about 2 miles from town, I believe. I was working on the Wellington, about 2½ miles from town, up French Gulch. There was a Dutchman there who had frozen both feet and they had amputated his toes. I was told that the winter they were snowed in for some 70 days he had, in a drunken brawl, just about killed someone. To scare him, some of the boys said the victim was dead. He started hiking for Leadville, and frozen feet was the price.

The WELLINGTON had been tunneled in seven or eight hundred feet, and through greediness stoped to grass roots, then an incline shaft had been run toward the entrance at perhaps a 60 degree angle. It was I believe about 700 feet down, and they'd stope up wet and hot. So we'd strip to overalls and jumpers, digging boots and slickers. They very near lost the shaft while I was there. It caved in to surface, and we could stand near the gallows Frame and see the moon through the hole. Forming a chain gang we cribbed up and saved the place. I hope these things will help you to get a picture of these places.

The Kingsleys
Laura & Clifford

Some letters that were sent me were in response to awakened memories in the minds of the writers, while others sought answers to questions. I could not be of any assistance to Donald S. Dressen whose letter follows, but it shows the mobility of people during the mining era, as well as the hazards and difficulties surrounding them in their daily occupations.

March 5, 1954

I . . . read your book and enjoyed it. Particularly so in that I was searching for some mention of my grandfather who died in that region about August of 1880.

David Worth Roberts was born March 18, 1849, in North Carolina and apparently came to Texas in the early seventies about the time of the death of his first wife. In 1875 he remarried in Williamson county, Texas, to Frances Elizabeth Wells who was my grandmother. He lived in that vicinity for several years, going to Colorado in the late 1870's. He is said to have been a geologist and to have invested in real estate and mining claims in a small way. However, his primary interest was in sawmills in the mining region, and he built more than one prior to the coming of the railroad into that area. I remember seeing years ago photographs of one of the sawmills and I believe President Grant was shown in the picture. My grandmother, who apparently did not live in Colorado, told me of the great difficulties that he had in transporting heavy boilers from the railroad's end to the sawmill site. At any rate, just as this enterprise was beginning to operate, he was severely injured one day when one of the belts at a sawmill broke and struck him in the head. He lived for several days thereafter and was taken by railroad to one of the larger towns for medical treatment. He died it is believed between August 22 and 28, 1880, and was buried near Breckenridge, Colo. (Another source gives the date of his death as Sept. 17, 1884, at Leadville, but this date is believed to be much too late. My mother who was born on April 28, 1881, declares she was born months after the death of her father).

If you can suggest any source that would enable me to confirm the elements of this biography, I should be most grateful.

Mrs. E. Cramer's letter (Jan. 19, 1954) contains the reminiscences of a woman familiar with several mining localities:

I hope everyone has enjoyed your book (Stampede to Timberline) as much as I have. Having lived in five of the Camps, you will understand why. I lived in Pandora—Telluride—Cripple Creek—Breckenridge and Montezuma. I rode over the new R.R. from Durango to Silverton in July of 1882.

I wonder who the old lady was who touched you on the arm and said she knew the names of the old fire engines when you were sketching them in Breckenridge. Wonder if I remember her as a little girl. They told you about Father Dyer up in that country, wonder why they did not recall Mr. Passmore. He was such an earnest Soul-Saver in that area and well known. The old lady you talked to in Montezuma—wonder if I used to know her when she was small. You speak of Mr. Bradley preaching in Telluride. I had entirely forgotten him, but he came out of the mist of memory—also remember Bishop Spalding, in his Episcopal robes, sitting up on the little platform in the old school house.

You speak of that bad place on the road to the Tomboy Mine. Wonder if it is in the same place where we got off and led the horses—and that zig-zag trail that fascinated you—I used to watch the burro-trains go and come along there. And remember when the poor mail-carrier was swept away by the snow slide and they thought he had skipped the country. I think it is wonderful that you have recalled the old towns. Too bad so many have become "ghost towns" and the little homes that were so nice are gone or left to ruin.

I surely wish I could have seen as many of them as you have but I think I would have felt safer on a horse—to think of a car traveling thru all those mountains!

Whenever I needed accurate information about Breckenridge, I turned to Helen Rich, author, historian, and friend of many years. I had been told that Father Dyer's Methodist church in Breckenridge was the first Protestant church built on the Western Slope. Residents of Lake City, Colorado, made the same claim for their Presbyterian church established by Alexander M. Darley, missionary pastor, in 1876. Which was correct? Helen Rich would know, and here is her answer to my question (March 4, 1955):

I am not able to find that there was any church, a building with a steeple and all, in Breckenridge prior to 1876—the date of the Lake City church. Services were held in houses, saloons and such places.

Father Dyer's church was built in 1880. I thought it might have been begun in 1879 but I have just taken a look at the deed for the land which he made to the trustees of the church. This was a quit-claim deed and made the "17th day of June, one thousand eight hundred and eighty." His book "Snow-Shoe Itinerant" pp. 332-3 says, "I gave half my lot to the trustees to build a church on" and although he does not give a precise date I feel safe in fixing one because of that deed which is on record in the courthouse. He goes on to say, "We carried a subscription paper till I got enough to start on; and went to the sawmills, got all the lumber I could, and we went to work and put up a house twenty-five by fifty feet, Posts sixteen feet high, and inclosed it. I nailed the first shingle and did more work on it than any other man.

"While I went to conference (this was in the fall of 1880, I am sure. HR) the friends finished the roof and put the floor down; and the next Sunday we had a service in the first church on the Western Slope in our conference, with a good organ."

The conference, of course, was the Methodist Episcopal Church conference so this church, where you saw his portrait, is the first church in the Methodist Conference on the Western Slope. Again Dyer does not give any date but that first service must have been after the conference that took place in the fall of 1880. Unfortunately the church has not kept proper records. The deed he gave was found in a litter of waste paper in the back of the church only two or three years ago.

I have not been able to establish when the Catholic Church was built here. People think it was between 1880-90, but no one knows. If you are at all worried I shall be glad to get in touch with the church at Leadville which should have the records. The only other church, St. John's Episcopal, was dedicated in 1892.

History is a worrisome thing, huh?

Lincoln City   Before leaving Breckenridge, I always drive up French Gulch east of the town past the Wellington mine property to Lincoln City. There's not much from the old days left to see, since the Wellington mine buildings have disappeared, and even the dredge near Lincoln City is falling apart and slowly settling deeper into the creek bed; still, Farncomb Hill remains, the source of the crystal gold found in the Wire Patch, which caused the City to boom in the 1880's.

On one of my first trips up French Gulch, I learned many things from a Mrs. Murphy who once lived in Lincoln City. "It is not as old as Breckenridge," she said. "But it started in the early sixties. At first the mail was carried across Webster

Pass from the Eastern Slope into Lincoln. After Farncomb Hill was opened up, miners swarmed in, and the population got as high as 1,200. You saw the Wellington mine at the lower end of French Gulch. It was the only big lode close to Breckenridge, and it was a good producer for twenty years. The ore was gold, lead, and zinc. Right across the gulch from it was the Country Boy mine. As you can see, the creek bed was dredged out for miles."

Louisa Ward Arps was introduced to Wapiti by Helen Rich on a trip over *Wapiti* the hill from Lincoln City to the valley of the Swan River. After the trip, she wrote me (Oct. 5, 1952): "From Lincoln City we angled up Farncomb Hill (where the wire gold that is on exhibit at the City Park museum in Denver came from) to Dogtown and Wapiti. Neither is impressive, but my husband took a record shot of both, chiefly for you."

I received another letter from a couple who had found their way to the isolated camp. Of this trip Mrs. Roy Fogle Dent, Jr. wrote (March 16, 1957):

> I have been meaning to write to you to ask if you have ever come across a community we discovered and which is not mentioned in "Stampede to Timberline."
>
> You start out from Linocln City and take the first road (a jeep trail) to the left up the side of the "wirepatch" on Farncomb Mt. . . . At the top we came right out above the "wirepatch," which is unmistakably what it is. Wapiti is farther along the road but still high on the other side of Farncomb Hill and if you keep on the road you go right by it.
>
> We came across two *very* well preserved cabins (with diggings on the side of the mountain well away from them). Both looked as if an attempt had been made to make them look "elegant" by 19th century standards. Some of the rooms had old wall-paper on the walls and ceiling (not matching) and other rooms had wood paneling on walls and ceiling. In the far house was a homemade chair, table and bedstead and a trap door with stairs into a basement (which we were afraid to enter). There was an opening in another room to an attic. In the near house there was also unmatching wall-paper and a sign on the door warning of explosives stored in the basement so we didn't investigate it quite so thoroughly. They (the houses) are set in a beautiful flat wooded parklike site and a bench-mark says 10,604'.
>
> From there we followed the road downhill on the opposite side of the mountain from the "wirepatch" to Parkville and Tiger, on the Swan River. In Breckenridge we stopped at the Columbine grocery and asked the name of the community. The grocer phoned someone who said it had been called "Wapiti."

Very truly,
Virginia Dent

The Swan River, which flows into the Blue a short distance north of Brecken- *Park City* ridge, has nearly as many dredge dumps as has the larger stream. Swan City near *or* its mouth is only a site, and Tiger farther upstream is fast becoming one, according *Parkville* to a friend, Dennis Smith, who visited it in 1973. He reported that "the buildings that you saw have burned—only the assay office is left, and it is being fixed up by the Historical Society."

The Swan forks about a mile above Tiger with its South Fork nearly clogged

North American Gold Dredging Co. dredge at work in Swan River, 1899 or 1900.
Courtesy Erl H. Ellis and George P. Hurst.

Aerial view of dredge rock piles at Swan and Blue Rivers junction, 1973.
(State Highway 9 [north-south] runs diagonally across lower left of picture.)
Courtesy Kucera & Associates, Inc., Denver.

with waste rock left by dredging activities. Under some of these stones is the site of Park City (or Parkville) of which George Feltner wrote:

> You can drive to the site of Park City, or one could the last time I visited Swandyke, about a year and a half ago. Nothing remains of Park City save some grave stones. As I recall, the Masons have erected a monument to the city, near the Swandyke fork in the road. In Park City the Masons held their first meeting in Colorado. It was once proposed as the capitol of Colorado— as what mining city wasn't. But it gave Breckenridge a run for its money in a fight for the county seat. What a wealth of history and adventure lies forgotten in those hills of ours.

In 1950 I saw the monument of which he speaks.

On the north fork of the Swan River is Rexford, which Ed. R. Lewandowski   *Rexford* describes as follows (Dec. 11, 1955):

> I trust you are still interested in "ghost towns," and it is in this vein I am writing. I am a member of the Colorado Mountain Club and have used your book . . . as a reference hundreds of times. You are correct in your assumption that the road goes just north of a saw-mill camp. I'm sorry I can't help you any further, but I hope my meager information will be of some help.
>
> On August 24, 1955, Margaret Bivans led a trip to Glacier Mountain, near Montezuma. We climbed to the top via Tiger, and on the way we passed the remains of about a dozen buildings. The location is just east of Garibaldi Gulch, on the North Fork of the Swan River. The buildings were in a fair state of preservation. There were a few 2-story dwellings and a number of cabins, obviously dwellings of some sort. . . . We found a few scraps of news-paper remaining in the structures, one scrap was dated 1889. One unique dwelling, roofless, had a 10-foot pine tree growing in the center of what had been the floor. The "town" can be reached by jeep only, a passenger car would never make it.
>
> U.S.G.S. Professional Paper 178—Geology and Ore Deposits of the Montezuma Quadrangle, Colo., by T. S. Lovering, 1935, gives more of Rex-ford's history:
>
> "The Rochester Queen or Arrastra Queen, as it was later known, is on the North Fork of the Swan River 3 miles from its mouth and about 3 and ¾ miles southwest of Montezuma. It was discovered by Daniel Patrick about 1880, and in 1883 it was developed by a 75-foot shaft, 325 feet of drifts, a tunnel 100 feet yong, and a 75-foot winze. The town of Rexford sprang up a few hundred yards below the mine chiefly because of the extensive work car-ried on at the Rochester Queen. A mill was built and connected with the por-tal of the mine by a wooden tram line, and ore was milled for many years. According to Burchard the mine produced $5,000 worth of ore a month for three months in 1881, and shipments were sent to the Argo smelter in 1881 and 1882, but no work was done in 1883. The ore contained iron, copper, galena, and some honeycombed gold-bearing quartz. According to the Colo-rado Mining Directory for 1883, the Rochester Queen had produced 500 tons of gold-silver ore prior to that year. The veins were said to be fissures from 6 to 10 inches wide containing decomposed iron-stained porphyry and quartz and had a fair tenor in both silver and gold, assaying when sorted about $50

a ton in these two metals. The mine was inaccessible in 1929, and the town of Rexford had long been abandoned."

The most recent report is from Dennis Smith of Littleton who wrote (Sept. 8, 1973):

I finally got to Rexford. We walked in for about four miles from a fork in the road beyond Tiger. The old store and boarding house were in good condition. Kathi and I were tired when we got back to Tiger. The road wasn't very steep except in a couple of places. It was truly a gem of a ghost town. All the nails were square.

Swan's Nest, 1950. Ben Stanley Revett's home, built in 1898.

Swan's Nest   The first time I drove up the Swan River I looked across the mounds of tailings near its mouth and caught a glimpse of a big red and white house seemingly buried by them. It was Swan's Nest, the home that Ben Stanley Revett built in 1898 for his bride, Mary Griffin. Revett was already prominent in the district, for as early as 1889 he had visited Breckenridge and by 1894 was definitely interested in developing the area. The first dredge that he built was operating on the Swan River by 1896, to be followed by heavier gold boats capable of digging deeper into the rich, submerged deposits.

In 1969 I received a letter (Sept. 29) from Mrs. Frances Revett Wallace, Revett's daughter, which describes "Touring the Breckenridge area and seeing the miles of tailings piles that are a memorial to Father's perseverance."

As the daughter of Ben Stanley Revett, who put in the first gold dredges near Breckenridge in 1896 (not 1907 as your book states), I am, of course, tremendously interested. When I first read the book, I was sure that the first gold dredges were put in at Breckenridge quite a bit prior to 1907—for I could remember following Dad around on the dredges . . . prior to that date.

In the "Colorado Magazine," published by the State Historical Society of Colorado in October, 1962, (Vol. XXXIX, number 4) there is an article, "Gold Boats on the Swan," the story of Ben Stanley Revett by Belle Turnbull, in which the date for "the first gold dredge to dig in Colorado ground was built and put into operation in *1896!*" It was the Risdon dredge, which was already a deserted relic in my earliest memories—and other "bigger" ones were then at work.

My son, Revett Wallace, and I have just returned from a trip to Denver, up to Dillon, and to Father's old house down on the Swan, 6 miles north of Breckenridge, where I spent every summer of a happy childhood from 1901 to 1918—in following Dad around on the dredges like a faithful puppy-dog.

It was wonderful to see again, after so long, the view of the Ten Mile Range, from the porch of our big house—which mother always called her "Ten Million Dollar" view—it is so beautiful!! The house is still in use, as a Lodge, and is amazingly well preserved, as it was built of California Redwood in 1898!

On our trip we took in Georgetown, Silver Plume, Dillon and, of course, Breckenridge—where we stayed at the most modern and comfortable "Breckenridge Inn." Then to Fairplay—where it was nice to see an old, deserted dredge still floating in its pond among the miles of tailings piles that interlace all that area. To Colorado Springs, and back to Denver—all in 3 days. It was a wonderful experience and my son was particularly pleased at his tour into "ancient family history."

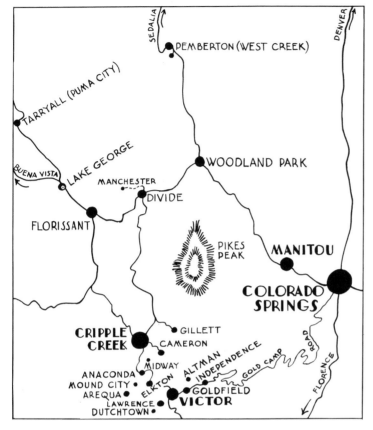

CHAPTER VI

Pikes Peak Country

Within sight of Pikes Peak the Cripple Creek Mining District, one of the richest in Colorado, remains an area of mines and mining interests. Its two cities, Cripple Creek and Victor, are quieter than in their boom days, but the residents enjoy the high, rolling hills, the bracing air, their mountain gardens, and the distant snow-capped barrier of the Sangre de Cristo Range to the west.

The area was not always like this, for from the time that the first trace of gold was found in 1874 by a member of Hayden's surveying corps, men began scratching at the surface of the ground until they laid bare the treasures hidden deep below. William W. Womack, who filed on a homestead where Cripple Creek now stands, wasn't interested in prospecting. He sold his land in 1884 to the Pikes Peak Cattle and Land Co., and they sold it to the cattle firm of Bennett and Myers. But Bob Womack, son of the original owner, who rode the range throughout the 1880's, was always on the lookout for float and kept whatever pieces of it he found. The Mt. Pisgah hoax in 1884 brought swarms of men to the mountains, but when they learned that the lure which brought them was a salted hole only ten feet deep, they left in rage.

In December 1890, Bob Womack showed his accumulation of gold samples to E. M. De la Vergne and F. F. Frisbee and persuaded them to inspect his claim. The following summer he discovered even richer float, and while celebrating in Colorado Springs he sold his mine for $500 in cash. This started the gold rush so that in no time the cattle land was staked out with claims. During 1892, the year the District was organized, it was mostly placer mining; the lode mines were discovered the following year.

After the demonetization of silver in 1893, hundreds of miners whose jobs were gone flocked to Cripple Creek to dig for gold in the hills that surrounded the townsite. The big mines, however—the Portland, Independence, Gold Coin, and Strong—were in or near the new camp of Victor, six miles away; and so sensational was the output of the many mines that two railroads were built into the District—the Midland Terminal from Divide and the Florence and Cripple Creek from Canon City. These transported the crowds arriving each day and handled the gold shipments that were sent out to be smelted. By 1900 another road was added from Colorado Springs, and to meet the needs of the growing population two electric trolley lines connected Cripple Creek with Victor.

Fires destroyed much of both cities: Cripple Creek in 1896 and Victor in 1899; and labor troubles and strikes closed down mines and wracked the entire District with violence in 1894 and 1903. In spite of ups and downs, the District has survived, and anyone visiting it cannot but be impressed by the magnitude of its mining days, as shown by its older buildings in the several towns and by the many headframes, mills, and especially the huge mine dumps.

As the rival cities of Cripple Creek and Victor prospered, small settlements sprang up near them, usually in close proximity to a single mine or group of properties. The lives of some were of short duration, while others lasted as long as the District continued to produce.

North of Cripple Creek was Gillett, called the Gateway City, as it was the first place the Midland Terminal railroad passed through as it entered the Cripple Creek Mining District. A special attraction was the race track, which provided weekend entertainment during the boom years. Of several small towns that clung to the hillsides south of Cripple Creek, one was Anaconda. For years a small, dilapidated stone jail has marked its site.

Victor like Cripple Creek was rimmed on three sides by smaller camps. To the northeast, at the foot of Battle Mt., was Goldfield; and farther on was Independence, whose railroad station was dynamited during the labor troubles of 1904.

Goldfield, 1942.

Directly north of Victor, on the hilltops, were Altman, once "the world's highest incorporated city," which by 1942 contained a shed, a few battered cabins, and a rusty fireplug; Midway, halfway to Cripple Creek; and Cameron, once a busy town with an amusement park at which the High Line trains deposited many of their passengers. In the twentieth century, according to Don Ripley, the Cameron mill was erected on the old townsite.

To the northwest of Victor lay Mound City, Elkton, and Arequa. Of Arequa, my friend Don Ripley wrote in 1942:

> On a 1902 map I notice a town was all surveyed, called "Arequa," right across from Elkton; the streets weren't named but I have been told that there was at one time a large settlement there.

Lawrence and Dutchtown

To the south of Victor were Dutchtown and Lawrence, mere sites today. Yet C. W. Kingsley, who lived in the area, recalls that Dutchtown consisted of a couple of fair-sized log buildings and a few board and batten miners' houses.

Lawrence, I knew, had been laid out by the Woods Investment Co., as had Victor. I'd also heard that the first chlorination mill which treated low-grade ores of the District had been built there. A few years ago, while assembling vital statistics about Joseph Raphael DeLamar, whose start in mining was in the Silver Cliff area of Colorado, I found his name linked with a large gold mill at Lawrence. Nellie Grater of Victor verified this information and expanded it as follows (Nov. 26, 1948):

> Bess and Don Ripley asked me to find out about the old mill at Lawrence for you. Not many people here remembered, but from Mr. Harry Nelson who was a surveyor here in the early days I got the following: It was built by Captain Delamar, a very wealthy man, and Ed Holden, a well known Mill man. In case it might interest you he also told me their head chemist, D. C. Jacklin, left here to go to Mercur, Utah, where he became interested in the big copper industry and later became one of the big shots in that industry.

A number of these smaller towns are further described by C. W. Kingsley who grew up in the Cripple Creek District but now lives in California. He wrote to me in 1975 at the age of ninety, still interested in hearing about Colorado and eager to discuss the exciting days which he remembers so well. The letter in which he reminisces about the satellite camps of the Cripple Creek District was, however, written in 1951:

> Dear Mrs. Wolle,
>
> I am glad you feel that you may gather enough incidents to make a supplement to your last book. The names I have given you are people like I hope I am, trying to get things accurate, or events, rather. . . .
>
> Mr. O'Brien of Newhall is so crippled that he will have to have some one write for him. . . . His daughter had much to do in getting the state to recognize NEWHALL (Calif.) as the spot where gold was first discovered, Jan. 23rd, 1848.
>
> I'm sure if you told her of the condition of the Catholic church (in Gillett) . . . she would have her dad tell her of the Bull fight in 1899, and the Monte Carlo Casino on the hill. The first time I saw it, it was much dilapidated, the town (Gillett) had some buildings, a grocery store and I believe the lumber

Gillett, 1942. Roman Catholic Church.

yard was running. Mrs. Perkins of Newhall, Calif. was christened in the Catholic church in Gillett. About the only time we kids got over there was when we left the High Line at the Black Sampler and started our hikes to Pikes Peak through Gillett.

Did you know that around 1903 there were somewhere around 50 trains a day in the district, with the Short Line, Midland, Florence & Cripple Creek, and High and Low lines?

When I was there the 6th Street Irish ran the town; if you were in with them life wasn't miserable for you. On pay days the Finns, Jacks, etc. were lined up for a block or more at the Postoffice, sending money home.

Anaconda was peopled by miners and its narrow main street had quite a few businesses. Arequa was what we called Arequa Savages, mostly mule-skinners and not very large. Elkton was a pleasant small town, then came Santa Rosa, where the S. O. Oil Company had its supply depot (with only a few houses).

Then of course, Victor, then up on the slope of Big Bull approx. a mile and a half from the center of Victor was Hollywood, where a few families lived, and by the way the only place in Colorado I believe a Leopard was ever shot. The place had probably 6 or 7 houses and one fall a man missing his chickens went out one night when he heard a disturbance, and shot this thing and found it was a leopard. I believe that was in Oct. 1908 and in August Goldfield had had a Circus so everyone supposed they lost him, but daren't admit it.

When my folks moved to Victor (that fall) some one tried to start Dances in a place called Altman. There were a few families there; anyhow the Arequa Savages were too rough. If memory serves me right, electricity had not got up there at that time, and the place was lit with big lamps running through the center of the Hall which I attended.

Then in the saddle between Big and Little Bull was a switchstation on the F.&C.C. called Alta Vista, three houses were the only houses (in my time), and one summer the Kingsley Families used one as a summer home, and walked back and forth to town. The F.&C.C. did all their make up of trains there to go to Florence.

Cameron had in the beginning of the 1900's an amusement park, and Short Line and Midland trains ran excursions. It had buffalo, deer, monkeys, and the things that make up that type of amusement. They had an overpass over the rail tracks and all in rustic style as well as the fence. I think it was real pretty and well built. I believe it was abandoned at the time of the strike.

At Cameron the school had mineral ground and the La Montaign brothers made it pay nice dividends, with a nice but small mine. Cameron was perhaps 30 families along about 1908; the C. C. Water pump station was located there, which I shall never forget as it was my haven of rest, when I walked from Colo. Springs, got lost, stayed out all night, followed a trail in the snow that led me to the railroad tracks near Clyde, and exhausted I was when I fell, upon coming in contact with the warm air of the boiler house. Course it was only February I believe. After postponing a few meals, and riding blind baggage from Chicago, Pop came out and got me, and you'd think I was quaranteed, the way I stuck to home for a while.

The Drilling contests were thrilling and much training went into those who entered—Double Jack, Single Jack, Hitch Cutting—all those drew big crowds. Here is an amusing phrase of the hand driller, to an inexperienced hammer man. The old timers holding the drill would say, "Let me skin awhile," meaning it was getting hard on his knuckles.

One time a chum and I got hold of a box of caps and were busting them one at a time on the sidewalk as we were walking down Third St. All at once the chum said, "Look, someone's finger on the window." Soon afterwards he realized it was his own.

Mr. Hugglett, Engineer for the Pueblo Mental Institute was a Boss on the Portland mine, also Sheriff.

Did anyone call your attention to Miners' Consumption? We had well ventilated mines in the Cripple Creek District, but men in blind stopes, before the water Lyner, suffered. A man afflicted would hardly cough when in the hole, but the agonizing time he'd have at home.

I must tell you of Dr. Elliott (Victor) who served in the Boer War, he was English. During the Flu epidemic, he was of course very busy, and through the work became ill himself. However, he got out of his sick bed and administered to the most needy patients. Some of them he told that "they'd get well but I'm going back to the Hospital to die." Which did happen. A *ruff piece* of *Pure Gold* was he.

C. W. Kingsley

Quite a few people sent me letters about their recollections of the Cripple Creek-Victor area. Since each adds bits of local color as well as information, it seems best to let these letters speak for themselves.

Victor   Virginia Prentice of San Bernardino, California (Dec. 12, 1949):

My mother went as a bride forty years ago to the Ajax Mine at Victor, where my father was then superintendent.

Victor, 1938. Portland mine.

Mrs. C. N. Clinesmith of Ellensburg, Washington (Nov. 22, 1949):

When my family moved to Cripple Creek in June, 1897, the District had 50,000 people. Cripple Creek claimed a population of 10,000 with 91 saloons. I saw Victor born and lived through the tragic Miners' strike. My father was President of the Engineers Union.

From Don Ripley's letter (1951), which answered questions of mine, I learned additional facts about Victor:

Jack Dempsey, who came from Manassa, Colo., worked in the Portland Mine, Victor. His brother was killed in the mine. His brother's name was "Jack." After the brother's death, the famous brother took the name of Jack.
Lowell Thomas graduated from the High School in Victor.
The Bull Fight at Gillett was spurred by rivalry between Cripple Creek and Gillett. It was intended to boom both towns. I've heard that Soapy Smith started the scheme. Soapy Smith would sell soap in Denver. He'd wrap a ten

dollar bill around a cake of soap in a basket. Then he'd sell the cakes for $1.00. No one ever found the ten dollar bill.

Mound City was near the El Paso mine. The Monte Rosa mine was under our house in Victor. Across the street by Ovren's Bookstore was the Gold Coin.

El Paso Tunnel might have been the Columbia Tunnel. I've been through it. The Roosevelt Tunnel was the first drainage tunnel, driven probably with old steam operated drills. This tunnel drains into Cripple Creek Canyon. 1908 was about the right date for this tunnel. The Carlton Tunnel was completed in 1941; I kept track of the footage of this tunnel (as I was bookkeeper at the Ajax mine). Laterals are now being driven from the Portland Carlton Tunnel level (3460′) to the Ajax and Vindicator Mines with the intention of *raising* the Ajax shaft.

The Woods Bros. built the big hotel in Victor. A. E. Carlton and L. G. Carlton, I am certain, were originators of the Golden Cycle Corp. A. E. had the bank, railroad and Colo. Trading & Transfer. L. G. was interested in the mines and mill.

Cripple Creek   From Marshall Sprague, Colorado Springs writer, I learned more about the Mt. Pisgah hoax which caused a rush to the region; but since the lure was a salted mine, the rush lasted but a day and a half and caused so much skepticism that the discovery of the district's rich ores was delayed several years. (Mt. Pisgah is just west of Cripple Creek.) Sprague wrote (Sept. 22, 1949):

> I am writing a history of Cripple Creek [*Money Mountain,* 1953]. . . . Your book contains an account of the Mt. Pisgah—so-called—fiasco of 1884 and I want to ask you where you got the information. The story I have does not coincide as to personalities involved but of course that is not surprising since so many were involved. My report of the discovery was the report made by Capt. De Coursey to the citizen's committee in Colorado Springs who sent him with two others to find out if it was fake or not. I have located the actual shaft and it is nowhere near Mt. Pisgah but on the north slope of McIntyre Mountain nearly nine miles west as the crow flies from the summit of Pisgah. However, McIntyre, I find, looks exactly like Pisgah from some angles and I am wondering if the name Pisgah wasn't attached to this boom of April 5-19, 1884, by mistake. Anyhow the location names posted on the Teller Shaft were S. J. Bradley, locator; and D. G. Miller, M. E., surveyor; I am wondering where this Chicken Bill comes in. I am beginning to get a suspicion that the hoax was promoted by Thomas Gross or Grose who owned the surrounding land as a homestead—a Gross or Grose whose address was Truro, a name you will find on the 1901 Interior map of the Pikes Peak Quadrangle. Truro is right at the head of the valley where the Teller shaft was. A Capt. H. B. Grose of Truro, who had an assay furnace at Alma, was one of the very early assayers giving glowing accounts of assays at Cripple Creek in early 1891. I am hoping to find a connection here to the effect that some Grosses or Groses were mixed up in the 1884 fake. The Dell family in Guffey remember the fake very well.

H. L. Standley of Colorado Springs grew up in Cripple Creek and remembers many incidents from his adolescent years:

Cripple Creek, 1903. Bennett Avenue. *Photo by G. B. Sanborn, Leadville.*

There has been no intention of ignoring your letter asking about any recollections I may have of the early days in Cripple Creek; but I fear I have little to offer. I went to Cripple Creek with my family in June of 1893 and was there until June of 1903, and during the last few years of that period did some photographic work among the mines and nearby mountains. When I left I disposed of most of my negatives—threw most of them away I think—something I have regretted many times of late years when they would have had a much greater degree of interest than at the time they were made.

My years in Cripple Creek were the ones between my 12th to 22nd birthdays and I fear that I was not all an observer of the life and events of what was occurring then—as a youngster I was just "growing up" and nothing particularly concerned me. Had a job in Colburn's Stationery Store for some years, later working part time so as to permit time for my photography. . . .

During the boom years I can remember of how crowded the sidewalks in Cripple Creek would be during the evenings—after the work day—of the many saloons, all prospering—and of the prospecting for ore that was going on all the time. Of course there were "shooting affairs" occasionally in the district. I witnessed none and think that they were much less frequent than most folks like to believe they were. When I went to Cripple Creek I was taken along as a passenger in a mule-drawn wagon starting at Pueblo, spent a bit more than 48 hours on the road and the road made its way along what was later the route of the first railroad in the district—The Florence & Cripple Creek Railroad. Also saw the arrival of the first trains on this road, THE first being a work train followed later in the day by a passenger train. Witnessed both of the fires in Cripple Creek and the Victor fire. The first fire in Cripple

Creek occasioned a hasty moving of the contents of the store I was working in, which were brought back within a short time, hours perhaps. The second fire which came a few days later occasioned the same sort of hasty move of goods, but this time there was no place to come back to.—At the time of rebuilding after the fires many of the buildings carried the names of their owners or builders—many of these were still to be seen at the time of my last visit, memorials to the men of the day and serve to impress upon one the rapidity of the flight of time. Probably none of these men are living today.

From Virginia Brown (Mrs. Neil) Curlee of Louisville, Kentucky, I learned another story about Cripple Creek:

July 21, 1960

My mother was the daughter of a Denver pioneer. . . . She was fascinated by the effect of the buffalo on the history of the west, and collected all the data possible on that subject. . . . Mother used to tell us about the miner in Cripple Creek who used to clear the streets on Saturday nights when he walked down the main street smoking a stick of pure dynamite (he hoped). Her uncle, Fred Reed of Cripple Creek, had told her that yarn when she was a child.

When Virgil A. Krewson of San Pedro, California, discovered *Stampede to Timberline* in the city's library, it awakened memories of his youth, which he was kind enough to share with me (Jan. 18, 1969):

Taking the book down I glanced through the table of contents, and one chapter inveigled me—Pikes Peak Country—because it was in Pikes Peak Country where I spent the earlier days of my youth.

I was born in 1893 at Colorado Springs during those turbulent times of the silver panic, as it was called, and the gold rush to Cripple Creek. . . . My father was a Carpenter, erected homes, buildings, etc. (my dad and Stratton worked together at times, they were good friends). The panic caused many hardships, no work, no money. My dad in the early spring of 1893 took off for the diggings at C. C. He built a cabin atop Bull Hill near Altman. When it was completed, my mother and her brood moved to the Cabin 11,000 ft. above the sea. The cabin, though, was very big and comfortable, as my dad was a good carpenter. She and Dad went through those hectic days of the strike. In the after years I heard again and again how my dad planted powder all around our cabin with notices of warning; you can be sure we were not molested by the deputies. As I was only an infant then, I know very little what really happened, only the telling of the Cripple Creek experiences from the older members of my family.

As a boy I was what was called on the Short Line R.R. a Tourist agent, in other words a Peanut Butcher. My brother Frank, also known as Steamy, had the concession on the three railroads, Short Line, the Colorado Midland Terminal and the Florence and C.C. I sold eatables, view postal cards, magazines, etc. The tourist Agent part of the job was telling the passengers all about the interesting sights and facts on their trip to that great Gold Camp of Cripple Creek. To this day I still remember the spiel I gave. As I made trips on all three R.R.'s . . . I have lots of stories about the Pikes Peak Country.

My mother was a pioneer; she crossed the prairies in a land schooner;

many the times she told of having to hunt for Buffalo chips, sometimes a great distance from the campsite—she told of her fears and hardships. It was chips or no hot meal.

In 1950 I received the following letters from Edith Stuart (Mrs. Lloyd R.) Jackson of Columbus, Ohio. She is a writer and is assembling a book about the Cripple Creek District where she once lived. In 1974 I met the energetic historian and found her most gracious and willing that I include the material she had written to me twenty-four years ago.

Feb. 18, 1950

Your book came yesterday morning and I must confess that the dishes remain undone, and very little housekeeping has gone on. . . . First, your book will always be unique and valuable in the annals of the west. . . . It could not be done by anyone else for you contacted the living who could tell you of the facts and most of them have long since crossed the Big Divide, no doubt. The manner in which you went about acquiring your information; and your persistence in the face of all the obstacles of such a pursuit can only be appreciated by those who have been Rocky Mountain hikers and climbers and know the immense difficulties, dangers, and distances which would have been interposed between you and your goals.

The sketches of places which I know fill me with the *mood* of the areas as no photograph could do. . . . I have always thought that one of the reasons for the old prospectors and habitual mining explorers (whose names one can trace from area to area), being so persistent in the high mountain areas was that they deeply appreciated the beauties of those inaccessible places. Another thing, too, which is often not taken into account is the fact that those who live in the high, sunny altitudes and become accustomed to them feel so buoyant and energetic and optimistic that no where else can they match that zestful feeling of health and vigor, and so what seem to us the horrible hardships those pioneers endured were rigorous, yes, but they didn't mind them, they gloried in being tough enough to take them.

I really struck paydirt first in the life story of my uncle who was a pioneer in the Cripple Creek District and has at this age still retained a most accurate memory of his early experiences there. In 1948 I spent some time in the District taking some old freighter and teamster drivers around with me to locate and discuss their memories of some of the great towns in the District.

I sense in your Pikes Peak area write-up, knowing the fierce local pride to resent any suggestions that the district might be a ghost area or ever become one, that you may not have had too much encouragement from the local sons toward your effort. . . . A second handicap you would have had would have been the shut-mouth policy dwellers there have adopted because of the bitter feelings left after the 1903 strike. These lie scarcely covered under the surface of all social life of the older generation, and are still vividly recalled by those who are younger as attitudes in which they grew up. Even I, whose parents lived there all their married lives, my father having come in Nov. 1891, as a child, having lived in Altman until his marriage in 1902, with all the contacts and friends my family and I have had there, have had a hard time digging into the memories people have there.

There are several points, which I would like to point out to you, as not jibing with what information I have. On page 476, under the description of Altman, the miner you quote as saying that there was no town of Altman in 1894 gave you the wrong impression. My father and his family and many others did live there as early as 1892, and the children were in schools practically from the first.

The El Paso County School system was well established and was set up in the District as soon as needed. The photograph I have of the first school, teachers and children is taken and dated as in January of 1893 and that was in Cripple Creek. Since the earliest paying mines were on Bull Hill and that vicinity, and families moved there to be near their work, the settlement must have been of some size very early, even though it was not an incorporated town. My mother and her brother attended the Cripple Creek school in the fall of 1892 and also in the spring of 1892, after their father had become interested in the district in Dec. 1891. They lived in a house built for them on what is now Eaton Avenue, and watched the camp grow into a sizeable community for a few months. In that time my grandfather became well-to-do because of his interest in a number of early mines and moved his family to Colorado Springs, where he joined them on weekends, but his sons were in the District with him learning about mining and promoting mines. My father was a sample boy and carried mail and newspapers all over the area from which the earliest history was developing in the mines and knew every property and could recognize rock from any of them and correctly identify it. He early worked in assay offices of various local assayers, and on local newspapers in both Cripple Creek and Altman. . . .

I wonder about your giving credit for the scene leading to the writing of the song: "A Hot Time in the Old Town Tonight" to another part of Colorado when all the written stories about it always seem to assign it to early Cripple Creek, Colo. . . . I have a story about it from a woman who lived in Cripple Creek town early in 1891—helped run a hotel there for years.

The Gillett bull fight story was stirred up last spring and summer in the Cripple Creek *Times-Record*. . . . My uncle in Indiana, the 1891'er, scolded me for what he read in the paper that I had sent in from one of the interviews I had taken from an old timer and proceeded to tell me such a detailed story about the entire affair that I was convinced as usual his almost photographic memory was superior to the other stories I had collected. I was interested to read in the Colorado Historical Society bulletin for Oct. 1949 the accounts about Arizona Charlie and his dramatic encounters with the Humane Society in the summer and fall of 1895 in Colorado in his rodeo shows. What was told about him there squared with the detail my Uncle had given me and I am inclined to believe his recollections are quite accurate. . . . His account does not quite agree with yours. . . .

Some years ago I spent some time looking up the records about the first settlements of the District. According to the plat books on file in the El Paso County Court House, the town plat of *Fremont,* Colorado, was filed November 1891 by Bennett and Myers and included the present business section of Bennett Avenue, Myers Avenue, and to the north of Bennett, Carr and Eaton Avenues from a block west of First to a block east of Fifth Avenue, in other words about from the Depot on Bennett to just beyond the Court House.

Apparently the settlement northeast of that area, which we now call Old Town, and where I lived most of my youth, and where the earliest house and school were located, did not become a platted subdivision until Feb. 1892, at least that is what I found in the El Paso records. In that same month, the Bennett and Myers group platted several other sections to add to their town. The Old Town one is described as the Hayden Placer Co., Subdivision of Cripple Creek, and was platted by F. W. Howbert and Samuel H. Knisley, in Feb. 1892. According to a letter I have from one of the men who actually laid out the Hayden Placer townsite, this is the story on the naming of the sections of what later became Cripple Creek:

"The naming of the town, that is Old Town, was a matter of much discussion. The name Fremont was suggested by the large contingent of men from Fremont County. Another suggested name was "Moreland." In the end, however, we who wanted it named Cripple Creek won out. The first elected officer of the new town was a Dr. Engleman, who was Mayor. Dr. Engleman had a drug store at the time. The very first house built in Cripple Creek was a log building on the west side of Main Street about two hundred yards from the gulch to the south and from the turn in the street coming up from the lower town.

"The lower townsite was part of the Broken Box ranch property and was owned by Bennett and Myers, real estate men from Denver. They employed a man with an old timey surveyor's compass to lay out that part of the town. They, being experienced real estate men, knew how to put over town lot schemes and the lower town simply walked away with Old Town in growth. One thing that killed Old Town was the fact that they did not want any saloons or dance halls or any of the vice mills that just seemed to go with a mining camp. The lower town organized a town government and the first Mayor there was Dr. John A. Whiting, who was a most excellent man and physician."

This same correspondent wrote concerning the early newspapers:

"I remember the *Cripple Creek Prospector,* but do not remember who had charge at the time it was started. However, there was a race between the *Prospector* and the *Cripple Creek Crusher,* which would get out first. The *Crusher* was run by a man named Pottinger. I just happened to be in town and Pottinger knowing I knew something about the printing business, got me to help him get the paper out. We beat the *Prospector* by only a few hours. This was in the late fall of 1891."

Now, according to the Compiled Ordinance of the City of Cripple Creek, Colorado, for 1902, which I have in my possession, the following is given:

"Cripple Creek was organized and incorporated as a town under the laws of the State of Colorado by the Board of County Commissioners of El Paso, May 31, 1892.

"Fremont was organized and incorporated as a town under County laws of the State of Colorado by the Board of County Commissioners in El Paso County Jan. 19, 1892.

"In accordance with the results of a special election held Feb. 23, 1893, the Town of Cripple Creek and the Town of Fremont were consolidated into one municipality and the name given the consolidated town was 'Cripple Creek'.

"City of West Cripple Creek was organized and incorporated as a city of the second class by the Board of County Commissioners of El Paso County, Feb. 1896.

"In accordance with the results of a special election held on the ninth day of March, A. D. 1897, the town of Cripple Creek and the city of West Cripple Creek were on the 19th day of April A. D. 1897 consolidated into one municipality under the name of the City of Cripple Creek."

The Compiled Ordinances, which I have quoted above, gives John Simmington, Mayor in 1892-93 for the Town of Cripple Creek, so it may be that the Dr. Engleman he mentions may have been a temporary mayor of the first organized camp before they were legally incorporated.

There is one subdivision, a small one, platted as East Fremont, just south of "Moreland" (or Cripple Creek Town). This seems to have been on the high hill just east of the Depot on Bennett Ave. This was first platted on March 24, 1892, by Geo. W. Willetts and F. P. Mannix. On Oct. 22, 1892, it was re-platted as a re-subdivision called Cripple Creek Heights. This probably took in the part who wanted the name Moreland attached to one of the infant towns. I think that this may have also taken in the Redlight section of Myers Avenue, as I picture this location in my mind, and that may have been the reason that the name was not acceptable to the folks who held out for Cripple Creek in the Hayden Placer town.

Altman was incorporated I know not when, but in the El Paso records I found the plat of Altman and the statement by S. J. Altman as President and of Mrs. Ella Altman as Secretary under date of Sept. 25, 1893, of the Free Coinage Gold Mining Company laying out their deed reservations, rights and privileges held to themselves from the city of Altman, which was thus officially mapped out at this time. Most of the people who lived there in the early days built cabins on the mining claims by permission of the mine owners, and owned only the houses. My father's mother moved up there in the very first of the families there and lived in a tent house. She was a widow with three small sons and went there to cook meals and in other honest ways to support them. The miners built a log cabin for the family on the North Burns mine claim. Years later my Dad sold it for $10.

In the Denver Public Library photograph collection of Cripple Creek District you have no doubt seen the pictures of the great fires of 1896. A map of the city of Cripple Creek at its greatest extent, about 1901, which hangs in two of the County offices in the Teller County Court House, shows the extent of those fires on the map. Two years ago I read an account of the first fire in an old Cripple Creek newspaper owned by the County Assessor, Mr. Tom Rolofson. My father was employed at that time on a Cripple Creek paper, a union organ, and was eating his mid-day meal when the fire was discovered. From all I could learn, the town was mostly a wooden town still at that time, and, while some brick buildings were destroyed, most of the places which went were wooden ones. This was all quickly rebuilt in more substantial fashion, and the business section was by ordinance zoned so that only brick buildings were permitted to be erected. Very quickly after that the entire burned area was built over again and the scars of the fire completely obliterated. This means that the foundations you saw and the wrecked looking appearance of the town were not due to that fire or those fires, but to the wrecking program which took place during World War One and immediately afterwards. Sadly enough the wars always practically close down the gold camps, and this means that the empty buildings and houses become such a tax burden in depression times that the owners often resort to wrecking to get something out of their property and to stop the tax burden. The terrific amount of places thus wrecked in the War One period was the cause for the bad appearance of Bennett Avenue and of the gradual disintegration and collapse of the entire south side of the street, and of much of the town.

Some owners, not quite so scrupulous, or prevented by city ordinance from further wrecking are believed to have started fires so that they could collect something. At any rate a few such fires in the remaining buildings here and there have caused their close neighbors to go to pieces and now, as of our last visit to the town, the entire south side of Bennett Avenue was about uninhabitable. Whether it has been or will be bolstered up now with the new mill renewing the prospects of the district I do not know. I only know that until 1948, we never felt that the town might become a ghost town, and then we were so shocked by the lower Bennett Avenue falling apart that we wondered if the town would long exist.

It was this widespread wrecking of houses and buildings when we were children that removed from the town most of the outstanding and expensive residences and the hotel and so on. The Grand Opera House on Myers Ave. burned in 1907; The big Odd Fellows Opera House burned in 1916, most people believing it of incendiary origin but nothing was ever proved about it. Gradually, as the population so sharply declined, the churches could not carry on and had to consolidate or give up entirely. Thus Baptists and Presbyterians and Congregationalists united and kept the brick Congregational Church which had been erected on the site of the huge tent building which was the first church in the community, and sold off their other frame structures. Some of the school buildings were similarly wrecked through the years, only one being destroyed by fire when we were of Junior High school age. The final consolidation of the school buildings took place some years ago in the former high school building, and the last of the grade schools was torn down. It had occupied a prominent position high on the big hill on which the Catholic Church and the reservoir are located, and its profile missing from the scene has always seemed strange to those of us who were accustomed to it.

One who has not lived in Cripple Creek cannot realize that the absence of mines in the city limits did not mean that the city was not the center of mining. With the splendid street car system which ran every fifteen minutes, the miners could live in the comfort of one of the most modern cities of its time, with pure water, a fine sewer system, electrically lighted and well served by three train systems. This soon robbed the smaller communities of population, which desired to live at the lower altitude where the schools were better equipped and the stores handy. But I can remember seeing streams of miners walking down from the Altman and Midway area along the miners' trails to their homes in Cripple Creek. They used a peculiar loose-jointed knee action on the steep trails which enabled them to walk numbers of miles daily without too much fatigue. Being good walkers was the pride and common practice of people in the district, particularly those who lived there from earliest times. My father was one who went fishing on Sunday for years during the season, often walking twenty-five miles during the course of his expedition, coming home with a fresh bouquet of wild flowers in one hand, a creel of trout on his hip and a tin bucket of wild berries in season. He also gathered mushrooms when he could find them, and was an expert on birds, flowers, trees and rocks of the territory he knew so well. Until the age when automobiles had really taken possession of the towns of the district, this was not an uncommon kind of habit because the streetcars were wrecked about 1919, as I remember, and that meant one had to use the fewer trains or shank's ponies. Since I was in high school after 1919 I learned to glory in being a good hiker, and mountain climber too, and my Camp Fire Group often hiked twenty-five miles on our Saturday excursions around the District. I have gone on many expeditions up Pikes Peak from our side of the range. This trip up Pikes Peak was a sort of local hobby, and in my father's day the young men of his Altman hometown often would take a hike to the Peak and be back in a few hours time, thinking nothing of it. . . .

There was always great rivalry between the various communities and much of it not friendly. There were gangs of boys in each town who made it their business to try to conquer those of the other towns. Those from each town had a nickname and had to be tough to take what came or else be woeful indeed, when they swung down off the street car or train on their visits into Cripple Creek or Victor. The Victor gang was called the Red-necks. I suppose the Altman boys were the toughest since their fathers were the union men and they had a tradition of toughness to maintain toward the lowlanders. Cripple Creekers looked down their noses at the other towns as a result of some of these practices. There was strong policy toward not permitting foreign born workers to come to the district, that is those from southern Europe. It was always felt that there were more foreign born people in the smaller communities than in Cripple Creek, and that may have led to some of the attitudes of the towns toward each other. Cripple Creekers very strongly desired to live as normal a life as if they had been in Denver or Colorado Springs, Kansas City or Chicago, and made every effort to do so, so that any stories that tended to make the town sound like the typical tough mining town were much resented, and all Cripple Creekers are sensitive to any derogatory remarks about the town on these lines, even those who lived away for generations in other parts of the U.S. This accounts for the strong annual picnics in California, Denver, Colorado Springs and elsewhere where old timers get together each year to be homesick together for the dear dead days in Cripple Creek district. Recent years have caused much better cooperation between Victor and Cripple Creek and perhaps some day the schools will unite, a possibility which would have caused a minor war, in my period there. . . .

Until the past few years, perhaps since the World War II period, Cripple Creek did not suffer but rather benefitted from its distance from the biggest mines, and now that the mill is being built at what was formerly Elkton, this condition will continue to be favorable to those who do not want to live with mine dumps in their back yards. However, I think that the lineal descendants of the rock miners have always preferred to live close to their leases and work, and so perhaps there are more of the very old time miners left in that area than at Cripple Creek. At any rate, most of the surviving childhood friends of my father live on the other end of the district. . . .

The World War II Veterans seem to have moved in on the control of local affairs in Cripple Creek at present and they seem to be making plans to bring tourist trade on a large scale, much as Central City and Aspen have succeeded in doing. Their biggest problem no doubt has been the old time dislike of notoriety or playing up the boom time history of the camp. A ski course was established several years ago by one of the veteran citizens of the District, Mr. Jess Vetters, who also owns the townsite of Gillett, where he has ranched for many years and where the former school house is now incorporated into his comfortable summer home on the ranch. Mr. Vetters has handled most of the freight hauling business of the district for many years. His people were early in the district, and an uncle was one of the city officials in the palmy days of Cripple Creek.

The Imperial Hotel, long the property of a story book family, the Longs, was sold several years ago and the new proprietors instituted the summer theater in their basement and special dinners, which started a train of visitors to the District and has contributed greatly in these lean years, I am told. However, they have had a hard row to hoe, for the old timer citizens of the city have not wanted them to play up the dance halls and the shady characters who once resided on Myers Avenue and they have had to be careful not to give offense in that direction. . . . At any rate, at one time, there were a lot of well to do people living in Cripple Creek who later made Colorado Springs and Denver and other points their homes who wanted no remarks made about the town in which they lived which would keep away the sober business type of people they wanted to attract to Cripple Creek. This accounts for the fact that Julian Street so infuriated everyone. [Julian Street was a freelance magazine writer from the East.] At the time he came there there were only a handful of the notorious characters left in the entire district because the war had made the place have a depression and there wasn't any loose money around any more. The one poor soul he interviewed was a half demented creature who was a sort of local joke, but who for reasons of age and lack of support was unable to make arrangements to leave the place and so clung precariously to some sort of existence. To the proud and dignified members of the various fraternal orders, the ladies' social and educational clubs, the active church members, the local city officials, the county officials, in short to everyone, the painting of their quiet respectable community was a blow below the belt, uncalled for and unmerited, and I suppose if Julian Street's reputation as a writer remains alive it will be only because the intense contempt of the Cripple Creekers was so publicized that he will thus be remembered.

Are you familiar with the study of "The Labor History of the Cripple Creek District" made in 1906 as a Ph.D. thesis of Dr. Benj. McKee Ractall(?). This account is the best thing of its kind that has ever been done about the district. It is accurate in every detail, . . . and completely unbiased. I first ran into it at Colorado College in a reference in a M.A. thesis I had borrowed on the History of the First Fifteen Years in the District. (It was published by the University of Wisconsin). (The author was given an Hon. Doctorate by

Colorado College in 1949). His accounts of the events of the two strikes are right and should be the source for consultation about the labor events in the district. The newspapers of the time were very biased and often incorrectly informed by one side or the other. . . . The newspaper men of Cripple Creek proudly boasted that they made the famous camp. Certain it is that millions of dollars from school teachers, lawyers, doctors and clerks were secured because of the publicity they sent out, and that the thousands of pock marks on the hillsides were often the token of their misplaced faith in outright fraudulent mining stock companies. . . .

Artists who have made paintings of Cripple Creek . . . do not grasp that the thing which is Cripple Creek is *space*—its streets were broad, if hilly, for the most part they were comparatively level in its greatest extent, in great contrast to the precarious perches in other towns of the district. Its homes were neat, well fenced, many full of flower gardens, it had miles of wooden sidewalks in its boomier days, and well kept up streets. The sight it presented at dusk as the great blocks of city lights came on stretched far out over the bowl formed by the rim of mountains, and punctuated at the west by the cone of Mt. Pisgah, was one of the fairest and most memorable sights that my memory holds. This view from the highline drive is the Cripple Creek that all natives love and hold tight in their fond recollection. There is always a sense of excitement coming to one as he tops Tenderfoot Mountain and there below suddenly lies the city, laid out block after block, substantial, miles of brick buildings, and a home of the finest kind of people. That song "Dear Hearts and Gentle People" makes me think of the way I feel toward those I have known and still know in that city that was. It is a gallant city, a one-legged veteran still self-supporting and stumping about grimly independent, if you please! Consult Colorado's records for war bond support, Red Cross support, kindness to and entertainment of the homesick soldiers stationed in the vicinity in War II, and remember that the Gold Order was a death sentence to the community, yet they did these things and hung on! Such faith and loyalty may be considered soft-headed by more sophisticated moderns in other parts of this U.S.A. Perhaps they are the last of their breed, the final frontiersmen of the west. But they *are Cripple Creek*. This is the spirit the artist must catch—the gardens of the Columbine specialists growing in the cultivated foundations of what was once no doubt a bawdy house on Myers Avenue, the small city parks and playgrounds, they are set up in what were once disheartening holes left by ruthless wrecking operations; the swimming pool, heated and supervised, in the highschool grounds to which bus loads of swimmers are brought all summer long from all over the district; the school band which has taken honors all over the state; the impressive and remarkable records of numbers of children sent to college from these little towns; the churches too poor to support ministers for a generation which nevertheless have had an active Sunday School program through the years which had given a far better religious training than any city church I have ever attended; the love of nature these people all have and practice—they choose to live here so that they may freely flick the trout streams in summer and hunt for deer and grouse in the fall, and cultivate their small but precious salad patches and flower gardens and lawns in the brief summer of two miles high.

Cripple Creek, 1938.

Above all they love the views—the distant Sangre de Cristos, the colorful trembling aspens and the wild flowers they can see as they drive about,
the side of Pikes Peak which they consider superior to any other view of
that noble monarch of the plain. You will forgive me for sounding off—but
Cripple Creek must be an emotion felt before one can paint it well. . . .

Sincerely and cordially yours,

Edith Stuart Jackson

In a letter just one week later Mrs. Jackson continued her interesting recollections:

As nearly as I can discover, . . . there were probably as many as four different little settlements in early Cripple Creek and this probably accounts for
the diversity of information on who were the first grocers, mayors and so on.
They had to locate near springs and sources of water, of course; this meant
Old Town at first because there was the Cripple Creek which then had quite
a flow, and a spring about where the Midland Terminal Trainshed was. There
must have been water also of some kind out in the area of west Cripple Creek
where most of the early placer mining went on. Then down just south of
Bennett Avenue there was another spring, which was enlarged into a well by
an old timer and from it water was sold by the bucket and barrel. Then up
Poverty Gulch, since it was here there were some very early mines; and on the
hill north of it which became the "Cripple Creek Heights" there was a well
or spring very early and must have been some population around it.

This meant that each little community had reasons for wanting to be the
center of things and different preferences for names. Old Town, however, was

the first place chosen for the first family to live there and their house was the first built for a family, according to a woman who was a member of that family and is still alive in the town. Her Dad also built a hotel on Main Street which was one of the first; the early ministers lived there and so on. By the time he had done that, however, there were some buildings up and down about the spring near Bennett Avenue in that part called Fremont, and a large population which lived over on toward Mt. Pisgah which later became West Cripple Creek. This part had a street named for Pete Hettig, and an old map shows his store location way out near where the County Hospital now is, I think, the extreme northwest corner of the city. He seems to have had his place behind what was later the large outfitting store of Work company just below Bennett Avenue. Later, I have been told, he went to Altman and had a store, too; some people say he had the first store in the Cripple Creek town. I think he had the first in Fremont but that Mr. Gowdy did in Old Town, and maybe Hettig later went to West Cripple Creek and held a small store there. One woman told me that he and her Dad were pardners in a number of the very earliest prospects there and were "assessed" out of their interests when they tried to get outside capital to help them develop their mines or were unable to do so and had to sell out for lack of funds for small sums. I have not yet been able to check this information but intend to do so at the land office in Pueblo where the first claims were filed. . . .

I have a particular interest in Peter Hettig. . . . He was the second husband of the mother of the woman who married one of my mother's brothers. This sounds involved, but there it is. I am anxious to learn all I can about Hettig's life and death because it might help me to trace my cousins, whose father died when they were small and their mother and grandmother took them out of Colorado and our family has no idea what happened to them and would like very much to know. I have a picture postcard of an old prospector and his cabin which is supposed to be Hettig but could instead have been Mr. Gowdy, I think, who really did handle the mail and have the grocery store in Old Town (Cripple Creek Town) in the earliest months. He was from Colorado Springs and borrowed thousands while he was trying to collect from the Government for his services as first voluntary postmaster. Mr. Howell's account in "Seeing Cripple Creek" in 1906 tells about him and confirms what my Uncle had said about him, since he knew him, being fellow residents from Colorado Springs.

The two newspapers which were first set up seemed to have been divided on their choice of location, too, one being in Old Town, or Cripple Creek Town, and one in Fremont.

Another thing which I note: the very earliest people into the district came via the Florissant route, over the Bear Hills (or Bare Hills) route from the west, or up the canyons from Colorado Springs, such as Bear Creek and Cheyenne, and there was also a foot trail right over the flank of Pikes Peak which the walkers took. I do not know how soon they came up the Phantom Canyon way, but some probably came in that way also. In other words, when one is afoot or on horseback, as the first ones had to be, they came up canyon routes as the natural way to travel and these in turn became the future roads and railroad routes, except from the west. I think this one may have come through the Box Canyon which has always had severe washouts and never was a dependable road but probably connected with the Shelf Road which led from South Park to Canon City, and was a traveled road by the ranchers of the Park who went to Canon for provisions and safety in the Indian raids. It was from these ranchers and from Canon that the very first people came who responded to the Womack discovery. Most of them had been prospecting and mining in the South Park area and were experienced in that way. The people who came from Colorado Springs and who later controlled the mines were the real tenderfeet, but they were business men and they grasped the possibilities of making something out of the camp which the prospector did not have the resources or influence to do. Then too, the wealthy mining men from Leadville and other older areas soon came in to put their money in the baby mines and take control of them. Moffat was such a one. So was Irving Howbert of Colorado Springs, and others. Many of the experienced mining managers who later came were from other big districts and many had already had unpleasant experiences in the labor relations in the older camps. This probably accounts for both miners and managers so early getting into difficulties in the Cripple Creek history.

I have just received a letter from a man who was a boy with my Dad in Altman which confirms my surety that quite a few people lived there early. He drew me a map showing and naming the names of those who lived there from early days. He says that they built their school in the winter of 1893-94 when he first lived there, but he came in quite a long time after my folks lived there, and school may have been held in a rented place just as it was at first in Cripple Creek, where the Elkhorn restaurant in Old Town, the Iron Clad Dance Hall in Fremont, the old Methodist Tent and so on served at various times for schools until buildings were built.

Sincerely yours,
Edith Stuart Jackson

With the Cripple Creek Mining District such a bonanza, it was to be expected that the countryside beyond that area should be explored and prospected in all directions in the hope of discovering similarly paying mineral deposits. A few of the small mining camps outside the immediate District are included in this chapter, the first of which is Manchester.

It was from my friend, Richard M. Pearl, professor of geology at Colorado *Manchester* College, minerologist, and author, that I heard about Manchester. Without his specific directions I would not have found it. One summer afternoon my driver, Russell Olin, and I, armed with Richard's two brief letters, drove west from Divide (west of Colorado Springs) until we were positive that we had reached the ghost site. We had carefully followed Richard's directions:

Aug. 15, 1950

I have just checked the Manchester directions again with the folks who own the land there and drove us there:

Turn north at Divide. Roads branch at gateway marked "Ute Park"; turn left there, continue to cattleguard, then you are in Pikes Peak National Forest. Continue in the same direction to a very dim sign which says "Manchester Creek—Wagon Road only." Turn to left there, go down a steep little incline, and follow 2 miles to the ruins. Holes, foundations and corral. Total about

8½ miles from Divide. If you pass the ghost town, you will later pass the series of beaver dams I wrote about, and you will know you have been too far. Manchester is on Manchester Creek but I don't think the road comes out any special place at the other end, so the Divide entrance is best.

Best of luck.

Sincerely,
Richard M. Pearl

Six months earlier he had sent me information about the place:

Feb. 25, 1950

Manchester is a town whose foundations are still visible; it is about 8 miles north and west of Divide, along Little West Creek in what was the Little West Creek mining district. There was a post office and general store, plus a saw mill there, and of course some houses. The claims were patented there in 1885; neighbors of ours still own a piece of the original patented land, isolated in the midst of the Pike National Forest.

No doubt you have investigated the postal records for your research. The postmaster in Denver, who handles all paychecks for distribution to postmasters throughout Colorado (rather than from Washington), told my father-in-law that there was once a total of over 900 post offices in the state, whereas there are now fewer than 600. Some names have no doubt been changed, but the rest of the discontinued names must largely represent ghost towns.

Cordially,
Richard M. Pearl

There was little left to indicate the site except a few small dumps and prospect holes and some rotting lumber where cabins once had stood. Grazing cattle stopped to stare at us, and as we started back toward Divide, a rabbit bounced ahead of us up the road.

West Creek Mining District

Another section that caused a flurry of excitement and attracted hundreds of persons in 1895 was the West Creek District, halfway between Woodland Park and Deckers. Letters from two men who were attracted to the new gold field explain how it started and why it fizzled out so fast. Stanley Bush of Gering, Nebraska, writes (Feb. 26, 1951):

I will be happy to supply you with any information I can about the Ghost camps of the Pikes Peak Region. Here in Gering I have very little material, but when I am in Green Mountain Falls next summer I'll dig it out and see what additional information I can furnish. There are many old settlements . . . in the Ute Pass Area, for it was carefully explored in connection with the gold rush at Cripple Creek. . . .

Near West Creek, the boom town that was originally started when two old prospectors salted a hillside, there is an old cemetery that is worth study. It is just two miles above the town of West Creek, on the North side of the road.

Then southwest of West Creek, on the Trail Creek Road, is the Gem Mine. Operated by a hermit that lived on bread and honey, it is still a mystery of the area, for the hermit never disclosed the source of his Turquoise and smokey quartz hexagons.

The Gold Standard Mine that I mentioned is on the Rampart Range, just overlooking the town of Green Mountain Falls, and used to be a mining camp employing some 60 workers. Gold was found, an assay office constructed, and some work begun, but the difficulty of removing the ore from its precarious mountain site halted operations. . . .

Then across the valley, above the town of Green Mountain Falls, stands an old powerhouse that supplied the town's lights and power before lines were strung from Colorado Springs. Deserted now, and partially destroyed by rock slides, its old machinery and control wires make a historic scene.

Further down the valley, above Manitou, Colorado, a road that I have never traveled winds its way to a small valley and a true ghost town that was a mining camp of the early days. Called Crystal Park, the camp can be reached on foot, but far more information could be obtained from Ralph Brenton, 4004 Chestnut, Kansas City, Missouri, for I believe that he once lived in the camp for a short while—or has papers and early pictures of the buildings.

Back on top of the Rampart Range, and some distance North of the Gold Standard, is one of the earliest cabins in the area. The homestead home of Henry Brockhurst, early rancher, it is a typical log house that was used for years as a home, and then as a stopping point for overnight pack trips and steak frys. Henry is still living, just above Green Mountain Falls.

Northwest of Woodland Park is the Skelton Ranch, an old dude ranch that operated during the time when guests arrived in Denver by train, and then took the overland stage to the ranch, coming up Jarre Canyon out of Sedalia. I was there while all the buildings were standing, but now they have been destroyed by the US Forest Service. The Service may have some information about these buildings. There was a large meeting hall, many small cabins, a large dancing pavilion, bank, post office, and barn occupied nearly half a block. I've explored them all, but was too young at the time to think of getting pictures or records. Now, I hear that they have all been destroyed.

Theodore H. Proske of Denver describes his experiences in West Creek as follows (Dec. 5, 1951):

In 1895 when the boom started at West Creek, I was induced to go there as the report was that all thru the mountains around West Creek were mineral lodes open for the prospector. I met an old timer who wanted me to join with him in locating claims, there was plenty of granite with streaks of iron rust that were supposed to carry gold. Like many others we located two claims, did the required amount of work, and waited for someone to come and buy our claims, but when the time came to survey these claims, we found that later prospectors had also made locations, and we found 8 discovery shafts inside our claim lines, a pair of men had secured orders from people in Denver to locate for them Claims, then dig these location shafts and collect $100.00 for each claim located, get them surveyed and send the certificate to their customers in Denver, who would then feel they were mine owners. This was in

1895; in 1859 no doubt this practice was started, if paying ore was found, next came law suits.

In the West Creek District the real estate boomers located and surveyed some 24 townsites and sold lots in coming great mining towns. At the height of the West Creek Boom about 10,000 people came in. When this boom busted, Puma City was started, and like many others, I went there, but this boom was not near so big as West Creek.

Puma City or Tarryall

In 1947 I met Otto Groening, eighty-three years old, and one of the original residents of Puma City. He was sitting in front of the false-fronted general store, and he seemed pleased to find someone interested in hearing about the past. As nearly as I can reconstruct from my hastily scribbled notes, this is what he said:

"Puma City was started in 1896 by Gilman Bros. of Denver, real estate agents, when the Cripple Creek boom was leveling off. It was a good location, for the Colorado & Midland ran through here. I came here in 1896. At that time there were several small camps between here and Lake George, including Gold City and O'Brien. O'Brien was started by a Leadville man. Hayman was laid out in 1898. Today the only building left at the site of Hayman is the white schoolhouse.

"The place had five saloons, three dance halls, three livery stables, a Wells Fargo Express Company, three stages from Jefferson, one from Cripple Creek, and the main one from Lake George. Both Coors' and Zang's breweries freighted in beer; Gilman Bros. ran the general store. The boom lasted a year and a half. Then people from Colorado City became interested and helped with the development. As more people came in a lot of boardinghouses were built. The place became crowded. There were eight to twelve men on each property. By 1900 the town was dead.

"From the first, the surrounding country was ranching. The ranchers hated the miners. The prospectors killed cattle, so the ranchers blamed all the miners. Only one mine near O'Brien shipped ore—the Boomer which was lead and zinc. The other mines were lead and silver.

"The name of Puma City was changed in 1900 to Tarryall after the U.S. Post Office found another Puma City. Long before that date the camp of Tarryall near Fairplay had disappeared.

"There was a big trial here that I remember. It was recorded at Fairplay at the courthouse. Two saloon keepers, Craig and Cox, each kept dance hall girls. Cox's place also provided gambling. Craig's girls went to Cox's place and gambled. Craig told Cox if he induced the girls to his place that he'd kill him. The girls continued to go there. One night there was a square dance with three sets of people. Craig came in, went up to Cox and said, 'I'm going to kill you, Cox.' He had his Winchester on his shoulder. Cox tried to get away, but Craig shot him. The girls went out through the windows. The trial was at Fairplay. Craig went to a lawyer who was a big shot who said he would clear Craig but that he'd take everything that Craig had and more. He did, and Craig was cleared. Mrs. Cox was left with nothing. She left Puma City in 1897."

"What did you do after the town died?" I asked Mr. Groening.

"I was trained as a harness maker, so when the boom was over I advertised my trade, and the ranchers brought me their work. Later on I bought land and sold it as ranches to dude ranchers and tourists. The last seven or eight years I've owned the whole townsite.

"You say you've been to Pemberton and West Creek. I knew Pemberton. The camp was named for him, since he started the place. After he sold out and came over here that camp was called West Creek. Pemberton ran a stage line and a livery stable here in Tarryall. We had daily mail service then."

Two years later I received a note from Marshall Sprague:

> I had a long talk with Otto Groening at Puma City some weeks ago and he remembers talking to a lady some time back that must have been you. I noticed the name Sterritt Ave. on his plat and found this was Matt Sterritt from Florissant; Matt was the man who located the Deerhorn in Cripple Creek and had his claim jumped by Clint Roudebush, the Leadville promoter (crook?)—but the miners rose in wrath and chased Roudebush out of town. Matt took a lot of money he made in the Deerhorn and sunk it in the Puma City foolishness; he lost another $20,000 in a Denver bank failure.

Balfour, which was about fifteen miles from Hartsel, was another mining camp in South Park (like Tarryall) of brief duration. According to Carl F. Mathews of Colorado Springs: *Balfour*

> I do not think it lasted longer than 1895, as when the prospectors found no gold and began to reason why, it was evident the ground had been "salted," as was the case with West Creek-Pemberton, north of Woodland Park. Twenty-five years ago there was absolutely nothing there except for a few depressions showing where basements had been excavated.

CHAPTER VII

Leadville, the Cloud City of the Rockies

Leadville, Colorado's rich silver camp, to which thousands stampeded in 1878 and 1879, is still a mining town. Occupying a gentle slope in the Arkansas River valley, with the Mosquito Range rising above its eastern boundary and Mount Massive, the state's second highest peak, forming a backdrop to the west, its cross streets lead up to the mines from which the city's wealth was extracted. The first prospectors who washed the sands of California Gulch in 1860 were searching for gold and they found it hard to separate the bright particles from the heavy black sand that clogged the riffles of their sluice boxes. By the end of 1861 the gulch was worked out and miners left, searching for better diggings.

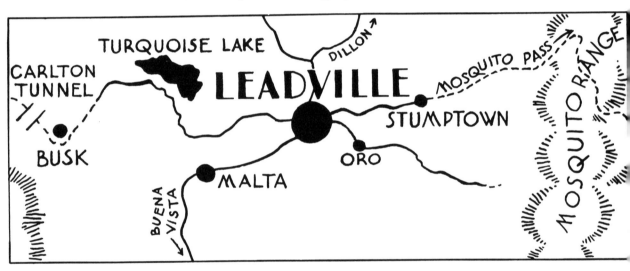

It was 1878 before the black sand which earlier miners had cursed was found to be rich in carbonate of lead carrying silver. That same year the first discovery of importance was made by George Fryer and his partner. Not long afterward, the Little Pittsburgh, from which H. A. W. Tabor made his fortune, was uncovered. Even before John Campion's Little Jonny mine made him a multi-millionaire, Leadville's streets were teeming with milling crowds all bent on making their fortunes before the rich deposits were exhausted. The boom lasted until the demonitization of silver in 1893. Then labor troubles harassed the camp; mines that were closed down filled with water, and not until old properties were worked for zinc and lead in the early part of the twentieth century did Leadville resume its position as a sound mining community. In recent years the Climax Molybdenum Company, whose mine at Fremont Pass has an elevation of 11,318 feet, found that working and living at this altitude was detrimental to its employees and their families. The company therefore built a suburb at Leadville (only 10,000 feet high) to which it moved its people and from which the mine workers commute to their shifts.

Leadville is proud of its past as well as its present, and I was not surprised that many people wrote me their recollections of Leadville in its early boom days as well as in the early twentieth century. No one described arrival in that roaring camp more graphically than my good friend, Mrs. Clara Gaw Norton, now deceased. She was six years old when she made the trip and her brother was sixteen.

After her parents lost their home in Idaho Springs by fire in 1878, her father rode horseback to Leadville to prepare a new place for his family. Shortly thereafter, her parents drove across the Divide in a light wagon drawn by fast horses, while a freighter followed more slowly with a load of household goods. The trip took two weeks. As there was no hotel in the mining camp when her parents reached there and their new home was not completed, they gratefully spread their mattresses on the floor of a friend's cabin. As soon as practicable the mother sent for her children, who were to drive over alone in a wagon with the rest of the furniture. This was in March when roads were muddy, and the heavily loaded wagon sank more than once to its hubs. Even the pole, used as a fulcrum, was not heavy enough to free the wheels; so Clara's brother made her add her weight by sitting on the tongue. Each time the wagon lurched ahead she fell off into the mud, much to her annoyance.

While crossing South Park, they became lost in a snowstorm but relied upon the horses to keep to the road. Late that evening the weary beasts floundered through a ranch gate and stopped before a house in Hamilton. Half frozen with cold, little Clara remembered the door opening and a woman's voice saying in amazement, "Why, it's a little girl." They spent the night at the ranch house, and by morning the storm was over and they set out again. Before long other freight outfits caught up with them, and together they began the grueling haul over Mosquito Pass, 13,188 feet above sea level. A steady procession of vehicles struggled up to the summit, and whenever one got stuck all the freighters helped to extricate it and get it started again. Even a stage carrying several prostitutes to Leadville's State Street became bogged down. Mrs. Norton remembered this incident vividly, for one of the girls stole and ate the lunch that had been packed for her by the ranch house woman, and Clara was furious.

Once past the summit, the trip was less harrowing. While they were still some miles from Leadville, they heard a rig approaching at a swift pace, and her brother thought he recognized its sound. He was right, for the driver was their father who, worried by their delayed arrival, had come out to meet them on the trail.

When John B. Stevens of Oneida, Illinois, wrote of his father's part in the Leadville discoveries, I was glad to be able to refer him to a source which I had used in an earlier book. His letter read (April 5, 1951):

> My father, George Stevens, tells how he and my uncle, I. N. Rafferty, discovered and panned gold where Leadville, Colo., now stands some time in the fifties when there was nothing there but sagebrush. Both have passed on many years ago and I am an old man but I still remember as a lad hearing them tell of their experiences for a number of years throughout the west in the fifties. They were quite successful in their mining operations and came to Illinois and invested in farm land. Is there any account of their discovery of gold in your book?

In my booklet *Cloud Cities of Colorado,* published in 1934, I quoted from *Reminiscenses of W. P. Jones* (a pamphlet at the Historical Archives of the Society of Leadville Pioneers):

> During the winter of 1859 there were rumors a plenty . . . of gold in abundance in the Arkansas River across the range. In March we made a party of four, E. Johnson, Old Man Boon, Sailor Jones and Myself. . . . As we were crossing the South Park our little old wagon broke down. . . . Then we were caught in a blizzard . . . so that the Stevens party had gotten ahead and gone up California Gulch.

From John R. Pitts of Leadville (March 23, 1968) I found out where the camp called Bromley stood. He also mentioned Alexander Chisolm who spoke of his father's and his uncle's mining interests:

> Your sketches bring back memories of my boyhood. I well remember the Montview House, and your sketch of Independence causes me to recall the days when my father on Sunday outings would show me the old shop in Independence where he had worked as a blacksmith.
>
> Do you have the book: "Gleams of Underground" by Alexander Chisolm? It is an autobiography of a man named Alton Lang who was born in Bromley at the foot of Independence Pass. According to Chisolm, "Up near the head of Lake Creek I found no traces of the former camp of Bromley."
>
> Chisolm himself informed me that he came to Leadville in 1896 when he was four years old. His uncle came to Leadville in 1877, and in 1886 he found a small mine near Independence. He slightly knew George Fryer, who made the first important discovery in 1878, of lead-silver ore on what came to be known as Fryer Hill. Chisolm's father, Kenneth, came to Leadville in 1896. As a teen age boy Alexander Chisolm spent his summers in Leadville working on leases held by his uncle and his father.

C. W. Kingsley, who wrote me several letters full of anecdotes and information gleaned from many mining camps in the state, provided the following item:

> Did you ever hear of this story of State Street in Leadville? A young fellow, 17 or 18 years old, came into Leadville on the bus. This was in or around 1909 or '10. He hit State St. and asked for something to do as he was broke. The street had him splitting wood and washing windows. He evidently in a few days felt happy that he was getting the necessities of life, for as he was washing windows he began to sing. It was not long before his audience were many. The girls on the row said it was a shame a boy with such a voice should be down there. Well, as always, the Scarlet Ladies did something about it. They collected enough money to send him to Denver and I believe to start him looking for a voice teacher.

As new Leadville mines were opened and developed, not only within the townsite but up every gulch and mountainside to the east, small settlements sprang up at certain of the new properties along Jonny and Breece Hills and in Big Evans Gulch. Northeast of the rapidly growing silver camp, beyond Fremont Pass, Robinson and Kokomo became centers of activity. Several of these suburbs are mentioned in the letters that follow.

Leadville, 1933. Mines and Mosquito Range beyond city.

Leadville, 1933. Jonny Hill looking west.

Malta,
Evansville,
Finntown,
Cletemore,
Graham Park,
and
Adelaide

George Feltner of Littleton (Feb. 28, 1950) added to my knowledge of Malta, the railroad junction for the branch line to Leadville, and pinpointed the location of Evansville, Adelaide, and other small camps:

There is Malta, of which practically nothing is left. But there is a cemetery in the trees on a sedimentary hill to the south of there. A part of that cemetery was moved from Oro City when the excitement there died down. Some houses were moved. Some stores, too. And even the dead went with the living to the new camp. There Tabor ran a store, and in that store a friend of mine used to while away his time with his brothers and sisters while his mother worked. Augusta Tabor took care of them.

. . . You are right about Evansville. It was in Big Evans Gulch, about at the foot of Breece or Jonny Hill. All that was left of it when I was big enough to roam about the hills around Leadville was the schoolhouse. It remained there, intact and shuttered for a good many years while weeds, grass, willows and evergreens crept over the disappearing foundations of the houses that once stood near it.

I never did learn much about that settlement. It must have been very shortlived. . . . There was another little settlement just about at the entrance to Strayhorse Gulch, where Finntown lies. It was called Cletemore or Cleatamore. I have forgotten the spelling of it, but it was named for a place in England where most of the inhabitants had come from. It may really have been a part of Leadville, one of the tentacles, so to speak. Such as Jacktown was in the southwest part of Leadville. Anyhow, it was a rough-and-tumble town, a dangerous place where fights and killings were frequent, as I learned.

Graham Park was above Finntown and to the south of it. There was a hotel or two in the place. A hotel burned down there once with some loss of life, one of its former inhabitants once told me. There were the inevitable saloons, of course, and a store or two and a school. About 15 or 20 of the buildings there were still standing in the middle of the 20's. Bootlegging was carried on in one of them. To the south of that house was another, almost fully furnished and in good condition. At one time several hundred people lived there. Several families lived there during World War I and for a few years after that. And I believe school children attended the school there instead of in Leadville. People living there worked at the Wolftone, Greenback, Louisville and other mines in the vicinity.

Adelaide was on the Jonny Hill road, beautifully situated in a little park. It was inhabited until some time in the twenties. By the thirties only foundations of the school, church, townhall and so forth were left. At one time, in the 90's, my father owned a house in that settlement. He worked at the Jonny mines. A friend of mine, a teacher I went to school to at the Central School in 1924 or '25, used to teach school in Adelaide. She also taught in Oro City and at Soda Springs, west of Leadville. She was born in one of the little towns near Leadville and has spent all her life there. . . . In about 1922 my folks rented a house to some people that had been burned out of their home in Adelaide. At that time a number of Finns lived there, maybe because Finntown was overcrowded.

Those Finns used to have a great time up there—more drinking and fighting. But they had socials, too, that were well and enthusiastically attended.

I attended a few in the late twenties while in High School. The affair opened with a play in Finnish. There was a big hall in the town (now moved to Leadville) and in the east end was a big stage. The play, as I remember, was pretty good. A Finn friend interpreted for me as it progressed. There were some good comedy scenes which brought laughs because of the action to us who could not understand the language and because of both the dialog and action to those who could.

After the play there was a dance. The music was a piano, a violin and an accordian. Many of the dances were native to Finland, but they also catered to the younger generation. Following the dance refreshments were served. There must have been a dozen cakes, sweet rolls by the dozens, cookies, cupcakes, coffee and tea. And I believe there were some little sandwiches. These socials, I was told, had been held since some time in the early 90's, and had changed very little if at all. I remember the admission price very well. It was 25¢. I snowshoed back to Leadville at about 1:30 or 2:00 a.m. Where, in these days could one get a bargain like that?

Olive Eckhardt Freudenberger of Salt Lake City describes Leadville as she knew it (Feb. 9, 1954):

I spent all my childhood and the early part of my adult life in Leadville. I can well remember back in 1905 when Jonny Hill was a mass of lights at night—the mines were all working a twenty-four hour shift. Some down-town mines were working at this time, too. They used to haul high grade ore past our house and the mule skinner, who used to handle a team of six to eight horses, used to brake the heavy wagon while the man beside him held a gun and two men on top of the load had guns across their laps. It always seemed funny to me because no one was ever held up. Down in my heart I used to wish some one would.

I have been through the Busk "Ivanhoe" Tunnel many times but I know it hasn't been opened for years. We used to take the old Midland train to Glenwood Springs for our summer vacation. It was a steep climb along the side of the mountain up to the tunnel and sometimes the train would stop and the passengers used to scramble off the train to toss snow balls at each other in June. When the train went through the tunnel, the windows were shut tight but the smoke and cinders oozed into the car and made the coach so smoky that we could hardly see the people in front of us. Once through the tunnel the train seemed to fly down hill to Glenwood.

Ivan C. Crawford, now deceased, whom I knew as a professor at the University of Colorado and who later became Dean of the College of Engineering at the University of Michigan, commented on my illustration of the interior of the Pioneer Saloon in Leadville (Sept. 6, 1949):

I was born and raised in Leadville, Colo., and my parents arrived in the state at a fairly early date—my father walked from Denver to Leadville in 1879 and mother came to town in a stagecoach in 1880. . . .

I must confess that the sight of stools before the bar in the Pioneer Saloon caused me to quiver a bit until I stopped to realize that you made your sketches in very recent years. As a paper boy on the *Leadville Herald Demo-*

Leadville, 1942. False front.

crat, I left the paper at the Pioneer Saloon for a number of years. At that period I am sure no one could have found any excuse for furniture of this type in front of a State Street bar, or any other bar for that matter.

Nearly a year apart I received letters from two men, each of whom asked the same questions about a mechanically operated musical instrument that I had sketched in a Leadville cafe. I answered each as best I could, although I could not guarantee that the building which housed it was still standing. Later I discovered that it had been razed. Their letters explain themselves:

Dec. 17, 1966

I am writing this letter regarding something mentioned in your book "Stampede to Timberline." This past week I borrowed a copy from a friend because I am interested in the old ghost towns and mining camps of the west. Also I love to collect the old coin-operated pianos that were popular during the 1920's. In reading your book, you mention that while in Leadville, Colorado, you stepped into the Crystal Palace Hotel Cafe, then operated by a John Bernat. You describe a fantastic "Wurlitzer automatic player" in the back of the room, with stained glass panels, etc. You also mention sketching this machine. I would greatly appreciate it if you would send me a copy of your sketch of the outside and inside of it and any information you might be able to obtain regarding its present location or existence.

Very sincerely,
Donald Rand

March 7, 1967

I am the owner of the large Wurlitzer Orchestrion which used to be located in the Crystal Palace Hotel in Leadville. I read about it in your book "Stampede to Timberline". . . .

My friend advised me also that you had a sketch of this particular machine when it was set up in operation. Would it be possible to purchase the original sketch from you so that I may frame it? Or, could I obtain a copy of the sketch? I am quite interested in the history of these machines and would like to know all that you know about it.

Is the old Crystal Palace Hotel still standing? I was told that my particular instrument came from the Silver Dollar Bar or the Silver Dollar Saloon which was in the hotel. Is this true?

Sincerely yours,
Q. David Bowers

Although I am not a railroad buff, many of my correspondents are, and to them I am indebted for facts and tales about railroading in the mountains and through the tunnels that pierced the Continental Divide. First let me correct a mistake in *Stampede to Timberline* that was brought to my attention by T. B. Aldridge of Denver (Sept. 14, 1950):

At the top of page 153 in your book: *Stampede to Timberline,* you state: "The South Park never did reach Leadville . . ."etc. As careful as you were otherwise with this most wonderful book, I wonder just how this error happened? I would very much appreciate a letter from you, to insert at said page in the book, that it is in error.

According to a quotation from the LEADVILLE DAILY HERALD of Feb. 6, 1884, in Mac Poor's DENVER, SOUTH PARK AND PACIFIC: "The completion of the South Park railroad to this city, which was shown by the running of the first locomotive over the new line into Leadville yesterday, is an event of no small importance."

One of the places about which I received quite a number of answers to my inquiry as to its whereabouts was Busk, which turned out to be a construction camp at the eastern portal of the Busk-Ivanhoe tunnel which burrowed through Mt. Massive to provide a railroad shortcut to Aspen. Lt. Com. Lester C. Harlow U.S.N., of Alexandria, Virginia, included in his letter information about other railroads in the state (Oct. 3, 1958):

On August 12th, you spoke to a group of radio engineers at Estes Park who were visiting out in Boulder that week at the Bureau of Standards. Prior to your talk you had dinner up on the hill above the Stanley Hotel. During that dinner . . . I sat beside you and you expressed an interest in the old narrow gauge railroads of Colorado. We discussed the Alpine Tunnel and also the fact that my mother made one of the last trips on the Denver & Salt Lake Railroad up over the top at Corona.

Before leaving Boulder to return east I obtained a copy of your book. When I reached page 535 I noticed two towns were mentioned that you had been unable to locate at the time of writing—Busk and Birdseye. Here is where my hobby came to play. Sometime in the past I managed to obtain some employee's timetables for certain railroads out there. Just in case you still did not find Busk, . . . perhaps you might locate it by driving your car a given distance.

As far as the Colorado Midland was concerned, Leadville was not on its main line. It was served by a 3.8 mile stub from a place called Arkansas Junction. Busk was 8.2 miles west of Arkansas Junction, or 12 miles from Leadville, on the main line. It was 2.0 miles east of another mainline point known as Ivanhoe. It was 8.7 miles east of a place known as Hell-gate, a place featured in some timetables with pictures. (A classic one was some elephants pushing a circus train over the steep part of the tracks.)

With regards to Birdseye, I am enclosing a copy of one page of an employee's timetable which may also serve to give you an idea of railroading in this area in 1902. You will notice it is 144.8 miles from Denver. It has a passing track or siding 480 feet long. I am sure you have located it by now.

Now let me add one to your list. Your book did not mention Woodstock. Perhaps you were by it and never knew it. It was located three miles west of the Alpine Tunnel at a point where the railroad makes a loop. It was built in 1880 when the railroad came through. The builders ignored the fact that it was in a snowslide area. Four years later, on March 30, 1884, the residents of this tiny town settled down for the evening. Nearby some prospectors heard a roar, a crash, and cries of agony. They sent one of their number to Pitkin to summon help and they went to the rescue. They rescued a Mrs. Doyle and her daughter alive, but there were thirteen killed. I heard that after that only the station remained there until 1910 when all operations over the Alpine Tunnel route stopped, ending an interesting era of railroading. The rails were removed in 1918 during World War I.

This gets a little off the track, but there used to be stories about the curse on the Alpine Tunnel. Seems as though the Indians cursed the railroad through there. (A possible excuse for the Woodstock tragedy.) In addition there were seven years (about 1888 to 1895) in which the railroad did not operate due to various causes. Then when they did decide to operate, six crewmen suffocated to death when an engine stopped in the tunnel and the fire burned up the oxygen. One young man survived by crawling on his knees back to the entrance. I later talked to a man who had visited the tunnel in 1940. He said it still looked as good then as when built. However, I have since learned that in recent years it has collapsed. I wanted to get up into that area, but was never able to.

Speaking of railroads, I understand that up around Corona there is still a Denver and Salt Lake engine at the bottom of a ravine where it was blown off a trestle many years ago. Perhaps I will get over to that area again sometime.

Busk   Each of the other seven who wrote about Busk added bits of data that were new to me. Since the closing of the Hagerman and Carlton tunnels through the Divide, and with the completion of the Ruedi Dam and the enlarging of Turquoise Lake into a reservoir, certain of the detailed directions in the following letters are no doubt obsolete, but those who did travel by train, or later by car over the roadbed of the Colorado Midland railroad, will never forget their trips.

Eight years earlier than Lt. Com. Harlow's letter, I had received the following from Howard K. Morgan of Kansas City:

Busk is very easy to get to. As to the town, there is one house on the hillside. Below it a few dozen feet away is another shed with an operating waterwheel running merrily. Another house completes the picture. All these houses are immediately adjacent to the old R.R., just south of the right of way a few yards.

The road is extremely good except for rough ridges where water crosses it from local streamlets. It is two cars wide and relatively smooth. Ask in Leadville how to get to Turquoise Lake. But you HAVE been there. You were on the old grade which is a wonderful road now. Stay on the south side of the Lake Fork creek canyon which feeds Turquoise Lake. Then a canyon opens from the left of SW and this is Busk fork. Still a wide valley and the RR is climbing steadily up it. . . . At the end of the canyon the road turns to the west and within a few hundred feet is the Carlton Tunnel entrance and you can drive within feet of it. The lumberwork has fallen outward and the mouth gapes from beneath it. Here are the remains of the old toll gate for Autos with the information that it's a dollar for a car to go through. Up to about 3 or 4 years ago cars drove through. There is the telephone which connected the two ends to let cars pass through one way at a time. Older than the Carlton tunnel is the Hagerman tunnel which the Colorado Midland first used. The railroad can be seen stretching from the Carlton tunnel entrance back along the Busk creek climbing slowly. It then comes back through some other loops above and went through the mountain almost above the Carlton tunnel. It was a long grind to this upper tunnel. The Carlton tunnel was built by a private concern on speculation for some million and a quarter. They then rented it to the RR at a very high price. The RR got mad and once more returned to the Hagerman tunnel and its steep pull on both sides. The Carlton was then used after the RR disappeared as a toll road for cars. Since the Midland is a wide gauge affair the road is much better now than the St. Elmo road (old roadbed of a narrowgauge).

From Boulder's neighboring city of Longmont, B. L. Boyles contributed this bit (Feb. 17, 1953):

You mentioned that you had started out to see the Busk-Ivanhoe (Carlton) tunnel, but never reached your destination. I was up there with friends in September of last year.

When you do go up there, don't fail to explore the Hagerman tunnel (on up the hill from the Carlton) and also the remains of the old construction camp used while the Hagerman was being bored. You can easily drive as far as the Carlton in an ordinary car, and a Jeep will probably make the trip to the Hagerman if you follow the right-of-way.

From Chattanooga, Tennessee, Alfred P. Rogers added other details (Sept. 14, 1968):

In your closing chapter you mentioned looking for the settlement of Busk. As you probably know, it was the eastern portal of the lower (Busk-Ivanhoe) tunnel. Since you stated that, at the time you were blocked by snow on the road that you were looking *down* the mountain for Busk, you may have been on the old Colorado Midland Railroad grade leading to the upper, or Hagerman Tunnel. It was the original route over the Divide, but because of severe weather, steep grades, and sharp curves, it was abandoned after a few years, when a Mr. Busk, financier of New York, built the lower tunnel and leased it to the Railway. The financial arrangement contributed to the demise of the Colorado Midland, in 1918, I believe. During one money hassle the Hagerman Tunnel was reopened and the Busk remained idle for two years. Today the Hagerman is choked with ice the year round.

The facts about the tunnels of the Colorado Midland were carefully put together by my friend Louisa Ward Arps, writer and historian. Here is the summary of what she found:

The Hagerman Tunnel, built within 500 feet of the top of the range, was sometimes called the Saguache Tunnel since it went through the Saguache Range. J. J. Hagerman of Colorado Springs was the President of the Colorado Midland R. R.

The Hagerman Tunnel was about a mile from the town of Busk on the eastern slope and 1½ from the western slope town of Ivanhoe, on Loch Ivanhoe. (This does not necessarily mean that these towns were there when the Hagerman Tunnel was built).

In 1890 the Colorado Midland decided it would like a lower tunnel but could not afford to build it. So the Busk Railway Co. was organized with J. R. Busk president (he was on the Board of Directors of the Colorado Midland). The day after this company was incorporated the Colorado Midland leased the rights to the tunnel. The tunnel company, in an effort to keep it clean, stipulated that coke was to be used by west bound trains; east bound trains were to coast through with no fuel being consumed.

It took three years with numerous changes in construction methods and companies, and with both day and night labor, to build the 3-mile long tunnel. I forgot to say that the towns of Busk and Ivanhoe, on either side of the Busk Tunnel, were thriving construction camps while the tunnel was being built. It was a major undertaking and the camps were live.

The "hole" through came in Oct. 1893. Trains ran through it that December. It saved 7 miles of ascent (I think on the eastern side). The tunnel was 15 feet wide and 21 feet high in the clear.

When the Colorado Midland was sold at foreclosure in 1897, the bond holders of the Busk Tunnel refused to accept the agreement with the new company. So the road re-opened the Hagerman Tunnel, building expensive snowsheds on either side, which were definitely not effective. It took them 30 days to dynamite the 2400 feet of solid ice in the tunnel. They used it for 18 months until June 1899 when the Colorado Midland bought the Busk Tunnel Company.

In 1917 the Colorado Midland went out under the auctioneer's gavel. A good many people bid for it, including a junk company, but Carlton of the Cresson Mine at Cripple Creek wanted it bad enough to throw away a lot of money on it. He had plans to build the road to Utah, and make it pay.

It was finally junked, and in July 1924 opening ceremonies for the Carlton Highway on its old road bed over the Continental Divide were held. I think the Carlton Company (the old man was dead) gave the right of way to the state but I could not find that in print.

Anyway, the Carlton people reserved the water rights in the tunnel, (and if anyone went through the tunnel, water flowed along side the car in the irrigation ditch, on the road under the car, and dripped from the roof) and raised the level of Loch Ivanhoe on the western slope 500 feet so the water would flow through the tunnel and down to Turquoise Lake to be used by the CF&I people in Pueblo. The CF&I people still use the Turquoise Lake water.

The road bed on the western side was under water from this lake-raising project. The article does not state whether the town of Ivanhoe was also submerged.

SOURCES: 1. Graves, Carl F. The Colorado Midland Railway & Locomotive Historical Society, Bulletin #36, 1935.
2. Article on dedication of Carlton Highway in Colorado Highways, Sept. 1925.

James A. Norton (deceased), who traveled widely through the west and shared his experiences through lecturing to a variety of audiences, wrote me (April 2, 1952) this graphic account of his trip through the Carlton Tunnel:

I first visited Colorado in 1940, and had a route mapped out for us by a friend. The route eventually took us to Leadville and the Tabor mine. Leaving Leadville we had the good luck to have chosen the road to Basalt, through the Carlton Tunnel. To judge from your book you only got there after the tunnel was closed to traffic and the road was in impassable shape. Here I shall try to pass along my experience to you about this drive.

We approached the east entrance of the tunnel where, as I recall, there was a white frame house, a small shack beside the road, and a huge sign which read, "S T O P". In slightly smaller type was an admonition to check with the caretaker (who lived in the frame house) before entering the tunnel. The reason was made clear when it was further stated that the tunnel was not wide enough to accommodate two cars side-by-side. Fortunately, the tunnel was clear, and without delay we received the green light to enter. Immediately the light of day was gone and even with headlights on we had difficulty seeing our way. As the car bumped over the ties and splashed into pools of water, the roof of the tunnel dripped water so that the wipers were needed to keep the dim vision from being entirely obscured. Meanwhile . . . a stream of water ran along side the roadbed back toward the east entrance, for as I recall the tunnel was slightly upgrade from east to west.

After bumping and splashing along for what seemed like an interminable time, we noticed what seemed to be a white spot somewhere ahead, as if an electric light were in the middle of the tunnel. The spot grew larger and larger, and at length we realized that it was the west portal of the tunnel. Finally we drove out into the sunshine again, and we were curious to discover how westbound traffic was regulated. Investigation showed that there was a weatherproofed (more or less) telephone here and the eastbound motorist was supposed to phone the other end of the tunnel for the okay before proceeding, as the huge sign indicated.

From here on to Basalt the road followed the old roadbed, generally, except that the old trestle sites were bypassed but even then their existence in the past was very evident; indeed, some of the old structures were partially in existence even then. Although the line through the tunnel was standard gauge, the tunnel seemed plenty narrow for an automobile, to say nothing of a standard gauge boxcar. . . . Possibly a sense of claustrophobia made the tunnel seem smaller than it actually was.

My husband confirms the accuracy of Mr. Norton's description of the rough and dripping ride, for he too had used the Carlton Tunnel during the 1930's as a shortcut to the Frying Pan area with two fishermen friends. When they returned in mid-September, the high western entrance was blocked by a truckload of potatoes, stuck on the thick ice that had formed at the portal. With everyone pushing, the truck was shoved forward until it reached the unfrozen, gushing water inside. After three miles of suspense, both truck and car reached the eastern exit, with their occupants soaked but unharmed.

As recently as 1969, Bob Hay of Evanston, Illinois, sent me the following letter (July 27):

> You wrote me and said you were going to Busk this August. Near Busk there are two other ghost towns, Douglas City and Rich City. There is not much left of Busk. There is more to see in Douglas City. I did not get to see Rich City. There is a man in Leadville named Dick Anderson who can take you to these ghost towns. You will have no trouble reaching Busk. Douglas City can be reached with trouble.

Hay wrote to Anderson and received the following reply which he shared with me:

Nov. 19, 1969

Dear Mr. Hay,

The only Rich City I know of was connected with the Colorado Midland. I believe you said you had a copy of Cafky's *Colorado Midland*—the location is shown on the map in the back of the book—"East and West Approaches to Hagerman Pass" as a group of cabins just north of Rogers spur and south of Hagerman Lake. Some of the cabins are still there, most of them in the same shape as those in Douglas City.

I'm sorry I don't know how long it was in existence and I don't know if it was a construction camp or if it was mining. You know at first there was an alternate route up the pass that could have included another tunnel—maybe it had something to do with that. If I find out more I will write.

Sincerely,

Dick Anderson

CHAPTER VIII

The Arkansas Valley and Monarch Pass

The Arkansas Valley

Hard rock mining requires water, so it is natural to find mining towns located beside streams or where creeks flow into larger bodies of water. Several such camps sprang up south of Leadville along the Arkansas River or on the gulches whose creeks drained into it. Most of them have disappeared. Those with a few cabins serve as summer homes for city dwellers; and places such as Buena Vista and Granite supply food, lodging, and gasoline to the steady stream of motorists who skim along the highway, and furnish supplies and shipping facilities for their stock to the ranchers who live in the area.

After a road was built to Twin Lakes and later completed across Independence Pass, other small settlements were laid out.

Matthew Oblock, Jr. of Aspen wrote (July 12, 1951):   *Brumley*

Two years ago I purchased a copy of *Stampede to Timberline*. . . . I would like to make a few comments about some of the places you mentioned and perhaps add a few items that might be of interest to you.

Going up Lake Creek from Twin Lakes towards Independence Pass you mentioned Everett. Driving on up the road to a little past the place where the Twin Lakes Diversion Tunnel's East Portal is located, there used to be a few cabins which was the site of Brumley. I believe that Brumley was either a stage station or else a mining camp. I am inclined to believe it was a mining camp. Going on up the highway and instead of crossing the bridge above which the road starts to climb the knife edge road over Independence Pass, take the old road to the right. This goes up Lackawanna Gulch to a group of mines. The best known was the Mt. Champion. It was located away up on the hill above timberline. A tramway used to run down to the bottom of the gulch where there used to be a large mill. This mill was razed during the 1930's. Half Moon Gulch was over on the other side of the Mt. Champion and could be reached that way from Leadville. These places are shown on the older Forest Service maps.

I never pass through Granite, eighteen miles south of Leadville on U.S. High-   *Granite* way 24, without thinking of Sewell Thomas, a consulting mining engineer from Denver who was identified with mining properties all over the West as well as in Mexico. He told me that he once spent two weeks walking from Granite to Pueblo testing the water of the Arkansas River for its pollution content; the information gathered was used as evidence in a law suit. He added with a wry smile, "The case was lost."

113

The Clear Creek road one mile south of Granite leads west off the highway to   *Vicksburg*
the remains of several small mining camps, of which Vicksburg is one. It is unique
for the canopy of aspen trees which arch over its main street. On my first trip to
Vicksburg I found no signs of life. On my second, a car stood beside a cabin. When
I knocked on its door, an old man opened it and invited me in. He was Edward J.
Levin, who had lived and mined in Vicksburg for many years. As I left, filled with
information, I took his picture and later I sent it to him. During the winter I received an answer from him:

Vicksburgh, 1950.
Main Street.

Edward J. Levin,
Vicksburgh, 1950.

My dear Muriel S. Wolle

I am the Man that Planted the Trees at Vicksburg.

I Got the Pictures you Took of Me in Vicksburg and they were Real Good and I want to Thank you for them.

Well Dear I Finally Finished my Lodge in Vicksburg.

I now have a 12 Room Lodge, Right up to the minute Fire Place.

Bath Room with Plenty of Hot water.

An 1100 Foot Private water Line with Lots of Pressure and the Finest Kind of Spring water.

I also Finished up my Work on My 24 Mining Claims and stayed until the first of Dec. and had 30 inches of snow Fall and the cold was 20 Below Cold, so I Got out while the Going was Good.

I Got a Nice Elk and had some Good Feeds, wishing you were at the Lodge to enjoy some of the Feed with me.

It sure seemed Nice to Get Back out of the snow and into the sunshine and Flowers For the winter [in California].

I will stop Here till about the First of May. Then I will go Back to the Lodge when Fishing opens.   Best Wishes

from yours

Edward J. Levin

Harvard City   About six miles south of Buena Vista, at the junction of South and Middle Cottonwood Creeks, the placer deposits that were uncovered in the late 1860's and early 1870's caused the town of Harvard City to be laid out. As soon as the Cottonwood Pass toll road was built over the Divide into the Gunnison country, the new settlement became a major outfitting and supply center for both miners and freighters. The discovery of lode mines on the nearby mountain slopes brought more men into the area, and the town flourished until the rush to Aspen and its silver mines drew off most of the population. The final blow came in 1882, when Independence Pass was completed and provided a more direct route to the rich silver fields.

A letter from Edgar G. Dicus (Oct. 6, 1974) of Ranchos de Taos, New Mexico, describes the trip west made by two of his wife's uncles who found good mines on the slopes above Harvard City:

> You might be interested in the following. My wife's name was Bogue and she came from the little (but very lovely) town of Vermont, Illinois, which lies about due west of Havanna and due south of Table Grove. Her two uncles Rob and Charles Bogue (brothers) left Vermont early in 1879 and made their way to Colorado Springs presumably by train. From there on they traveled on foot following the old Ute Trail around Pike's Peak and thence to the west. U.S. 24 rather follows the old trail, more or less to the top of Wilkerson Pass. When they arrived there they decided to make camp for the night above the trail right at the crest of the pass, and the last time we went through there the big rocks behind which they camped were still there.
>
> In the morning they followed the Ute Trail down into South Park. This trail turned to the left and dropped very steeply to the floor of the Park. Later the road followed the trail, but then the highway was changed—turning and descending to the right from the top of the Pass. They told us that the floor of the Park was filled with buffalo, deer and antelope who stopped their grazing long enough to watch the strange creatures walk past. The original Trail

Harvard City, 1910. Log cabin with sod roof, one of six cabins, three on each side of road. *Photo by Will Collins, Boulder.*

bore north and crossed into the Arkansas Valley at the Buffalo Peaks instead of turning south and crossing at Trout Creek Pass. They crossed the Arkansas and made their way to a big clump of cottonwoods on a creek where they camped for some time.

Since they were interested in prospecting they went up the valley to Leadville and spent the summer working for some mining outfit. In the fall they came back down the valley to find that the town of Buena Vista had started in the clump of cottonwoods in which they had camped.

I am not clear about the timing but I do know that they ended up in a spot above Cottonwood Lake on South Cottonwood Creek. A man by the name of Asay Fox had the Cora Belle mine site high up on the south side of the gulch. They swamped a road up the mountain and built a crusher-separator. A flume was built from the mill pond created by damming the creek. We knew this as the mill pond, but "outsiders" called it Fox Lake. Incidentally, Fox mountain was named after Asay Fox.

Charlie had a good strike high up on the south side of the gulch, and Rob had one on the north up under the saddle of Sheep Mountain. . . .

As you know, the mining town of Harvard was on the north side of the gulch about up from the foot of Cottonwood Lake.

Chalk Creek

The entrance to Chalk Creek Canyon, with its several old towns, is five miles south of Buena Vista at Nathrop.

Chalk Creek is a dramatic canyon from its mouth at Nathrop to the high valley beyond Hancock, close to timberline. Mineral deposits on Mt. Antero and on Mt. Princeton, which flank the canyon, brought prospectors to the region by the late 1860's, and the development of a number of mines required some means for cheap shipment of ore. The keen rivalry of the railroads to reach the Pacific coast, coupled with increasing pressure from the mining interests for low freight rates, hastened the construction of the narrow-gauge Denver, South Park & Pacific Railroad up Chalk Creek. Before long trains were puffing up the grade to above timberline, ducking into the Alpine Tunnel through the Continental Divide, and looping in wide curves down to Pitkin on the Western Slope.

After the railroad was abandoned, the right of way was made into an auto road, up which passenger cars climb easily as far as St. Elmo; those with four-wheel drive can continue through Romley to Hancock.

Near the mouth of the canyon, at the foot of Mt. Princeton, are the Chalk Cliffs, towering white formations whose sculptured pinnacles change in appearance hour by hour as shadows slide across them.

Those who explore the canyon today will not see the four-story hotel at Mt. Princeton Hot Springs with its mansard roof, cupolas, and verandas. It was built about 1890 at the base of Mt. Antero, just beyond the Hot Springs' bathhouse and pool, and was torn down in 1950 for its lumber. It was a ghost hotel when I saw it but still intact and imposing. Charles A. Page of Gunnison, a local historian, who wrote to ask a few questions, mentions the once popular hotel (Aug. 23, 1971):

> So you'll know I'm not a flat-lander (though there is nothing wrong with that), I was born in Buena Vista of a mother who waited table at Mt. Princeton Hot Springs in the 20's and at the Bay Hotel in Buenie about that same time, and of a father who worked around the mines and fired old No. 268 from Salida to Montrose. My paternal grandfather is buried in the Leadville cemetery and hauled ice blocks for the Ice Palace, while my grandmother went to Leadville in a wagon over Mosquito Pass. My mother's father ended up in the mine explosion at Newcastle, while my grandmother is a Gilchrist from the early Alpine days.

I have always been interested in the variety of places that Westerners have come from and the diverse occupations and professions they worked at or mastered in response to the needs of the new communities in which they found themselves. The paragraph above from Page's letter shows what I mean.

Alpine   Alpine is the first townsite passed on the way up the canyon, although little is left to identify it. The settlement, which was on a meadow below the railroad grade, still contains the old log stage station, built in 1889—now a recreation center. The smelter, whose chimney still stands, was close to the railroad in a grove of aspen trees. Helen J. Wright of Waverly, Iowa, wrote me of her interest in the place (Nov. 9, 1949):

> Mr. and Mrs. H. O. Kullman, of Nathrop, Colorado, bought the townsite of Alpine. I bought the assay office site, next the Alpine smelter chimney and built a cabin there.

Mt. Princeton Hot Springs, 1942. Hotel Antero.

Alpine, 1942. Stage station.

The Kullman resort among the trees is still very popular, but in 1972 C. E. Rathbun, the Chaffee County assessor, wrote me that the Kullmans are no longer owners of the place. In 1950 Miss Wright wrote again asking for information I could not give her (Sept. 10):

> I'm sorry we didn't get to talk to you, because I have a question to ask you: About a mile above Alpine, on the Stage Coach road, there is a small, circular stone structure, most of it below the surface of the ground. It is about four feet deep and three feet in diameter. There are signs of a tiny gabled roof over the entrance and a flat roof over the circular part and locks on the doors. No cabin near. What could have been its function?

Iron City

Her letter continues with information about Iron City, one of the smaller communities in Chalk Creek. It is scarcely a mile below St. Elmo, the largest of the towns in the canyon, and it contains St. Elmo's cemetery. Miss Wright says:

> The building with three holes and spillway on the north side of the old coach road is the power house in Iron City. I think it supplied electricity for the Ohio Lode (whose mill was in St. Elmo) during the first world war. I don't know the original use of the building on the flats near the stream; when I was first there, about five years ago, there were quite a few tools, a sleigh, wheels, etc., so maybe it was a tack room and repair shop for the stage coach line. Mrs. Carson, postmistress at Nathrop, Colo., might be able to give you more information about Iron City since she lived in Alpine in its last days.

St. Elmo

St. Elmo still contains many of the original buildings, although the wooden sidewalks have crumbled and most of the false fronts have disappeared. When I visited the town in the 1940's and made a number of sketches, Miss Stark and her two brothers Tony and Roy were living, and their store was the center of activity. The last time I saw the place, the schoolhouse across the creek was closed but seemed well preserved, and the Stark Bros. store was boarded up because the estate was still in litigation.

My book *Stampede to Timberline* was published in July 1949, and that same month an illustrated article of mine on ghost towns appeared in *Ford Times*. One of the watercolors showed the main street of St. Elmo, including the Stark Bros. store. Shortly thereafter and for a number of years tales drifted back to me from strangers who had had odd experiences in the place, all in connection with the Stark store and its owners. A friend who was camping above St. Elmo over a weekend went to the store on a Sunday morning to telephone home, and when Miss Stark looked out the window and saw that the woman's car bore a Boulder County license, she locked the store door and asked my friend if she knew "that Boulder sketcher named Wolle, who called St. Elmo a ghost town and so was ruining her trade because people were afraid to visit the town." It took my friend some time to mollify her and to get the door unlocked.

In August, Kathleen (Mrs. Malcolm E.) Collier added another episode to the Stark-Wolle feud (Aug. 15, 1949):

> We had an interesting experience two weeks ago when we were at St. Elmo. I went into the Starks General Store about noon on a Monday. The door was locked, and it was with reluctance that a sister of Mr. Stark (I don't know what her name is) opened the door. After I made my purchase of some hot

St. Elmo, 1942.

cocacola, I asked her rather timidly, for she had about scared me out of my wits, if she had seen the book with a reference to St. Elmo in it. She immediately bristled up and said, "Them dirty buggers." Then she scurried around and found a Ford magazine with a picture of St. Elmo in it by you. She said that St. Elmo was not a ghost town. I tried to explain that you had not classed it as such but she said that it was driving their business off as people wouldn't come to the end of the line as they believed it to be uninhabited. After smelling the smells in there for that brief time I wasn't sure that you were entirely to blame for the business falling off. She said a so and so painter had been sitting across the street all summer painting their place and she was afraid to cross the street for fear he'd put her in the picture. She said you'd made the picture in the Ford magazine with the chimney falling off their store when they were using the chimney every day. She said you had left off the post office sign and the western sign, and the sidewalk was all wavy and ghostly looking. Then all out of breath she demanded, "Do you have the book in your car?"

I went out and got it and she called in her brother. The Mr. Stark you wrote about has since died. The two then started out to laboriously read the part about St. Elmo, spelling out the words slowly. My husband who was getting impatient gave the horn a toot, and she said, "Tell him to keep his shirt on." After she had read a bit of it, she seemed to change her mind somewhat about you. I think she liked you a little more.

Then she told me I was very sweet to bring the book in for them to see, and I really was glad I had shown it to them. I thought you might be interested in their reaction.

Mrs. Collier's efforts were not lasting, for others had unpleasant encounters. Later that year, Wilfred L. Raynor, Jr. wrote from New York (Nov. 14, 1949):

> At St. Elmo, while in the process of taking pictures, we were told by an irate-looking gentleman, standing on the porch of his house on the mountainside, that "we'd make a couple of good pot shots down there." We weren't sure whether he meant for "trigger happy" deer hunters (which we'd been reading about) or himself. But we figured we'd better remove ourselves to a safer locality, which we did.

One young man described an amusing, if annoying, episode that occurred in the early 1950's when he went to the store to buy a tin of tobacco. Miss Stark sold him a can, and he drove back to his campsite. Filling his pipe from the tin, he began to smoke, but something was wrong with the odor of the mixture. Looking at the can, he read the date on it: 1920!

In 1881, when the Denver, South Park & Pacific Railroad was being constructed, both St. Elmo and Romley, four miles above it, served as railheads as the track layers forged ahead toward the Continental Divide. The next two letters from women who knew St. Elmo in its prime recall incidents that give the feel of the place:

> March 23, 1956
>
> I was born and raised in St. Elmo in 1884. I was married there in a little house which still stands, and in 1904 and in 1905 my first babies were born there. My father came there in 1879. In 1880 he sent for mother and 2 children; another four children were born in St. Elmo. My mother-in-law, Mrs. Emma Launder, ran the American House (which is the square building in the center of your sketch) from 1894 to about 1911. Also we went to school there with Tonie Stark, his sister Annie and brother Roy, and believe it or not, I graduated from High School in St. Elmo. The reason for this was that the school in Buena Vista was full and as there were 2 pupils in St. Elmo ready for high school, the County Superintendent permitted the teacher at St. Elmo to teach those grades and after passing the examinations we were given our diplomas. My Grandfather, Grandmother, Father and 2 brothers are buried in that beautiful spot at the foot of Sugar Loaf Mountain at Iron City, just below St. Elmo.
>
> My father, Daniel Clark, could have given you many authentic details of the town. He owned, operated and sold the Iron Chest mine, for which he got $4000.00 and I don't believe it ever produced a dime. He also owned several claims at Hancock. He served as Mayor and Councilman of St. Elmo, was also County Commissioner and at one time a member of the Republican Central Committee.
>
> Mrs. Christine C. Wyse

The second letter, from Joan Hellman of Martinez, California, refers to her mother-in-law, who also knew St. Elmo intimately:

> [no date]
>
> Mrs. Hellman went to St. Elmo when she was about 16 to work in the hotel for her aunt, Mrs. Dell LaPlant. Her aunt was sort of a Klondike Kate, exceptionally tall and good looking and well known in Salida, and at 80 still interested in Men and Mines!

St. Elmo, c. 1885. Main Street, looking east; railroad grade to Romley and Hancock just above buildings on left. *Courtesy Denver Public Library, Western History Department.*

St. Elmo, 1885. Murphy Mine Silver Cornet Band (named for the Mary Murphy mine above Romley). *Courtesy Denver Public Library, Western History Department.*

After living in Pine I guess nothing could shock Mrs. H. (except) an incident at Romley of which you may have heard. The camp cook, a Charles Chaplin (that spelling isn't right), served a couple of tables food that was spiced with cock roach powder, killing the boss and 14 others, then immediately married the owner's wife a great many years his senior. She was buried in Buena Vista. . . .

The house where my husband was born, in 1913, is the first house on the right after you cross the bridge. But the family had been in that house at least since 1900, as an older boy went to that little school as did Mr. Hellman, the father. Always admired a grown man who quit the mines for a year to learn to read and write English.

Did you know that in those days Romley had an M.D.? Must have had an interesting past to be way up there. . . . Rose Richards, the Post Mistress of Buena Vista, 1953 or so, had lived at St. Elmo, and you would never believe the beautiful cut glass and dishes she had had at St. Elmo, and how they ever reached there intact! She has since died. In 1949, the last time we were at St. Elmo, we took her, and she and Annie Stark were talking about an apparently wealthy man, a Mr. Hudson, who came from the East to erect a plaque in memory of his Uncle who had struck it rich there and been buried in an avalanche and he wanted to know the site. He interviewed both these ladies, and Annie said that she had been the one that found him! What I wouldn't give to get in that store by myself. Those old newspapers!

Joan Hellman

Romley   The mine that made Romley was the Mary Murphy in Pomeroy Gulch, so high above the railroad grade that all ore had to be trammed to the waiting cars for shipment down the canyon to the smelters. The mine office stood beside the track across from the diminutive station, while the post office was one of the buildings below on the flat. The school and its "teacherage" were not far away but were hidden among tall trees. Ora (Mrs. E. A.) Kehn recalls that (Aug. 20, 1950):

As a girl I went to the old school in Romley and met my future husband right at the old depot! . . . My father freighted the machinery from St. Elmo across Tin Cup Pass for the Halsey Mill, and we spent one summer in Garfield. Came back in early September in a raging blizzard over the Pass. I'll never forget it. My father also freighted over Independence Pass from Buena Vista to Aspen. I spent all of my girlhood (almost) in Buena Vista.

Some of the buildings at Romley are of more recent date than others, especially those veneered with tar paper. The aerial tram that crosses the highway to the lower dumps is from the Mary Murphy property. Helen Wright says (Sept. 10, 1950):

The big dump, tailing, etc. between St. Elmo and Romley are from the Mary Murphy. The buildings, mostly planks covered with tar paper, not the log type, property of the Mineral Recovery Co. I've been told that this was a "gyp outfit" which used the aim of recovering the mineral in the Mary Murphy and other dumps as a front.

Many unexpected requests come to me about the location of specific mining claims or mines, or about whether I have heard about a certain relative or business

Romley, 1950. Post office.

Romley, 1950. Schoolhouse.

partner. Because most of these I am unable to answer, it was a relief to learn from Orin G. Peterson (Dec. 9, 1960) that my book had helped:

My father, Emanuel Peterson, used to refer to his having worked in the Pat Murphy Mine near St. Elmo and having teamed in that area. I have always had an interest in seeing that area which was vastly increased when we found a quit claim deed to "an undivided ⅓ interest in and to the following described lots Viz: Lots No. one and No. two in block No. sixty-nine in the town of Virginia City. Said lots each being 25 feet wide by 105 feet deep;

and being situated on the south side of Washington Avenue; each of which said lots has a log business building upon it."

John W. Scott

Notarized: John KinKaid

Tin Cup

This deed certified to be filed for record in Gunnison County at 9 a.m., Sept. 20th 1881 and recorded in book 35, page 312.

J. R. Hinkle, recorder

Fees $1.50

This was a formal printed document which heightened our interest in what was already an interesting phase or epoch in my father's life. Incidentally the consideration was five hundred dollars for share in lots. Of course I'm quite sure he gave up title to this property before he left Colorado, but we had a consuming desire to see where my father had been.

This last fall we made a trek to the St. Elmo and Hancock area and knew we were near the site of the Mary Murphy Mine. No one we saw seemed to know of a Virginia City. There were a few men near the Hancock area, and I told of my interest in learning about the site of Virginia City.

One of the men said, "I have a book at home that would tell. The name of it is *Stampede to Timberline* by Wolle." We obtained a copy at the Lincoln Public Library . . . and I learned what we were so much interested in. That there was both a Mary Murphy and a Pat Murphy Mine and that what was Virginia City had been incorporated into Tin Cup.

Hancock   Beyond Romley and below the townsite of Hancock are loading stations for two important mines—the Allie Bell and the Flora Bell. The sturdy log ore-storage house of the Allie Bell was breaking in half in 1972 when I last drove to Hancock, and the trestle over the road at the Flora Bell was gone. The water tank at Hancock disappeared years ago, and its stone foundation is choked with weeds. I'm certain that the buildings described in Wauhillau La Hay's letter (Jan. 10, 1950) were those of Hancock:

When I lived in Colorado, we used to weekend a lot at Mt. Princeton Hot Springs in Chalk Creek Canyon behind Buena Vista. We'd always drive up the canyon to St. Elmo and, weather providing, we'd turn left there and go past the mines whose names, as I recall, were the Allie Bell and the Something-or-other Bell. At the end of the road, just below a long ridge of Mt. Shavano, we came across a few buildings and on one, we could barely make out the word "Comstock." That may or may not have been the settlement's name. As late as 12 years ago, some company had started working one of the mines up there. We heard it was an eastern syndicate but no one was ever around when we were. We used to pick wild strawberries and mountain iris up there. A railroad ran to the town during St. Elmo's boom days.

Alpine Tunnel   The Alpine railroad tunnel above Hancock is closed now, but hikers and those with four-wheel-drive vehicles can make it to the east portal. From three persons who reached the tunnel entrance, I received letters describing road conditions and giving warnings as well as encouragement to those wishing to try it.

Jan. 5, 1953

In regard to driving a jeep from Hancock to the Alpine Tunnel, we do not believe such a trip would be possible because, in taking up the rails, a large number of broken spikes were left in the ties. We feel that these would soon rip the tires to shreds. In addition there have been a number of small rock and land slides across the road bed, as well as eroded places. However, there is a nice trail along side the old road bed, and it is not at all a difficult hike.

M. C. Poor in his book, "Denver South Park and Pacific," shows the distance from Hancock to Atlantic Siding which is at the portal of the tunnel to be 2.9 miles, and we judge the distance is around three miles. We find Crofutt shows the distance between Hancock and the Tunnel to be a mile, but after walking it, we are sure he was wrong. We hiked to the Tunnel and feel it is well worth while even if one isn't interested in old railroads or Colorado history as the wild flowers growing along the right-of-way are very beautiful in the early part of August in the high altitudes.

Guy and Else Herstrom

Dec. 8, 1958

I have enclosed some pictures I took this summer of the old Alpine tunnel and the surrounding area. . . . We drove a jeep to within three miles of the tunnel and hiked on to it, passing through the deserted ghost town of Atlantic. We also approached the tunnel from the St. Elmo side by following the old mercantile road that supplies were carried over during the time the tunnel was being built.

Our family has a cabin at Tin Cup and we all enjoy exploring old mines and ghost towns. We have been to Dorchester, Bowman, Gothic, Pitkin, Ohio City, St. Elmo and Atlantic and to such mines as the Enterprise, Mary Murphy, Star, Forest Hill and Jimmy Mack. A jeep is necessary to reach most of the old deserted places.

Jerol J. Grenawalt

July 26, 1963

I wanted to let you know we made it to the Alpine Tunnel. . . . The mouth of the tunnel is choked with debris—rock, dirt, ice and timbers.

I thought you'd be interested in the present conditions of the road to Hancock since that was one of the things we discussed with you. The road all the way to Hancock had appearances of having been improved in the fairly recent past. The road can be negotiated in an automobile and in many places it is wide enough to allow passing. The old railroad ties have either been removed or covered up. We even drove several hundred yards over the ties beyond Hancock with our Rambler wagon but had to stop at the first cut because of mud and water.

I believe a skillful driver in a Jeep could make it within less than a quarter of a mile of the tunnel. At this point a huge rock slide completely blocks the way of any vehicle. . . . Anyone trying the Jeep trip, should take along a good pick and shovel and an axe, since there are a few places where even a Jeep might meet more than its match.

Donald L. Harms

Chalk Creek
cont'd.

No letter that I received while writing *Stampede to Timberline* provided more first-hand information about Chalk Creek and the railroad than that from W. C. Rupley, a railroad man (May 22, 1946):

We lived in Denver 1900-1909 during which I was chief train dispatcher for the Colorado and Southern railroad, office in the old union depot. My territory included all of the South Park division of the Colorado and Southern, a line from Denver to Leadville, crossing the Continental Divide at Boreas, a branch of it a few miles from Denver at Sheridan Jct. running to Morrison.

At Como, about 90 miles from Denver, the Gunnison line branched off and that line holds most of the glamour for me and mine. 25 or 30 miles from Como was Schwanders where a 2 or 3 mile branch took off for Buena Vista. Schwanders was at the mouth of a canyon (in those days we spelled it canon) and high water often ripped the track out of that canyon. Then came several little stations and on the right as you go up was Chalk Mountain, a formidable peak, apparently composed of chalk, for through the centuries it had washed off from the ruins and its chalky sediment had spread out fan-like over a wide area. That little narrow gauge was built right over that flow of chalk and the telegraph poles, probably never more than 18 feet high above ground, possibly 22 feet, were set in that bed of solidified chalk dust. Years went by and more and more chalk swept down, making it necessary for the section men to form a cut with side walls of chalk, which walls grew until they gained considerable height, maybe twice the height of a man, or as high as a box car. Meantime the chalk also rose up higher and higher on the poles until in 1907 or 1908, I forget the dates, the poles were only 4, 5 or 6 feet high above the ground. A man might have jumped over them in places. I am probably telling you a lot that you already know.

A place not far below [toward Denver] St. Elmo was a station called Cascade and the stream surely made a beautiful cascade there over the rocks.

In the summer of 1902, when our boy was 6 months old, I took my wife and son up there. They stayed at St. Elmo 2 or 3 days with the Stark family. Mr. Stark was section foreman for the railroad, quite a rough type, but Mrs. Stark, a little light-weight woman, made up for it in hospitality and sincerity. They ran the store, or a store, had a daughter 16 or 17 at that time and two boys younger, Roy and Tony, I think. The girl's name was Annie, a pretty mountain girl. My wife was not rugged, but Tony the younger boy urged her to climb some of those mountains that leaned over the town and she tried. Tony had a beautiful collie dog and he and the dog would literally run up the steep declivities and, when they had gained a few rods ahead, Tony would call back to my wife, "Be sure to come on up; you will never regret it." He was about ten or eleven years old. She climbed on and she never ceased to marvel at what she saw. Mrs. Stark had laid down two kitchen chairs to fence in our baby while she went on with her work, looking out for him in the meantime. . . .

The Starks also had two burros; everybody had a burro or two. I saw Tony carry out a bundle of hay and drop it on the ground. The burros went for it. A couple of minutes later Tony came out with a large double arm full of paper to burn—he was cleaning out the store. You know what the burros did—they left the hay and went over a couple of rods and began eating the paper. That wasn't funny to Tony, he had grown up with them. He got my wife on one and he mounted the other to take the mail up to the Mary Murphy mine at Romley. Tony started ahead; my wife's burro went a couple of rods and then all at once changed his mind and went around to the back kitchen door and she had just as much control over the burro as a kitten would. The burro knew it.

On west of Romley comes Hancock where there was a water tank for the R.R. engines, and next was the tunnel, 1800 feet long, down hill both ways from the middle of the tunnel, that being the Continental Divide. Between that tunnel and Hancock were the worst snow blockades I have ever heard of. I have photos of box cars in the deep snow cuts, a man standing on top of a box car reaching up as high as he could with a shovel extended above him and unable to reach the top. It wasn't railroading in the winter, it was just fighting snow. Couple four engines together at the tunnel mouth, two on one end headed down grade and two on the other headed the opposite way, so that when they got through (if they did) they could again back up through the snow that rolled back in. Once in a while these engines were crowded off by the snow on the high side and they rolled down the mountain. It was not unusual for these engines, starting from the tunnel at high speed, to tunnel through entirely under the snow for a long distance like a mole and, if they stopped, the fireman had to get busy quickly to open a vent over the smokestack so the fumes would not get back into the cab and suffocate them. All windows were boarded up. Slides on two occasions came down from the hill and carried cars right out of the middle of the train down the hill. The telegraph wires on the high side were buried much of the way with snow, and why the wires carried messages I do not know; it seemed it should have grounded the wires, but it did not; some thought because the snow was dry.

On the west side of the tunnel was Alpine Tunnel station and an eighth of a mile on west was "THE POINT," since the railroad turned sharply to the left there, and from that point one got a magnificent view of the broad valley in the immediate foreground as well as the snow capped peaks to the west in Utah. My wife went down to "The Point" ahead of me and she was so excited when that scene burst on her view that she called back to me to come on down as she could see the Wasatch mountains over in *Egypt*. She had been a teacher of geography. Sitting on the bank at the point and watching a four-engine train plodding up the 4% grade straight across that enormous valley was a sight. It seemed scarcely to move, but the steam was pouring out of all the stack exhausts in evidence of the struggle. Once in a while the lead engine would whistle. One could see the issue of steam and then wonder if that sound would EVER reach him; finally a faint squeak would waft in to confirm what had been seen. A woman is here from Chicago. Her father helped to lay water pipe for the R.R. at Parlins and Pitkin further down the west slope in the 1880's. . . .

We, around the railroad offices, always referred to the elevation of Alpine Tunnel, the highest railroad point in North America, as 11,596 feet. I notice you give a lesser figure. The engineer's office in Denver could give you the correct figure. On one trip with the superintendent in his private car we stayed all night in Gunnison, left at day-light to continue on to the Baldwin coal mine, I think about 14 miles further on. While we were washing up for breakfast some one told us, "Look quick," and there I was treated to the sight of a coyote chasing a jack rabbit, and that needs no comment. It was the only such exhibition I have ever seen.

I am not informed as to when the tunnel was abandoned, but I thought they continued to operate trains through it for at least a few years after 1909 when I left; you are probably right, however. The tunnel frequently caved in in a small way 1900 to 1909, but never any great quantity, probably a few tons at most, unless for one or two exceptions a little more.

By the way, Mrs. Stark told us that years before 1900 the railroad on one occasion had 110 men there digging snow between Hancock and the tunnel to keep the line open for trains. She alone fed all of them for twelve days straight, during which time she hardly closed her eyes. She had to stay up all night baking bread and doing such work. The men filed in in groups eating off the dishes their predecessors had eaten off of, without them being washed. She had 2 or 3 little children besides.

She was this kind of a good Catholic. Seeing my wife burn the hair from her comb she chided her severely. She never did that as she was aware that after she died she would have to go all over creation gathering up every hair. Her husband beat her, but she could stand that for it would lighten her term in Purgatory. She visited us in Denver, staying at our house probably two days, but then she went to spend the remainder of the time with the Sisters because they let her pray all the time in the chapel.

I was about to apologize for this long letter, but why should I? I might have written much more. . . .

One more item: Did you know of old Bill McKee, section foreman at Fairplay, on a branch that led off from Garo, 8 miles west of Como? He had a burro he called Napoleon. He would have the burro pull the "push car" up the hill, or the railroad grade with a load of ties or a few rails, and when they came down, the burro would ride down with the rest of the crew. Bill swore that burro would not pass a broken rail. More than one national magazine had articles about Bill and Napoleon with photos 40 to 50 years ago.

I retired from railroad work seven years ago.

The Salida Area and Monarch Pass

Back in 1942, before I had published *Stampede to Timberline,* I realized that there must be mining camps of which I had never heard, as well as certain remote ones that I had visited but about which I could find little information. Letters written to postmasters, librarians, and Chambers of Commerce in an area in which I was interested brought helpful replies and enabled me to plan more complete itineraries in a given vicinity.

I had heard of Cleora, Turret, Calumet, and Whitehorn as places that could be reached from Salida, and of Shavano above Maysville. The secretary of the Salida Chamber of Commerce to whom I wrote, Wilbur B. Foshay (now deceased), supplied the following (June 29, 1942):

Cleora   The town of Cleora was started by Santa Fe engineers thinking that would be the end of the road. Then in the settlement of differences with the Rio Grande this branch was taken over by the latter company, and the town was started here at Salida, originally called Arkansas City, and the businesses from Cleora moved up here.

The stock yards are now at Cleora. There is an old cemetery there, but I don't think there are any of the old buildings. Up at Turret there are still some. Those at Whitehorn are all gone. As to Maysville, I think there are some of the old buildings, and perhaps at Garfield. As to Brown's Canyon, it led up into the hills, and into some mining activities, but I don't think there were any old buildings down at the mouth of Brown's Canyon that are still standing. Then up at Shavano, which means a seven mile hike, there are some old buildings, and of course up at St. Elmo, quite a few, for that matter much of the town still stands, and at Alpine there are still some buildings. Near Cottonwood Lake I know of no old buildings and do not think there was ever a settlement. There are some old mines up there above the lake.

When I wanted to know more about Turret, after I had visited it, I turned to my friend Prof. Richard M. Pearl at Colorado College. He not only answered my questions but introduced me to Minneapolis, of which I'd never heard. "It's not worth sketching except for the historical record," he wrote (Aug. 15, 1950). "It is on the road between Salida and Turret, or perhaps I should say between Turret and the famous Calumet mine. To reach it I have only the following information: *Minneapolis*

> 0.0 Salida. Go north on Colo. 291 toward U.S. 285.
> 1.1 Cross Arkansas River.
> 1.2 Do not enter Spiral Drive. Continue ahead on gravel road.
> 8.9 Take left-hand road to Turret. The ghost town is along here somewhere, on left of road, as I recall. That road up Harrington Hill is a steep one, but I do not know of any way else.
> Turret, by the way, had exactly one person when I found it, and the old gentleman has since died, but I hear that more activity is in store for the place."

Another source of information on the area was F. E. Gimlett (now deceased), hermit of Arbor-Villa. He was a picturesque character, of uncouth appearance, with full flowing beard and massive mane of unkempt hair. He was startling to behold in the days when other men were cleanly shaved and their hair closely trimmed. He was elderly and had spent much of his life prospecting the nearby mountains and gulches, watching many of the smaller camps develop. His keen memory and insatiable interest in each mining area caused him to write numerous pamphlets containing brief histories of the places, interlarded with short articles of homely philosophy, political opinions, and his distrust of women in general. I therefore wrote to him with some trepidation. His answers to my letters, however, were both gracious and helpful. *Turret*

Nov. 15, 1942

Turret was founded in 1899. It was short lived with a population at its best of about 1000. Calumet was but a camp lasting only a few years while the C.F.&I. were mining iron there. Whitehorn also was active about this time with a population not to exceed 500.

Yours Truly,
The Old Prospector; Hermit
F. E. Gimlett

Turret, 1942. Main Street.

Turret, 1942. False-front store, originally a laundry.

Turret, 1942. Gregory Hotel.

While on a trip to the Calumet mine near Whitehorn in 1950, I met Carl F. Mathews of Colorado Springs, a photographer and history buff. Although we had never met we knew of each other's work. We talked about Turret, and he promised to send me some data about the place. Shortly after this encounter I received the following note from him (Nov. 12, 1950):

Turret was started about 1898, as a December, 1902, issue of the "Turret Gold Belt" is listed as Volume IV. The townsite was platted and named by Robert Denham, taking its name from nearby Turret Mountain. The Colorado Business Directory of 1902 lists the following business houses:

E. Becker, saloon; Chambers & Co., groceries; Wm. Clifton, barber; the Gregory Hotel (H. V. and V. V. Gregory, proprietors; oldest house in the city); E. G. Holman, grocery and saloon; Hotel Turret (E. G. Holman, proprietor); O. O. Larson, Justice of the peace; Charles Roberts, Gen. merchandise; A. J. Robinson, postmaster & Justice of the peace; P. M. Schlosser, Justice of the peace. (Why so many justices of the peace?)

In 1903 Becker's saloon was gone, but Fred Windeate and Zook & Gilbert both had saloons; Frank L. Hall was listed as a painter and S. S. Hewitt was constable; J. H. McCullough had a general merchandise store and Charles Roberts was still in business; the Gregory and Turret Hotels were still going; G. McKinnon, blacksmith, was a newcomer; P. J. (P. M.?) Schlosser had now become a notary public and civil engineer.

Population in 1902 was given as 195.

Richard Stratton of Boulder lived in Turret from 1915 to 1917 and attended school there when a small boy. When I talked to him in 1953 he recalled that he used to sled down to the main street from above the schoolhouse. At one time the teacher, Miss Eva Corlett, had but two pupils—his sister and himself. Another schoolhouse on top of the hill outside the town was built purposely halfway between Turret and Klondyke, a smaller settlement a short distance below, where the M. & O. (the Company) mine was located. He remembers it, because as a child he occasionally rode to Salida in its delivery wagon.

There were a number of mines close to the town, but so much water from streams and springs seeped into them that "they were very wet mines." The main street ran up Cat Gulch to its head, where the Gold Bug and Vivandiere properties were located. During World War I, Stratton's father worked at the Independence mine and saw to it that a man with a bunch of mules hauled its copper ore to Salida for shipment. Mr. Stratton, Sr. also opened up a granite quarry near Turret and later disposed of it for $55,000.

Stratton's father came to Turret in the early 1900's. He was married there, and the couple's first home was a lumber shack; their second was the Gregory Hotel at the end of Main Street. A small one-story saloon adjoined it on the northwest and was used by his parents, after its closure, as their kitchen. Dick remembers the big range that used a sixty-gallon barrel as a hot water tank and the twelve-foot well behind the building. "Wells only had to be that deep," he told me. "We never had so much room again," he added. "While we lived in the hotel, my sister and I each had a room of our own as well as a playroom. Close by was another empty saloon which had a good wooden floor, and that's where the two of us roller skated."

In its early days the town was connected with Salida by a buckboard stage driven by Merle Craig. He used "green broke" horses that in their frenzied dash down the mountain not infrequently knocked the vehicle to pieces. By 1930 a bi-weekly auto bus replaced the buckboard. Mail was delivered to storekeeper Briggs, who was also postmaster. On one occasion when delivery was late, he stormed into the Salida office saying, "I am Postmaster Briggs from Turret and I demand the mail." Although each year the population dwindled—to 17 in 1917—it was 1930 before the local office was officially closed.

Dick mentioned some of the men he had known in Turret, a few of whom he saw on later visits to the place. The big house on a knoll was that of Peter M. Schlosser, who bought Briggs' store and succeeded him as postmaster. He was a hermit, but had a wife and daughter living on the west coast. When radio was a new thing, he became greatly interested in it and was able to get stations at great distances from Colorado on his set. One day while listening to a California broadcast, he happened to hear his own daughter singing. In 1930 he was a Representative in the State Legislature. Old man Clark kept pet squirrels; he lived in a shack near Schlosser's house. Al Click was a mining man, and so was Dave Austin. On schoolhouse hill stood a house of similar design to that of the Schlosser residence, where another hermit lived, Emil Becker. He was quite a taxidermist and also tanned hides. To keep abreast of the times he subscribed to eastern newspapers. He occasionally shot an elk, but whenever he was close to starving he would get work in the granite quarries. June Denham, so named because he was born in June, owned a house at the lower end of the town surrounded by a big vegetable garden, which he irrigated by piping water from the old ore mills.

"The last time I was in Turret," Dick told me recently when I saw him in Boulder where he has a ranch, "the hotel where we lived was gone—hauled out or torn down. That hurt me. I'd have bought it if I'd known that was going to happen. It doesn't seem like Turret without it."

Every trip across Monarch Pass recalls one that I made in 1942 accompanied by Elliott and Elizabeth Evans. Before continuing west on U.S. Highway 50, after leaving Salida, we drove into Poncha Springs, chiefly to see once more the two-story Jackson Hotel, set back from the street, surrounded by trees, and bearing on its facade a painted sign with the dates 1878-1902. It was open, and the elderly lady who ran it, Mrs. Jackson as I recall, showed us through the rooms on the main floor, most of which were furnished in the period of the seventies. In the parlor was a square piano, and Elliott delighted her by playing several selections on its tinkling keys. The hotel seemed empty except for the lady, who was completely content in the quiet, comfortable building filled with memories.

Poncha Springs

In 1950 I learned from Mrs. Gordon Barker that the place by that time was known as the Poncha Springs Hotel and was owned and run by a Mr. Gennow. According to what he told her:

> After Mr. Jackson died in 1902, Mrs. Jackson had the hotel painted, and the painter recorded the date the building was built (1878) and then added the current date of 1902 as a sort of signature. The Gennow's copied those dates before they found out what they meant, but expect to drop the 1902 the next time they paint the place.

The building is still standing (1976) and contains a small restaurant and bar, but the trees that framed it have either died or been cut down, and the exterior lacks the simple charm of its heyday.

My first visit to Shavano was not until 1945, although I knew there was such a place high in the mountains and that it was reached via the Brown's Canyon road just west of Maysville. My best sources of information about the remote camp were F. E. Gimlett and a young man whom I never met. My letter of inquiry to the hermit brought the following answer (Nov. 15, 1942):

Shavano

> The town of Shavano was created about 1879, was short lived, and became in reality a ghost town in 1882. Outside of one murder and hanging, there were no outstanding events. Lots were given free as well as water and firewood with a proviso that each holder should grade his own street to the extent of his 25 foot front. The town was built of logs, and not even one frame, of these only a few tumbledown structures remain. The town is reached by following up the North Fork of the Arkansas 10 miles from Maysville and the population was about 500.

The next year I received two letters from Pfc. George H. Fricke, who was with the ski training unit north of Leadville:

Nov. 21, 1943

> Being stationed at Camp Hale, I have become very interested in the history of the mining towns in Colorado. . . . Where can I find data on the history of Shavano? This is a town about fifteen miles from Salida, at the base of Shavano Mountain. I don't think it ever had a population of more than 500. At the present time I am the sole occupant of the town as I bought some land

at its edge. I would be very much obliged if you could give me a few clues (about the place).

In answer to my letter Private Fricke replied:

Dec. 3, 1943

I agree that no hobby is more fascinating than learning of these ghost camps. The army is a limiting factor for me. Nevertheless I have come in contact with many of them around here on "G.I." hikes. Much to the disgust of my officers I go wild when I get in one and they usually have to root me out of some decaying house before they go on.

There is little information I can give you about Shavano. I have been there quite a few times. . . . This much I can tell you: There are about 10 buildings still standing, the majority of them without roofs. There are three which are in a fairly good state of repair, with roof intact and even a stove. All were log cabins except the mill, which was made of excellent lumber and frame. Unfortunately, most of the mill has been torn down for the lumber by the residents in Salida—the framework and machinery remain. The mill is by far the largest building in the town. It is close to three stories high. The only other building that can be recognized, is what appears to be a general store on the main street. . . . It has long shelves on one side.

Shavano is not very difficult to reach now. It is about 9 miles from Maysville. The road has recently been repaired by the county to within four miles of the town. Had the prospector who was mining up there made good it would have been continued to the town to facilitate removing his lead ore. The mine folded up, however. The last four miles of the road is in a sad state of repair. Beaver dams have washed out large portions of it. But the walk is not difficult from where you leave the car. So far as I know, the railroad that went over Marshall Pass did not approach near enough to the town to be of any use. Incidentally, Shavano town is 11,000 feet above sea level.

Another year passed, and in the summer of 1945 a friend and I headed up Brown's Canyon in my Ford and at the county bridge started up the steep grade in search of Shavano. After about six miles we passed a group of weathered cabins but doubted that they were the ones for which we were looking. Another two miles of climbing brought us to timberline and onto a narrow meadow. In the distance, near the summit of the mountain, was a dump and the stone-faced portal of a mine. Surely *it* wasn't all that was left of Shavano. Our road had petered out, and the meadow was far too rocky to drive across, so with great disappointment we turned around and headed back to Maysville. We stopped to investigate the cluster of cabins we had passed on the way up. Near the creek stood the skeleton of what must have been the mill. I entered one cabin, and, while I waited for my eyes to adjust to its dark interior, I heard a rustling in the farthest corner. Something was alive, and I backed out hastily. Peering through the doorway, I saw a large porcupine, disturbed by my entrance and lumbering toward me. We left Shavano.

After this trip I again wrote to Mr. Gimlett to identify what we had seen. As before, his response was most helpful:

The cluster of cabins you write about was Shavano. There was at one time a store mid these cabins, the mill (now razed) was across the river, a little creek (Cyclone Creek) supplied the town with water.

Shavano was a short lived town 3 years (1878-1881). The mine you speak about was the Billin's Tunnel, the portal built with cut stone by master artists. T'was not profitable and abandoned after investment of 1 million dollars.

P.S. The mine you saw from Monarch Pass was the Lilly-New York & Garfield, still producing.

Garfield

Garfield, some miles beyond Maysville on the Monarch Pass road, has few of its older buildings left, other than the schoolhouse south of the highway, which is now a private dwelling. Yet years ago, during the mining era, a number of properties were staked and developed on the mountain slopes above it. According to Dr. Cornelia M. Thompson of St. Louis (June 30, 1965):

My son, T. Chiles Thompson, owns five mines on Taylor Mt., which is west of Garfield—Tabor, Geo. Washington, Iron Monitor, Brighton and Patch. My grandfather, Joseph G. Marriott, employed Mr. Gimlett to manage them for him.

Hartville

Hartville, on the eastern slope below the summit of Monarch Pass, still eludes me, in spite of the fact that Mrs. Grace Young of Hastings, Nebraska, wrote me in 1966 that she had at one time camped there with her husband for eleven days. This was in 1932, while the old road was still in use. In the vicinity they found a cemetery of twelve or fifteen graves. Later she sent me a map marked with the location of Hartville and with a quotation from her husband's diary:

Camped last night 4 miles below Monarch Pass. After a good night's sleep we hiked across our little valley and across the highway to an old cemetery. It must have been all of sixty years old. Most of the graves were of babies or old people. Most of them were fenced in, some had the markers still in place. Other markers were piled in a pile under a tree. Most of the graves were covered with rocks, no doubt to keep wild animals from disturbing the graves.

CHAPTER IX

West of the Divide

The area west of the Continental Divide, fanning out from the foot of Monarch Pass, is full of cattle ranches and fishing streams, the largest of which is the Gunnison River. Tucked away between the creases of the sage-covered hills or hidden on lonely mountainsides covered with timber are also many of the old mining camps. Some are quietly disintegrating; others, like Tin Cup, revive each year when summer brings back those who anticipate a season spent in Colorado's high country.

North of Sargents are White Pine and North Star, and the sites of vanished Cosden, Tomichi, and Bowerman, while beyond Waunita Pass are Pitkin and Tin Cup. Nearer Gunnison, north of Parlin, is Ohio City. Southwest of Gunnison were

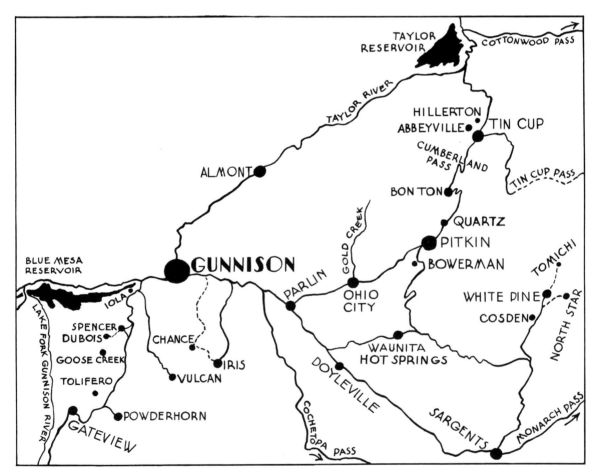

Cosden, 1941.

North Star, 1944.

White Pine, 1941. Main Street.

a number of small camps, most of them mere sites today. DuBois, Spencer, Tollifero, Powderhorn, and Vulcan were laid out when gold was discovered in the area in the 1890's. Although these camps' lives were short, the "Gunnison Gold Belt" drew many prospectors to the sage-covered hillsides.

Mining Camps Northwest of Monarch Pass

On the west side of Monarch Pass, just before the highway levels off at its foot, a county road to the right, or north, meanders up a valley for ten miles, passing through the site of Cosden and climbing a gulch to White Pine, North Star, and Tomichi, which was obliterated years ago by a snowslide. White Pine is the best-preserved and can be easily reached. The other two require four-wheel drive vehicles. In addition, it is wise to inquire locally at Sargents as to road conditions.

When these camps were flourishing, a small smelter stood at Cosden at the lower end of the valley. I was barely able to identify the slag heap as I passed the site in 1972. Letters from David H. Campbell and his wife, Mary, gave firsthand descriptions of both Cosden and Tomichi.

Cosden, White Pine, and Tomichi

Aug. 12, 1950

Referring to your letter of the 10th, to Mrs. Campbell inquiring about White Pine and Cosden, I was in Tomichi the winter of the big slide (1898-9).

I was assayer for the Granite Mountain Mining & Milling Co., that was exploring in the long Magna Charta tunnel under Granite Mt. at Tomichi.

That winter the snows were exceptionally heavy. In response to inquiry from our eastern office as to depth of snow, I snowshoed over the flat back of the office in the young pine trees that prevented the snow from drifting. With a snow shoe pole that was 7'4" long, I could not reach ground thru the snow in half the places I tried.

For the first 100 days of that year it snowed on 90 days. For much of the time it was difficult to get supplies into camp for there was no road into camp above White Pine. The first road into camp was dug on Sunday about May 16th, and we ploughed thru four foot drifts to do it.

Late in March it snowed six inches of very light snow, and before that had time to settle, one or two feet of very heavy snow came. The light snow made a very weak bond between the new heavy snow and the well settled snow beneath. This cleavage allowed the top snow to start sliding along the light snow, on the steep mountains above timberline or where the timber had been removed. For a few feet the slide was a few feet deep and did not reach ground, but then became heavier and took all the snow to the earth and rushed down taking all timber before it.

On March 3rd the old timers were afraid of a slide on Granite Mt. and we did not go to work or I would not have been writing this letter. The slide came down at 9:15 A.M., destroying all the mine buildings, filling the creek bottom 100 feet deep, and extending into the town wrecking four houses and burying six persons, of whom we got two out alive. It is of interest that there were slides on the west, north and east of us as far as we could see from town, the only place escaping being the town which was saved due to the very heavy growth of young pine that prevented the slide from getting a start above it. If that had come there would have been few survivors.

The Magna Charta tunnel shipped no ore for none of merchantable quantities was found. You state on page 172 that the tunnel reached for more than a mile, until it was under both White Pine and North Star. This is entirely incorrect.

North Star was one mile east of White Pine. The boys used to say it was a mile straight up, because the road was so steep. Tomichi where the tunnel was located was two miles north of White Pine, and the Magna Charta Tunnel was directed a little north of west of the town. Thus the tunnel was driven in the opposite direction from North Star.

Cosden, where our smelter was later located, and of which I was business manager and had charge of mines and all business except smelting the ore, was at the opening of the gulch about three miles below White Pine. There was nothing there when we started operations, and we built the smelter, office, bunk house, stables, two boarding houses and nearly a dozen small residences. When we were there two years ago there was no trace of the smelter and but few houses.

D. H. Campbell, Mining Engineer

His wife wrote:

March 29, 1950

For a year and a half I lived with my parents in Tomichi and White Pine. Also the first six months of my married life in 1900 were spent at the smelter at Cosden.

Feb. 16, 1951

Your letter came yesterday. My husband passed away Jan. 4, but I will tell you what I can of Tomichi, White Pine and Cosden.

Early in 1899 my father, Geo. C. Stephen, went to Tomichi from Denver to be superintendent of the Granite Mt. Silver mine. In June of that year my mother and I joined him and it was there that I met my husband . . . who had come out from Boston a year before to be chemist for the mine. My husband's father, Dr. Henry H. Campbell of Boston, was one of those instrumental in operating the mine and opening the smelter.

My folks and I lived in Tomichi that summer in a three room house which was covered with tin cans which had been flattened out, nailed to the sides of the building and painted red to resemble brick.

There was a store, assay office, mining office and a number of log cabins on either side of the road. The entrance to the mine was across the creek and a little above the town. It was above the opening of the tunnel that the snow-slide started. It went over the dump across the creek on into the town where it crushed several houses and buried several people. Two or three people were killed; I am not sure of the number. All who escaped worked frantically to dig out those who were entombed, and when they got them out they were put on sleds and taken to White Pine. All the way down they were fearful of other slides overtaking them. No one went back to Tomichi until the next spring when all danger of slides was past. The slide occurred in the winter of 1899 before I went to Tomichi.

The fall of 1899 the town moved to White Pine where there was less danger of snowslides. White Pine was somewhat larger than Tomichi as it had two stores and a very good boarding house kept by Mrs. Hammond, also a Post Office in Mr. Baley's store. I believe it was the spring of 1900 that the smelter started.

June 4, 1900, Mr. Campbell and I were married and started housekeeping in a little three-room log cabin on the bank of the little stream you cross just as you go into Cosden from Sargents. We stayed there until Christmas and then went to Denver and later to Boston. Not until 2 years ago did we see those places again when we drove all through that country, but everything was changed and we could not find one house in which we had lived.

. . . My father was in Leadville in the early seventies, lived all one winter in a tent while he and three others worked the Ballard mine.

Bowerman   The site of Bowerman, a small camp that boomed briefly in 1903 and collapsed as rapidly as it began, is in the Box Canyon District, three to four miles south of Pitkin. J. S. Bowerman found gold on a mountain top near Waunita Hot Springs, and for fear his strike might be discovered he kept in his cabin the sacks of ore he dug. His wife admired the shiny nuggets and carried a few around with her. One day while shopping in the Hot Springs store she showed them to the owner and to other customers, and when questioned said they came from a mountain near the Hot Springs. Next day twenty-five men swarmed over the mountaintop staking out claims. Within the week 600 prospectors were jostling for space in which to dig, and the Bowermans were helpless to protect their discovery.

A traveling salesman, James M. Diven, who visited Pitkin on his regular trips, describes Bowerman as he knew it.

Bowerman was actually just a suburb of Pitkin. It was only three or four miles, almost directly south. I made one call over there to see what was doing and took this picture. It was a celebration of the "Fourth," I think. You can see there are only half a dozen houses. It was Bowerman's brief day in the sun.

During the Bowerman activity there were a bunch of Eastern College people who apparently had invested some money in "Bowerman" and they stayed at the Scott Dickinson boarding house, an overgrown frame dwelling which stood just south of the creek (I think it was Ohio Creek) in Pitkin. It may still stand.

These Easterners remain fresh in my memory because I had to sleep with one of them one night. It was that or nestle out in the aspens, not a too comfortable alternative. As to the Dickinson hostelry—Mrs. D. and her handmaidens rassled the pots and pans (and I do mean pots) and Scott managed to stand the strain of taking in the money. Twin beds? I doubt if Scott could have found TWIN-BEDS in the dictionary. Of course, Mrs. Wolle, I am sure you have never had to sleep with a strange man during a Colorado gold rush but I can tell you—that altho the fellow I drew was a very decent chap, it was not, for me, an agreeable experience. Looking through my old films I find another one of Bowerman so am sending a print. Gives a better idea of the "town" in its hey-day.

Bowerman, 1903. *Photo by James M. Diven.*

Bowerman, 1903. *Photo by James M. Diven.*

Pitkin, 1880. Lower Main Street and Armstrong Street; Chicago Park in background. *Courtesy John G. Curtiss.*

Pitkin, 1880. Looking southwest; post office (#1); Board of Trade saloon on State Street (#2). *Courtesy John G. Curtiss.*

Pitkin can be reached most directly from Parlin on U.S. Highway 50. It *Pitkin* dates from the late 1870's. By April 1879 a small camp called Quartz was laid out along a creek of the same name, only to be renamed Pitkin later in the year in honor of Governor Frederick W. Pitkin. The ore that was uncovered was silver, and throughout the eighties and early nineties the camp ranked high among the state's mining camps. It is quieter today, yet many of its early buildings, such as the city hall and the stone hotel, exist, as does the railroad station, which upon the departure of the railroad became a private home.

Jessie Raber of Paonia wrote (Oct. 24, 1949):

> Mr. Horace Curtiss has lived in Pitkin and Tin Cup and knows the history well not only of these spots but many others which are now represented by only spots in which the sage is higher than the surrounding country. This indicates that the dirt-roof buildings have fallen, decayed, and the brush has a higher stand. . . . Mr. Curtiss spent considerable time in the early day development of these two towns. He is now retired.

Horace L. Curtiss and his brother John G. Curtiss (both deceased), who also lived in Pitkin during its prime, wrote the following letters and sent photographs of certain buildings, such as the two-story hotel on the main street called the Mason Hotel. It was also known as the Bon Ton Hotel, L. S. Mason, Prop., "The Most Modern and Best Equipped Mountain Hotel in Colorado."

Nov. 16, 1949

Dear Lady,

I have lived in Colorado since 1874, since 1877 on the west slope. Our family moved to Manitou in 1874, the year the Cliff House was built. Ed Nichols and I were kids together—in 1877 we moved to Lake City where I went to Sunday school in the first church on the western slope *but* when I tell that I feel sure no one who hears me knows that Breckenridge is in the western slope and had a church long before 1876.

. . . In the *Colorado Magazine* a couple of years or so ago were published some letters of Bishop Spaulding telling of services he held July 24 and 25, 1880, in Pitkin and Hillerton. He called the Pitkin Mission St. Bartholomews. My father and some others from Lake City located the Pitkin townsite as a placer, Feb. 22, 1879, and Father was the first postmaster, lasted till Cleveland was elected when he was ousted.

. . . Mr. Darley gave me a copy of his book "Pioneering in the San Juan." Some of the things the old timers had to tell you were enlarged on some—but no one knows any better—and it is a surprise to me that anyone could get so much so near to what it was as you have done. And it has been a pleasure to me to read your book. It's like a drink from the Fountain of Youth. It brings back so much.

Very Resp.
H. L. Curtiss

Jan. 24, 1956

I read your *Stampede to Timberline* first a few years ago and have just gone over a friend's copy and it has put a few ideas in my mind. I feel I am more or less a pioneer having come to Colorado as a boy in 1874, father having come to Colorado the year before. We first lived in Manitou until the fall of '77 when we moved by wagon to Lake City where we lived until June 1880.

Father was one of the locaters of the town of Pitkin (first Quartzville) and was the first postmaster, and served till June 1st, 1886, when he was one of the rascals turned out by the first Cleveland Administration. We lived there for twenty years and have lived here (Paonia) ever since. I know the Gunnison country quite well, particularly the Tin Cup-Pitkin sections.

. . . I am some acquainted around Buena Vista too, have been on Clear Creek up to Vicksburg and the place above and also to the top of Cottonwood Pass on the old B.V.-Aspen freight route. I have several old Pitkin-Tin Cup pictures that I would be pleased to send to you to look over and have copies, if you would care to keep them. I would like to meet you and perhaps "reminis" a little.

Very sincerely,
J. G. Curtiss (John)

John Curtiss did send me a number of photographs of Pitkin in its early days, and when I wrote him my thanks he responded (Feb. 11, 1956):

I wish you would keep any pictures that seem to appeal to you, for these days those who appreciate these old ones are very scarce, the little snaps of mine I don't care for because I have the films put away. Hope you get a lot of good from them.

. . . Now I don't want to take too much credit for having the pictures— my brother Horace and I were very close to each other all our lives and he had perhaps more opportunity to collect some of these pictures. He was some five years older than me, he passed away in 1951 at 86 years.

Pitkin, 1912. Main Street; the schoolhouse built in 1880 was later salvaged and used as the Town Hall. *Courtesy John G. Curtiss.*

. . . In Lake City we boys attended Geo. M. Darley's Sunday school at the Presbyterian Church and played with George and Ward Darley a lot. I don't remember their younger sister's name. We never saw any more of the family until while living here Horace and I called on Ward when he had the Pres. Pastorate in Delta.

I hope you Eastern Slope people are getting the much needed moisture as we have done in this section this winter. Of course all of us are very much interested in the Upper Colorado River Diversion project. I don't see how Congress can turn it down.

Bon Ton

A gravel road connects Pitkin and Tin Cup via Cumberland Pass, and high above Pitkin, close to the road, was the Bon Ton mining property with its assortment of frame buildings and shacks. The Curtiss brothers were familiar with it also. In 1949 Horace wrote (Nov. 16):

> The Bon Ton mine and mill came to life about 1910 to 1914—there had always been a lot of prospecting on the hill—but no silver or gold assays to encourage anyone—there was always what the assayer would say was base metal. Finally it was known that it was molybdenum; so a company got hold of the property we know of now as the Bon Ton, worked the ground and built a small mill which was a fizzle—it is all gone now for junk—the machinery was moved from a mill that had stood in Pitkin. The molybdenum ore at the mine ran 2 or 3 or 4 percent higher than the Climax ore. I doubt very much whether the mill they had could ever save what there was in the ore and the mine bust. Now for a watchman's story who was at the Bon Ton one winter after they shut down—he and his wife went down to Pitkin for a day or so at the Christmas time. There was running water in their house, but you had to let it run—and they did—but there must have been a small brush left on a windowsill above the sink and along came a mouse we will say, and anyway the brush was knocked into the sink, where it floated around and settled down over the drain—result an overflow. When the folks got back, everything in the house, even the piano, was frozen about a foot deep to the floor. Don't ask me to tell you another.
>
> P.S. That foot of ice on the floor at the Bon Ton is too much ice. Let's say everything was frozen to the floor and let's say a mountain rat knocked the nail brush into the sink, instead of a mouse—take it off the ice and add it to the mouse.

Garfield

It is only a short distance from the Bon Ton mine to the summit of the pass. Once over the top, the road drops directly into Tin Cup past the site of Garfield. It was Horace Curtiss' letter that described the place:

> What I want to tell you is that you missed a really worth mentioning place in the Tin Cup country. (I was working on the Tin Cup mine 60 years ago now). As you go over Cumberland Pass, down the pass and past the Blistered Horn Tunnel before you get to the wreck of the West Gold Hill Mill, about ¼ mile back up the road on the right hand was *"Garfield"*. They had a newspaper, "The Garfield Banner"—Joseph Cotter editor—you can see a copy of it in the State Historical Collection. Garfield never had a postoffice and the paper was mailed in Virginia City.
>
> I would say that when you drove down the fork of Willow Creek going

into Tin Cup from Cumberland Pass, that you drove right down the main street in Garfield. I could show you where it was, only an old foundation or so and a few piles of stones that one day were fireplaces in the cabins—maybe you could find an old iron fork or spoon and some pieces of broken crockery. I don't imagine they ever had more than 200 or so people. I don't know a soul left that could tell you about it—so you'll have to take my word for it, or pass up what I have told you. It would be fun for me to meet you and go over some of the things you have seen and tell about.

In John Curtiss' letter of a few years later John says that this Garfield must have antedated the better-known Garfield that is above Maysville on Monarch Pass.

Tin Cup

The first prospecting in the area was done by Ben and Charles Gray and James Taylor in 1860. The men searched for colors in the streambeds without success until the morning when Taylor set out to find their horses which had strayed into a neighboring gulch. From habit he tested the sand of a dry wash and took a sample back to camp in a tin cup. When he washed the gravel, he found gold flakes in the bottom of his pan. Throughout the summer the men worked their claim, until freezing cold and lack of water with which to wash the gravel forced them to leave. When they returned the following spring, they found Fred Carl Siegel and Fred Lottis placering much the same region.

Placering was carried on until 1878, when the Gold Cup mine was discovered on Gold Hill, three miles southeast of the present town of Tin Cup. With the new stampede to the district, Virginia City and Hillerton were laid out in 1879. Though the camps were rivals at first, it was Virginia City that absorbed the other settlements and was incorporated in 1882 as Tin Cup. The 1880 Census estimated the population of Virginia City and surrounding territory as 4,000.

The town lies on a wide meadow surrounded by hills with the Continental Divide east of it. The Community Hall, built in 1906 at the intersection of the two main streets, is the largest building and because of its cupola is often mistaken for a church. The rest of the settlement is made up of single houses and cabins, some new and some old, and a few stores.

On a trip to Tin Cup in the early 1950's, my friends and I arrived in the evening. Unfortunately it was a weekend, and all accommodations were rented to fishermen and vacationers. We'd had a long day and couldn't face another stretch of driving that night. "Haven't you anything?" I asked the owner of the auto camp. "Nope," he replied. "I do have one old cabin, but you wouldn't want it." "Why not?" asked my two companions in unison. "The roof leaks and there's no heat; just cots and a couple of chairs." "Show it to us," we urged. When we saw the interior, we agreed that his description had been an understatement, but we took it. The bedding he provided kept us warm, but the last things I saw before falling asleep were clusters of stars shining through holes in the roof, and next morning the water in our wash bucket had a film of ice on top. I couldn't help but wonder what Tin Cup had been like in its prime.

Interviews with residents and letters from people who knew Tin Cup when it had been a thriving community gave me some idea of its daily life:

I met Mr. and Mrs. Seeton, who owned a large house on the side of a hill near one of the cemeteries. According to Jessie C. Carroll, who wrote me in 1951, this was "the house of a wealthy English family in the 1880's and was elegantly furnished." It even had a tennis court.

Tin Cup, c. 1890. Looking west on Washington Avenue. *Courtesy John G. Curtiss.*

Tin cup, c. 1890. Frenchys Place; Frenchy was A. N. Perrault (lower right). *Courtesy John G. Curtiss.*

The Seetons told me that the name Virginia City was changed to Tin Cup by the post office department because of confusion over mail delivery, since there was a Virginia City in both Montana and Nevada. They pointed out the old wooden sidewalk in front of the Community Hall, and they indicated a cabin near it which had been a drug store, a shoemaker's shop, and a saloon. They spoke of the bullet holes in its roof after a man had been killed in front of the building and of the gun laid on a platform out in front. Access to the town from the Eastern Slope was formerly over Tin Cup Pass, and they described one trip which they made over the Divide from St. Elmo with a horse-drawn toboggan. The horse wore no snowshoes but it learned to walk in the snow-packed track of the narrow trail. With each storm, the well-tramped road grew higher and higher.

I showed the Seetons a sketch I had made of an emtpy two-story building, the floor strewn with all sorts of debris, including an ornate iron bedstead and a large coal range. It was not a store, they said, but a hotel. They also pointed out the site of the first slaughterhouse in the region and its big windlass used to hoist the animals. Naturally, it stood in Slaughterhouse Gulch. They mentioned the shut-down Blistered Horn mine on Cumberland Pass, which was developed by eastern capitalists and named by one of them who, while investigating the property, remarked, "I blistered my horn [nose] today." The mine was worked again briefly in 1916. The West Gold Hill mill that we passed upon entering the town was nothing but a promotion scheme. Besides these properties, there were also the Tin Cup, the Iron Bonnet, and the Bon Ton, the last of which was a producer of molybdenum.

Certain other mines in the Tin Cup District are mentioned by J. J. Grenawalt of Limon (Jan. 5, 1959):

> I have been to the Gold Cup mine numerous times. It is a large development with a sawmill and several shafts. The saw is run by an old car engine. Most of the buildings are in good shape. If you follow the road on up a ways, you will reach an old sheep fence, made of large logs.
>
> The Forest Hill mine is north of Tin Cup. You turn off the road that goes to Dorchester when you reach a sign that says, "Trail Creek-Forest Hill mine 2 mi." There are two mines but I believe the largest is the Forest Hill. It has shut down fairly recently and is in good shape. On top of the ore "hopper" you can find almost perfect crystals of iron pyrite. There are also some very old cabins near the mine.
>
> The Star Mine is north of the Forest Hill, almost 8 miles from the Dorchester Road. The country around it is very beautiful, most above timberline. The mine is worked by an old Norwegian and his wife. They were very friendly and gave us some ore samples. The buildings here are very large. The Enterprise Mine is across the valley on the Sawatch Range. It is a large mine with many buildings, and an aerial tramway down into the valley. The view from here is magnificent. You can see the Star Mine, way across the valley on the opposite ridge.
>
> There is nothing left at Dorchester except a ranger station. Bowman is nothing but a few foundations. If I can tell you anything about the mines in the Tin Cup area, I will be glad to do so.

A letter (Feb. 1, 1950) from Edith Stuart Jackson recalls an amusing incident in which her father participated:

Star mine, north of Taylor Reservoir, 1958. *Photo by J. J. Grenawalt.*

Star mine (arrow) as seen from Enterprise mine, 1958. *Photo by J. J. Grenawalt.*

My Dad lived as a small boy at Tin Cup when there were 13 families there working at a mill. Some of them were Scandinavian background and made skiis for my Dad and his brother both under school age, and they learned to ski. They thought it would be more fun if they iced up their slide, so they poured a barrel of water down the slope, and when my Dad went down it after it had frozen, he rasped off so much of his hide that he never forgot it.

Shortly after *Stampede to Timberline* was published, Mrs. Katherine Gallagher Young wrote me protesting statements I had made about her uncle Courtenius La Tourette. In subsequent letters she supplied the correct information, and I am glad to comply with her request to rectify any erroneous material.

July 18, 1950

I was born in Tin Cup in 1895. I am the daughter of Samuel Gallagher, a pioneer merchant of that place. . . . Later he purchased the grocery, dry-goods and hardware store of Mr. Freeman, who wanted to move to Denver and was a merchant there until his death, July 9, 1906. . . .

My great uncle was Courtenius (Charles) La Tourette. The information about him was all wrong and I was certainly shocked to read it. . . . My mother and father were married in Tin Cup, Dec. 1893 in Uncle Court's home. Uncle Court was a Lieut. in the Civil War and belonged to a fine Hugenot family and was an honest, honorable gentleman. I'll admit he did kill a man but it was done in self defense after he had been shot at five or six times. . . .

My mother was Annie B. Clickener who went to Tin Cup the spring of 1893 with her uncle Mr. La Tourette and wife. She married my father, Samuel Gallagher, the following December. He was a pioneer of Tin Cup having gone there in 1879, after spending five years in Leadville. He worked in the Gold Cup mine until he bought Mr. Freeman's store in Tin Cup. He was postmaster there in the 90's, also was town treasurer from the time he went into business until his death July 1906. He was the only merchant of Tin Cup listed with Dunn and Bradstreet, that I know of. Neiderhut did not go there until 1903 or 1904 and went in business with Bill Wool who had a meat market. As the city had built a city hall and were not using the Masonic building, they bought it and moved it from its location to the N.E. Corner of Washington St. and Grand Ave. Some of the Masons were very bitter about this as Wool had always been classed with the "rough element" and considered not honest or at least tricky.

Your informant about Tin Cup was wrong in most she gave you, I certainly wish Mother and I had known you were writing this book as we could have given you a lot more information.

Feb. 3, 1951

My mother is writing a biography of Uncle Court (Charles) La Tourette and when she comes home I will send you one. I am corresponding with a few old pioneers of Tin Cup and have hope of getting more information soon.

How I wish you had seen Alex Parent while up there: he could have told you more than any one else. My sister was there the fall of 1929 and he was living then, but think he has been dead several years now. His wife is buried in the cemetery in Tin Cup. They were only married a few weeks before she died. He used to be a mailcarrier from Tin Cup to St. Elmo.

The Railroad Magazine of June 1941 tells about the last run of the R.R. and has an article on "The Alpine Tunnel Route." I can remember Mother and me . . . sitting in a coach for hours while the engine and crew were gone to dig (snow) out of this tunnel. We were on our way back here on a visit and went by way of Gunnison and visited Uncle Court and Aunt Emma La Tourette.

They first went to Leadville, Tin Cup, St. Elmo and Gunnison and back and forth to Tin Cup. However Papa stayed through the depression there, winter '94-'95. He did a banking business for the mines after they did not have a bank there. After Papa died (1906) most of the mines got supplies from other towns, and it wasn't long until they all closed down.

April 15, 1951

If you ever have the book printed again would like you to let me know so some mistakes could be corrected. On page 185 you stated that Neiderhut kept an iron box in the cellar, etc. It was my father who did this instead of him. When Mr. Freeman had a store there in an early day he purchased a large iron safe, which cost him over $1000 to buy and install in his store. Then Papa bought him out in 1890 and continued to use it. He was robbed by holdup men in 1898 and forced to open it so they could take the money. After that he hid most of his money in the iron box in the cellar and got it out as he needed it. When the store burned there was $619.00 in the box or the safe (which fell into the cellar during the fire) and was burnt to a crisp. However, Mother sent it all to the First National Bank in Denver and they sent it to Washington, D. C. for her and she got it all. (I have their letter to her stating this). She also has a letter of condolence from the Struby Estabrook Mercantile Co. of Denver after his death stating that he "was always looked upon as one of the most straightforward business men in the state" etc. After the fire Mother sold the safe to Neiderhut for $25.00.

The Masonic Lodge, No. 52 A. F. & A. M., was organized in an early day, and the Masons built a large two-story building at a cost of $1500 I think. They occupied the upstairs and the downstairs was used for a town hall, dances, etc. It had a large stage and five sets of people could square dance in it with plenty of room. Some glorious times were had there. In later years most of the Masons left town, and as the town had decided to build a town hall, Neiderhut talked the leader into letting him have it for a few hundred dollars ($250) and it was moved a block and a half from its original location to the corner of Washington St. and Grand Ave.

Alex Parent was the last mayor of Tin Cup. In 1937 some Gunnison County officials went to Tin Cup and had Mr. Englebright, who was the last town treasurer, open the safe and they took all the money and town papers to the Gunnison Court House and by doing so closed the books on Tin Cup. There was approximately $2800 in the safe of town money and it had been kicked and rolled back and forth across the floor by the tourists for years. Mr. Louis Thomas was with them and told me about it.

How I wish part of the town money of $2800 could have been used to write a book about the town and pioneers with all the town records in it, also all old pictures. Wouldn't that have been a wonderful memorial for the pioneers?

Sincerely,
Katherine Young.

Jessie C. Carroll, who also wrote me in 1951, mentions Katherine Young's aunt (Feb. 16, 1951):

> I do not clearly remember Mrs. Emma La Tourette tho' I did meet her. She was considered the most ladylike of the town's people. I had an old book of the Women's Clubs of Colorado about 1890 or 1900. . . . It had the affiliated Club of Tin Cup as the "Tea Cup Circle" and had only Mrs. La Tourette and Mrs. Woll, who lived across the street, as members. . . .
>
> New cabins have been built in Tin Cup and old ones covered with "paper brick." A young soldier and his wife built the jail into a good cabin. An association of property owners was formed last summer (1950) and the town hall is being repainted. There are no town citizens or voters—all go out for the winter months and have their hopes for doing more digging in the spring.

A series of letters from James M. Diven of St. Louis are so full of personal incidents dating back to the early years of the twentieth century that I share them without deletion. They are centered in the greater Gunnison area, with occasional digressions to such spots as Granite, Twin Lakes, and the Wet Mountain Valley. Both Mr. Diven and his wife are deceased.

Jan. 14, 1958

From 1902 until 1918 I traveled in Colorado selling hardware for the old Simmons Hardware Company, St. Louis; at that time the largest hardware jobbing house in the United States. I made headquarters in Colorado Springs and my territory extended from Aspen, Glenwood Springs, on the northwest to Clayton, New Mexico, on the southeast, with side trips to the Gunnison, Crested Butte country, the Wet Mountain Valley, and Arkansas Valley down to Lamar. My visits to major points were spaced about every five weeks and to such places as Pitkin, Ohio City and Tin Cup as prospects for business indicated—Oh yes, and SNOW permitted.

As you can imagine with such a wide territory I supplied hardware for a wide assortment of industries of which hard rock mining was only one. I actually sold hardware some fifty years ago to Neiderhut's in Tin Cup, f'rinstance. It was Neiderhut and Woll back then. Mr. Woll pulled out later and went in business in Buena Vista.

As long as I have touched on Tin Cup I will continue there for the moment. The first flush of mining excitement was past but still some more or less sporadic activity as new Eastern money was attracted, and, of course there was always a core of old Prospectors like Uncle Dan Robley and "Coyote Bill" who were partners.

Back in my days in Colorado, vehicular traffic was animal propelled. Autos were almost objects of curiosity up until I left in 1918 and unthot of in the upper reaches of the mountainous country. I sometimes drove over to Tin Cup via the route you took, from Pitkin, but the "official" route was from St. Elmo. Daily stage with the mail when the "Pass" was open and balance of the year daily mail was carried on horse back. There were two mail carriers—One from St. Elmo to the top of the Range and the other from Tin Cup to the top. They met at the Top at a pre-arranged hour and exchanged mail sacks. Snow was always many feet on Top from about December until around July 1st, by which time it was down to where they could shovel the road open and

Tin Cup Gold Mining Co. dredge, which still contained machinery when picture was taken. *Photo by J. J. Grenawalt.*

wagons could get through. Supplies had to be brought in in late fall, to last through. A single, temperamental, phone wire led from St. Elmo.

I usually planned to go in to take my order so the goods would arrive in St. Elmo about the time the "Pass" would be open for wagons. One trip I made, the east slope was open to Timberline where the snow began but the west side was still such a mass of snow and ice hummocks that horse-back was needed. I phoned across and the mail carrier led an extra horse up for me. We drove the buckboard up the east side to where the snow began—leaving St. Elmo at two o'clock A.M. so as to meet the west-sider at the top at 6 o'clock. It was in middle of June. Snow was still very deep and we had to walk over it while still frozen or sink in too deep to travel.

Have you ever stood upon the backbone of the Continent at 6 o'clock of a bright, spring morning? Looking at that great landscape—mighty mountain peaks ranging off as far as the eye could see—I was moved to reverently remove my hat and say with the Psalmist—"Lord, when I consider the work of Thy fingers—What is man that Thou art mindful of him?"

Standing and looking, far down the west slope I detected something moving slowly over and around snow drifts. It proved to be a man and a woman and he was pushing (of all things) an old style wooden wheeled baby buggy with about a two year old youngster. The Tin Cup mail carrier said, "They left town at midnight. She is having trouble with her teeth and they are going *OUT* to a dentist." So much for Tin Cup.

I have spent two days between Pitkin and Buena Vista on a "South Park" train. Snowed in in Crested Butte—incidentally fallen off most of the hills around the "Butte" learning to ski. Snowed in on Rio Grande train on Marshall Pass. Many other vicissitudes common to we Travelers in those hectic days.

I started this essay with Aspen and will close with the same. I would leave Aspen at six o'clock in the evening on the Rio Grande going down to Glenwood Springs, there catching a main line train across to Buena Vista. About an hour's wait between trains. In Aspen, a Frenchman(?) named Vazey ran a lunch counter and served the most glorious, IMPERIAL SIZE, porterhouse broiled steaks. (Fifty cents as I recall—maybe 75). Down in Glenwood the White Front Saloon carried a big refrigerator full of *FREE* LUNCH, consisting of assorted cheeses, and such delicacies as pickled lambs' tongues and other tid-bits which I love. Each trip to Aspen, as evening grew near, my biggest problem was—Would I eat Vazey's steak, or wait for the *FREE* LUNCH. I simply couldn't do justice to both so close together.

Feb. 5, 1958

You ask about Crested Butte. During my years calling there, I think I may say the "Butte" was prosperous. The coal mines in that area were usually working full time and while the town did get some slight support from ore mining and cattle, coal had always been its main prop. The Elk Mountain Hotel (I understand) still stands. Chris Diehl was the first owner and operator in my time and he later sold it or leased it to George Hubbard. Of stores there were three general—the Company, Glick and Roth, and Martin Verzuh—a drug store, a candy and notion store, Robinson Lumber and Hardware, John Mikesker Hdwe. and the Bank. Of saloons, I think there were sufficient to irrigate the populace, at least, I never heard anyone complain about "dying" of thirst. I enclose an old print of a picture I took about 1910, from the bluff west of town, looking east. George Hubbard used this picture and one of the hotel (I took) on his stationery for years. Any mention of the "Butte" without including SNOW would leave it incomplete. I was pretty lucky, only getting held up a few days, but a dry goods salesman with 21 trunks of samples was marooned 21 days by snow. That is—he didn't get his trunks out for 21 days. I believe he did *ski out* about the 14th day. In all, I carry quite happy memories of my visits in Crested Butte—about 8 visits per year for 16 years. So much for the "Butte."

Riding the "South Park" R.R. was usually an experience. On my maiden trip it was spring, and the valley was practically a swamp. The R.R. ties were rotted and, leaving Gunnison for Pitkin, before we reached Parlins the cars sank thru the ties and we were marooned for several hours. Came noon and no place to go—I saw an half beef in the baggage car—with my pocket knife I cut off a slab. Built a small fire alongside the track and tried to broil it on a piece of wire. It was still mostly raw but well smoked when I ate it. No bread, nothing. Coming west at night on the "South Park" during snow time—the meadows were not fenced—cattle would roost on the tracks and the headlights on those engines didn't give enough light to see to spit—so the train would plough into a bunch of cattle and it was a 50-50 bet as to which would be derailed—the cows or the engine.

I judge from your book, Pitkin is only a remnant of its prominence during my time. Nice little town. Three pretty fair general stores—R. R. Williams, Scott Dickinson, and Crystal Bros. Mr. Williams was, probably, the leading man of the town, interested in local mining as well as his store. Scott Dickinson, besides his store, ran a big frame boarding house across the creek.

Crystal Bros. were oddities. Two old bachelors living in rooms in rear of their long one story log building. One of the brothers was a pansy specialist. Pansies flourished in that climate. I once mailed one to my wife packed in wet grass—as large as a good big saucer. I am glad you mentioned that Pitkin stately flag pole. It always impressed me.

The Bowerman strike occurred during my time. Just a flash in the pan, but did attract a number of easterners that one year. Caused the "South Park" to install a regular passenger train—one train each way from Gunnison to Como or Denver. The schedule took this train thru the Alpine tunnel about six P.M. so the R.R. established a rough lumber eating place a few yards south of the tunnel, with a grey haired couple to run the place. I happened to eat there one trip. As we filed out past the pay-spot, after our meal, the old lady stood beside her husband. I said "You folks are from Missouri." They exclaimed "Why, YES. BUT HOW DID YOU KNOW?" I said, "I knew by the way you fixed your mashed potatoes."

There used to be a large placer mine west of Granite which dumped its tailings into the Arkansas River thru a big flume right at Granite. Above there the river was clear, below, it was muddy all the way down. The State put an end to the dumping around 1915, as I recall it. While on the Arkansas I will mention the splendid auto road in the Canon from Parkdale west. I watched most of this built, from trains. Convict labor.

Coming on to the Wet Mountain Valley. The copper property I mentioned out of Hillside (I have forgotten its name) lies well up on the east slope of the Sangre de Christo. During one of its revivals I spent a night up there. A man named Comstock was manager for some "Boston" people. He told of how much copper they had "blocked out" and the mine's great future, but the supply of ready money ran out and it closed.

At Silver Cliff—the bluff you mention just north of town is the Geyser dump. I am surprised that someone didn't mention the *Geyser,* for it was the most important mine in that section in its day. It had seen its best before my time, but every so often some new eastern money would be available and new management would locate in the "Cliff" to keep things moving until the funds ran out. The last manager, an easterner, didn't like the primitive accommodations of the "Powell House," so had a long sewer dug and hot and cold running water and other niceties put in. Otherwise, I suppose the mine would have run longer.

Querida was still a good camp in my earlier years but finally died on the vine. I used to drive up there regularly and enjoyed a brisk trade in the two good general stores.

A bit of history regarding Rosita may be of interest. We are so used to seeing these large heads of Iceberg lettuce now-a-days that it may surprise you to learn that—so far as Colorado is concerned—they were unknown before 1915. The first, known, were grown in Rosita. That climate was ideal for lettuce. The man shipped his production by express to Denver CRATED IN BOARDS TORN FROM DESERTED RESIDENCES. One could tell by the various colored painted boards. The express went Rio Grande out of Westcliffe.

In my time a Rio Grande standard gauge line ran from Texas Creek on the main line to Westcliffe. The crew made headquarters in Westcliffe, leaving about 9 A.M., laying over at the "Creek" until about 4 P.M., thus catching passengers and mail off of main line trains from both east and west. My, how many hours I have spent in Texas Creek, between trains. Usually there were two or three of us "peddlers" and we would provide ahead, bread and a few other items, including coffee, and cook our lunch down by where the clear "Creek" ran into the river.

Just north of the Texas Creek depot was a neat little white painted school standing at the foot of a tall sharp pointed butte. (I don't know where the pupils came from; I don't recall ever seeing but one or two around there). One day, alone, to kill time, I decided to climb this butte and did. So steep it required hands as well as feet to reach the top. Did you ever roll a rock down a mountain? If not, I will reveal my simple mind by saying it is one of the most fascinating pastimes, to me, that I ever indulged in. Of course it is against Colorado law, as a rock the size of a football, once started, can end up in an avalanche, to the detriment of people or cattle below. I carefully started a rock down the opposite side from the school. BOY—it was sump'in!!! One called for another and each bigger, of course. In my excitement I had worked around and after I had started a buster—Holy Cats, it was headed straight for that school building. If it hit it—you could have driven a pick-up truck through the hole. But the Lord was with me—the rock missed the school—but I had enuf rock rolling for the day.

So much for this ball of string.

Feb. 28, 1958

I haven't been to Colorado since 1927 and have paid little attention to R.R. developments, so-o-one day I recently obtained a Denver and Rio Grande folder. What I discovered made me positively homesick. No rail line to Gunnison or Crested Butte and none to Westcliffe from Texas Creek on the main line. Back in my days, all of the main line Rio Grande trains routed via Pueblo and Salida and up to Leadville before turning west over Tennessee Pass. A train leaving Denver at 8 A.M. traveled some 300 miles and at evening dinner time was still only about 125 air line miles from Denver, at Leadville. The Moffat tunnel was just being completed and was not used by the Rio Grande until after 1920. Salida used to be a live railroad town.

Just here I want to mention a very large frame residence which stood on the point of land which divides one large lake into the two *Twin Lakes*, out of Granite. It was built by the Leadville mining man John Campion as sort of a summer home. One time when I drove to the village of Twin Lakes I was shown through the house by its caretaker. Finely furnished—bowling alleys—barn and carriage house equipped with fine harness and horse drawn vehicles. I recall that the caretaker was particularly impressed with an elaborately carved hall-tree from Germany which he said cost $1,500. It is possible that the building still stands.

To keep from making this epistle too long, I will end with the Wet Mountain Valley, my favorite spot in Colorado. I have stood on the east slope of the Valley and viewed the Sangre de Christo Range almost every hour of day and night and have never become tired of looking. The "Geyser" mine is the big dump just north of town (or so I always understood) and in my time was reputed to be the deepest mine in the state. My recollection is—the "Bull Domingo" was some distance north of the "Geyser." The "Geyser" was once a good payer but went the way of many other similar properties. Occasionally, some eastern capital would come in and revive it for a time. Never for long. There was a "copper" property up in the Range opposite Hillside which bobbed up about every so often.

Before closing, I want to mention the Schibles, Emma and Fred, who ran the Powell House in Silver Cliff and later moved the building down to West-cliffe. They would have been to you like a miner striking a "pocket" of nuggets. But I will stop here. I, too, would enjoy a chat with you. While I paid slight attention to "hard rock" properties, except as their activity affected the business of the towns, I did breathe that "hard rock" air and some of it did stay with me.

Cordially,
James M. Diven

The Gold Belt South of Gunnison

According to M. A. Deering of Gunnison, with whom I talked in the 1940's, Gunnison County contained "a mountain of iron ore" with one of the largest deposits of iron and manganese near Powderhorn.

The camp of Tolifero (Tallifero, Tolifaro, Tollifaro—the name was spelled in different ways) was near Powderhorn. Kermit Matthews wrote to see if I could pinpoint the place for him as his grandfather had mined there.

Tolifero

Sept. 7, 1950

I'm writing to see if I can get some information. You see my grandfather was a miner, sort of a businessman who got the "gold bug" and tried his luck at mining. When my mother was quite small he was mining near Tolifero. Since then my mother has tried to find out where that town was and to revisit the site. The only mention she can find of it at all is in your book. Have you any idea as to the direction Tolifero was from Iola, or of any present town it was near?

Your book really stimulated my interest in the early miners. My grandfather, too, helped to give me an idea of just who the miners were and where they went. He was first at Tolifero where he did not find gold. Then he became a guard at some of the mines in the Cripple Creek area during the time that the miners were stealing so much. His ladder was nearly sawed through one night, and finally he was run out of town altogether by the racketeers. He told me of how the miners were made to change clothes before leaving the mine, and how the washerwoman that washed the clothes strained the water and became quite wealthy. Judging from his life, your book has been complete and accurate in every detail of those areas.

Sept. 29, 1950

If you remember, I promised that if I found anything about the location of Tolifero that I would send it to you in return for the information that you sent me. I did get some from the sources that you mentioned. I'll just copy it word for word and then there won't be any mistakes in the transcription.

From the Librarian of Western State College: "Tallifero is located approximately six miles southwest of Spencer, Colorado, which is located on state highway 149 which runs from Iola to Lake City. State highway 149 junctions with U. S. highway 50 twelve miles west of Gunnison. On the map on which I was able to locate Tallifero there is a road that junctions with highway 149 seven miles south of Spencer. This road evidently goes over and junctions with the road which comes down the Lake Fork. Approximately four miles west of this road there is an old road which runs to Tallifero which is approximately four miles north. I have no information on the condition of the older roads mentioned as the map I have is seventeen years old. I also have a picture of Tallifero taken July 20, 1933. At that time a sheep camp was located on the old town site."

From the Powderhorn postmaster: "Tolifero was a mining camp about four miles from here, but was abandoned some years ago and the land was taken up by a man, who sold to a sheepman who has a summer home there and pastures and sheep pens. I think all the Tolifero buildings were sold and moved away. Tolifero was west of Powderhorn."

From the postmaster at Iola: "The old mining town of Tolifero is located about five miles west of Powderhorn, Colo. The Holeman ranch now is on the site of Tolifero."

Yours truly,
Kermit Matthews

From Neil Foreman, who knew the country intimately, I learned still more:

Dec. 8, 1950

I was born in Tin Cup in 1898. My home was in Lake City from 1913 until 1919. During that time we lived at Capitol City and three miles up North Henson Creek, father and I shipped some high grade silver ore out of that country, we did some prospecting around Rose's Cabin and the high country clear back North and East of Lake City. I then moved to Cripple Creek where I raised my family. I leased many years on the Cresson and Ajax and worked or leased on many other properties in that district. I left there in 1940 because the doctors told me that I had Cilicosis, but the army took me in 1942 and said that I was O.K. I was doing construction work at Camp Hale until I went to the army. . . . I have a sister at Powderhorn and that is where I make my headquarters. . . . I bought some gold property just over the ridge from Vulcan toward Spencer last summer and if I live long enough I will open a gold mine there. . . .

The main purpose of this letter is that I have located a large deposit of radio active mineral and when we get it producing it should cause as much of a stampede as gold ever did. Doing business with the government is awfully slow, so it will be some time before we get started, but if we do get a rush started there I would like to give you a personal invitation to come over and see history in the making.

Just an old hard rock miner,
Neil Foreman

Dec. 28, 1950

Tolifarro did set one mile west and about one mile north of Powderhorn. There is a road goes north about one mile west of here and when you top the hill you can see a farmhouse ahead of you which is the site of Tolifarro. It also sets in the upper end of Goose Creek. The Carpenter group was the only one that I know of that did any mining to amount to anything there.

In the lower end of Goose Creek there was a town by the name of Dubois, it had quite a history but never produced too much actual mineral as far as I know. . . . I think you would have to walk about four miles to get into Dubois, there was still some cabins there two years ago. I would enjoy making the trip with you if I am still here this summer.

I have not been to Rose's Cabin since 1919, but we had it fixed up and was living in it at that time. There are several good mines in that district that have good quantities of minerals and will produce sometime.

I was in Vulcan last summer and the road is in good shape but I know very little about it, in fact not any more than you have in your book.

Sincerely yours,

Neil Foreman

It is Anna R. (Mrs. Clyde) Jewell, however, whose memories of Dubois (Du Bois, Duboise), Spencer, Tolifero, and Gateview are most vivid. She shares through her letters many girlhood recollections. Of Gateview she writes (Jan. 31, 1950):

Dubois, Spencer, and Gateview

The place has been known both as Allen and as Barnum before being called GATEVIEW, and it had a postoffice. . . . There had been a school there until the local children grew up. The school for that district was then held at DUBOIS until that camp folded and ran out of children. A child was then borrowed from the Spencer district to live at Dubois with the teacher and be thoroughly tutored to "hold the district" until Jack Carr (present owner of Gateview Ranch) and I justified moving the school to that end of the district.

I recall the trip of my parents and Carrs to Dubois to see about moving the school equipment to a room in a cabin at Gateview, pending building a school on the river. Dubois then consisted largely of the secretary of the schoolboard, Theodore Renner (or some such) a mining man, and the Gabrielsons who ran the tiny money order postoffice and a general store. The Gabrielsons continued to live there until some time in the 30's when they took their first automobile ride which, courtesy of Assoc. Press, landed them a 3-inch item in the *New York Sunday Times*. Their ride was to Gunnison to retire.

The school equipment consisted of two seats and desks long enough to seat three children apiece done in plain jack-knifed pine boards and a third really fancy seat with a drawer under the desk and painted a dusty rose. The teacher's desk was a cabinet-maker class home-made table, complete with chair. The blackboard was of rather knotty pine planks painted black. The wide cracks between the planks got into the way of problems in addition. The eraser was king-sized, being cobbled from a one-by-four plus handle and a quantity of knit cotton underwear. There was a handsome, practically new cottage organ which had been bought for the Dubois school with proceeds from a "social" in the brief mining boom there. I wouldn't be too surprised, if it is still to be found in that little schoolhouse down the river from the Gateview ranch . . . until the mice got it.

You may recall that Walter Wilson, a Lake Fork rancher, was shot by deerhunters this past fall. Well, his kid-sister, Louise and I, second-graders, got much practice fitting long columns of figures to add neatly between the cracks on that blackboard. Louise commuted by train from the Wilson ranch and summer school at Gateview. She looked like a seven-year old version of the Dionne quintuplets, and the mosquito brigades practically ate her alive on evenings in late summer when the trains were habitually late. That is until mother arranged with Mrs. Wilson that Louise should come back to our house for the night if the train didn't arrive by sundown. Our teacher was a pleasant young woman whose home was "up Henson" where her brother had recently been killed in a snowslide near the Hidden Treasure (Mine). In winter she taught at Ames, Colo.

A year and a half later she wrote again at some length (July 4, 1951):

DUBOIS. Now about reaching Dubois: In my childhood we got there by going down the Lake Fork below (north of) Gateview to the old Charles Carr place (Madera), then turning right and going up over a hill and on. First you came to Dubois, then on to Spencer where "Old Doc" Collins (aged about 80 plus during the first decade of the century) lived. He dispensed homeopathic prescriptions on description of the ailment and only called on the patient in case of the most dire necessity. More of that later. I was never beyond Dubois as far as Spencer. But Dubois was certainly reachable from Spencer at that time. It was in the direction toward Gateview.

At that period, Lake Fork people reached a place on the Cebolla, where the Rosenbaums and the Hayneses lived, by way of Dubois. My second teacher, a middle-aged divorcee, taught the Rosenbaum and Haynes children, the Hockers, and one George Hawkins there during the winter and at Gateview (now Riverside school) in summer afterwards. I recall persons going to Vulcan, to Iola. Once we "went around by Taliferro" to avoid the "Powderhorn Hill" which was like going down a winding staircase in a vehicle. Altho I believe Taliferro was one of the later camps, when we went that way it was all moved away but one good-sized frame house.

I recall Mrs. Gabrielson's little general store which she kept in her small log house along with the money order office. She had a great many geraniums which bloomed gorgeously, and she gave Mother slips off them. The remote offspring of these slips Mother kept, and they all froze one March night in 1942 when a heavy wind blew the bay window in at the ranch we have near Dolores, Colorado. These slips must have been given 36 or more years previously. I recall the horehound stick candy, the peppermints, and what I called "sealingwax" candy Dad always got for me at Gabrielsons. (A sort of brown-sugar fudge).

This middle-aged teacher I mentioned deserves more mention. She and her older sister from Kansas happened somehow to come out to Colorado to teach in the little schools. What with summer school where children had to go farther and winter schools where they were closer, a rural or camp teacher could teach practically the year round. Our school, at least, was a snap, for only the children of three or four families could possibly be in walking distance. This middle-aged teacher had us do some ambitious things for our ages. Since she had to read Shakespeare's *The Tempest* for "reading circle" for teacher's examination in December, I had to read it with her, every word of it aloud to myself, and do my primary child best to comment on all of it to her. (I'm afraid it gave me an odd idea of *The Tempest*.)

Then this lady had quite a talent for drawing and painting, and we did that too. I still have a watercolor of the Riverside schoolhouse I made before the Shakespeare episode, and a full sized copy in water color of a calendar

Vulcan, 1946.

Gateview, 1974. Lake Fork of the Gunnison River flows between the cliffs.

picture of a small boy sitting against a white siding house-wall watching a bee sitting on an apple he'd been biting into. (No, I didn't trace it, and she didn't draw it for me. With her supervision I struggled through it myself).

But the odd part of the whole thing was that fifteen years later, the first year I was an instructor in the Western State College art department, this same former rural teacher was *my* art pupil in the required "public school drawing" course she had to take to get a standard state "life certificate." She was an unusual person in that line of endeavor, and she loved to tell the rest of the class how she gave me the first training I ever had. Of course she had much more ability than most, and naturally the course was tailored to the run-of-the-mill teacher-training student.

When I hear all about what wonders school reorganization will do for rural children, I still do not feel the least bit cheated in what I had, even if it was just 20 months all told before the eighth grade. What we did have was practically a private tutor. Incidentally, the schoolhouse where I took the eighth grade burned a week or so ago, and the people unanimously voted a bond issue to rebuild it while ardently refusing the offer of Cortez to let them haul the children in there for school. That has always been such a happy school where the pupils and the parents were so pleasant that they could get and keep the best teachers. To the children the Lakeview "last day of school" was on par with Christmas and the Fourth of July, and former pupils came back each year for it if within a day's travel. It is a largely "old timer" community and is a rich district, and it will probably resist consolidation to the last ditch.

I mentioned this ancient Dr. Collins who held forth at Spencer. The only time I saw him was when Jack Carr, who now runs the Gateview Ranch (where I lived as a small child) was in a desperate condition with a swelled grain of corn in his windpipe. He had some in his mouth, got to laughing, and got a grain lodged in his throat. His father snatched him and caught the evening train to Lake City where the doctor poked the grain of corn down his windpipe where it couldn't be reached. The corn swelled, almost shutting off the boy's breath. That was one occasion when Dr. Collins came to the patient and remained until the family was off to the hospital at Salida with the little boy. I remember the night my parents helped sit up with small, smothering Jack, who was about to drown in the secretions from his throat. The surgeon said later that the whole thing could have been avoided had his mother held him up by his feet and shaken him hard until the dry corn fell out.

And three weeks later (July 24, 1951):

I note that you say that C. A. Frederick moved his newspaper from Tin Cup to Spencer in the middle '90's. That explains to me, at this late date, how my father came to know the said Mr. Frederick so soon after we moved from Gateview to a place between Dolores and Cortez where Mr. Frederick had again moved his newspaper. I mostly knew him as a stern Seventh Day Adventist who had his troubles with the local schoolboard because he did not send his children to the local schools. He said they were taught at home when he did not have enough cash to send them to Seventh Day Adventist schools in California. Allegedly they learned to read and spell setting type by hand for his *Montezuma Journal.* His daughter and I roomed together for three months when we were both freshmen at Western . . . under the eagle eye of a stern

Seventh Day Adventist missionary. It turned out that poor Beth was no more a Seventh Day Adventist than I was, and I learned later she felt trapped.

SPENCER. Mr. Samuel Spencer, so long of the First National Bank at Gunnison. I knew his son and younger daughter well and see his widow frequently here in Grand Junction. Mrs. Spencer's older sister, Mrs. Mary Axtell Lawrence, taught her first school in the old schoolhouse across the road from the Gateview ranch house beside Indian Creek. When that generation of pupils had quit school, the building was divided into a carpenter shop on one end and a chicken house on the other. And the school for that district was moved to Dubois.

GATEVIEW. In the summer wagons loaded with fruit and vegetables would come up from "the Lower Country" (Delta and Montrose counties) peddling. There would be baskets of cherries and grapes and packed boxes of peaches and plums and pears. Sometimes there would be corn on the ear, cucumbers, watermelons and canteloupes. But mostly it was just fruit, and I knew so little about watermelons that I played that my doll family considered Baked(!) watermelon a treat. In the fall, the agent for a wholesale grocery house in Denver would call with his samples and take one's yearly staple grocery order. One's other orders were sent to Lake City and delivered to one's railway siding by train. It was surprisingly handy.

Lemonade was a summer treat, and none I ever get anymore tastes quite like it. We would take a quantity of quart beer bottles with stoppers to Charles Carr's at Madera and fill them at Carr's soda spring in the evening when it was cool and store them in the cellar or in the icehouse. Lemonade made with this water was my idea of delicious. There was an even better-tasting soda spring on the Powderhorn road, but we didn't get there complete with bottles so often. However, it was very good to drink even plain, and some thoughtful soul had chained a cup to a post beside it so that the public could drink and hold a germ exchange there.

In summer one might go to some once-burned-over area which had grown up to wild raspberries to pick. . . . They were marvelous and about 75% more flavorful than cultivated ones. We also picked gooseberries. But the neighbors also went for oregon grapes for jelly, and service berries and chokecherries. The Gateview ranch had so many currants and such, we had enough without.

Mrs. Clyde (Anna R.) Jewell

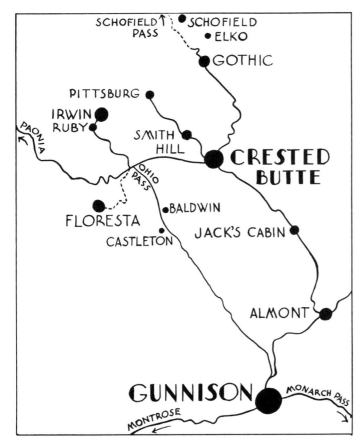

CHAPTER X

Gunnison and the Camps Northwest

Northwest of Gunnison both precious metals and coal were found in appreciable quantities and at approximately the same time. Crested Butte, the largest of the settlements, was laid out in 1879, and although gold, silver, and lead were found in the surrounding mountains and placers were developed on Slate and Coal Creeks, it was the coal deposits close by that brought the Colorado Fuel and Iron Company to the "Big Hill" overlooking the town. For years coal mining was the major industry, but when the C. F. & I. shut down in 1952, most of the miners left. Although those who did stay on faced lean years, they never gave up, with the result that Crested Butte retains much of its original quality. Some lead and zinc mining is still done on Red Lady Mountain at the Keystone and Standard properties, but the Butte's chief attraction today is its growing popularity as a ski resort.

During the coal mining era, nearby Anthracite (or Smith Hill as it was also called) grew up around the coal breaker. With a spur of the railroad from Gunnison reaching coal mines not only at Crested Butte but also in the next valley to the south, Castleton, Vidal's Spur, Teachout, Baldwin, and Floresta flourished throughout the 1880's. During the same decade, prospectors continued to search for gold and silver; and when they found it, the camps of Pittsburg, Gothic, Elko, Schofield, and Irwin sprang up.

Gunnison   On a trip to Gunnison in 1950, I stopped at the Wm. Hudson Garage, attracted by a sign in the window:

A Hudson Car
Driven by
Carroll M. Carter
of Ohio City
Was the First Car
To Cross This Pass
Sept. 22, 1911

The elderly man in the office told me that the car had indeed crossed Monarch Pass on the original wagon road, the present highway being the third to be built. The "Old Road" that is so marked today near the western summit is the second one. The man to whom I talked drove over the first road in 1915 in a car "that you drove up in low gear and hoped could make it, and you drove down in low and hoped it didn't run away."

Matthew Oblock, Jr. of Aspen in a letter written in 1951 reminisced about Gunnison:

> The town of Gunnison itself had quite a colorful past. Old Gunnison, which is west of the present town, was a rip snorting wild town in its early days and was started as a supply town to the various mines in the county.
>
> There is an old cemetery across the river from west Gunnison over by the palisades. There is just one stone marker there. The inscription, or I should say the epitaph, reads, "REST BROTHER REST." The story is that the man "died with his boots on" in some mining camp close to Gunnison.
>
> I do not know whether or not anyone called your attention to the masters theses on file in the library at Western State College at Gunnison. There are several there that deal with some of the early history of the mining towns in Gunnison County. . . . I read some of them while I was a student assistant during my Junior and Senior years at Western. One thesis dealt with the White Pine region and was written by Margaret Fassler whose father had business interests (I believe mining) in that region at one time. Another one which I remember was called, "The Rise and Fall of Irwin—A Ghost Town." Still another was on Journalism in Gunnison County by a Mrs. Spencer. If you would check through these and others on file there, I am sure that you could find much valuable material which you could use in case you ever bring out a sequel to *Stampede to Timberline*.

I did not linger long in Gunnison, for my goal was the mining camps northwest of the city; but as I drove up the Taylor River to Almont, I told the artist friend who accompanied me about La Veta Hotel in Gunnison and of how fortunate I was on an earlier trip to have seen it less than a year before it closed its doors to guests. A man wrote me, after he'd read *Stampede to Timberline,* to express surprise that I hadn't mentioned the building. Although I hadn't because it wasn't in a ghost town, I did know something of its history.

When Capt. Louden Mullins set out to build the finest hotel in the state, Gunnison had a population of only 5,000, with about half living in tents or shacks. The captain owned land on the west side of the town, and when the Denver & Rio Grande Railroad reached Gunnison in 1881 and continued laying its tracks west toward Salt Lake City, he decided to sell lots on his property, thereby promoting a new area to be called West Gunnison. It was in this section that he started to build his dream hotel, but by the summer of 1882 only the stone foundations were finished and Mullins' money exhausted. Next the property was acquired by Benjamin W. Lewis of St. Louis, and building was resumed. The Lewis Hotel and Improvement Company was formed with D. J. McCanne of Gunnison as manager and supervisor for the completion of the hotel, which by then was called La Veta.

Its massive exterior was brick with ornamental iron columns clustered around the main entrance and balcony. The first two floors contained dining rooms, parlors, a billiard room and bar which contained the largest plate glass mirror in the

Gunnison. La Veta Hotel. *Photo by G. B. Sanborn, Leadville.*

Gunnison, 1941. La Veta Hotel lobby.

state, a library with reading rooms, a bank, and a most unique feature—the Denver & Rio Grande ticket office and waiting room—on the south side of the building next to the railroad tracks. The hotel, with its 107 rooms, was completed and furnished by May 1883, but the official opening, attended by guests from all over the state, did not occur until April 15, 1884. Even so the citizens of Gunnison were not satisfied, and they organized a more formal event embracing participation by the entire Western Slope. This second fete was held in May 1884.

For many years the hotel ranked with the best in the state, but Mullins' original vision of Gunnison as a thriving industrial city like Pueblo failed to materialize, patronage fell off, and in 1944 the property brought only $8,350.00 at a sheriff's sale. A Colorado Springs salvage firm was the purchaser, and the ornate La Veta was razed for the value of its furnishings and building materials.

My visit to the famous hotel was made early one morning shortly before it was demolished, when three of us, who were starting back home from a short stay in the Crested Butte area, pulled up beside the huge four-story building.

"I want to see the interior," I said, "especially the lobby and what they call the General Grant suite, although I've been told he made his trip through Gunnison in 1880, two years before the hotel was even under construction."

The lobby with its domed glass ceiling was all that I'd hoped it would be. A large iron safe built into the wall behind the main desk was faced with a wide trim of carved woodwork, surmounted with a florid broken pediment of walnut. Next to the safe was a tall walnut cabinet with a clock in its center section and pigeon holes for mail on either side. The long panel of glass through which the clock's pendulum was visible contained the painted message, "Buy Your Tickets via Pueblo and take the Santa Fe Route to all points East." Bronze chandeliers, which no doubt had originally burned gas, still had their colored globes. An imposing staircase of walnut and ash led to the upper floors and completely dominated the far end of the lobby.

"I'd like to see the General Grant rooms," I said to a somewhat sleepy clerk behind the desk. "Just go on up," he replied. "The general isn't down yet." Startled by this banter, I climbed to the second floor and found two large rooms each with a black walnut Victorian bedroom suite, as well as settees, rocking chairs, and tables. These were definitely the original furnishings. After making three sketches, I rejoined my friends downstairs, and we continued our journey home.

The gentleman, James M. Diven, who wrote me in 1958 about omitting mention of the hotel, knew La Veta well. As a traveling hardware salesman, his territory included Gunnison. He wrote about the hotel (Feb. 5, 1958):

> Coming down to Gunnison—I again express my disappointment at your missing the old La Veta Hotel—the QUEEN of the Colorado Ghost Town hostelries—built in the "Black Walnut" and "Red Plush" era—when "All roads were down hill to Gunnison" and it was to be the "metropolis" of the state. Four floors high with foundation about three feet thick to support two additional floors "soon to be needed." Built by St. Louis capital. Bricks hauled from east of the Range. All visible woodwork including massive stairways of black walnut. All hardware solid BRONZE. Being "modern" it was gas lighted and with most elaborate bronze and enamel gas fixtures. Incidentally, while buying the gas fixtures they thot it wise to include sufficient to furnish the two "anticipated" additional floors. Those additional fixtures were still stored all

my years going there. It would be interesting to know what became of all those bronze gas fixtures when the building was razed—at least, I understand it was razed as I am not able to find anyone of the present generation who has seen it. It did not have private baths or toilets. It did have marble wash basins in all rooms with hot and cold water. Toilets and baths were on a back hall of each floor. To any new visitor the old "Gal"—grandeur in decay—was almost ghostly. Many of the rooms were in suites—a parlor large enuf to hold a basketball game with big bed room alcove. In my time two full size beds were, usually, installed in each bed room and we salesmen would often double for company. Once a new comer salesman was doubled with an oldtimer. Came bed time he asked where the toilet was and was directed to the rear hall. Coming back after a surprisingly short time he was asked, "Didn't you find the toilet?" He replied, "Yes, but I could be murdered back there and they wouldn't find my body for a week." This will give you an idea. And so we leave Gunnison.

The Crested Butte Area

Crested Butte lies at the end of a long valley surrounded on three sides by rugged mountains with sharp peaks. Roads and trails which radiate from the townsite provide access in summer to fishing streams as well as to mountain passes, ghost towns and sites, and an abundance of colorful mountain scenery. Deep snow in winter attracts skiers, who find accommodations either in town or in the lodges and condominiums near the ski slopes. Except for the ski area three miles north of the town and a few new homes, Crested Butte looks much as it must have in its boom days. The City Hall on Elk Ave., the old stone schoolhouse, the stone jail, the Emporium (formerly the C. F. & I. Commissary), St. Patrick's Church with its tall steeple, the Union Congregational Church with its bell from the church at Irwin, as well as most of the private homes date from the 1880's. Even the Denver & Rio Grande depot at the eastern end of Elk Ave. is, as of 1976, a museum.

Today few buildings remain on the Big Mine Hill south of the town, but an old photo shows the flat surface in front of the shaft entrance filled with sheds and chutes, with loading stations and tracks, a mule barn, a dormitory, and the superintendent's house. Until recently this last building has functioned as an excellent restaurant.

Gothic   Roads lead in all directions from the Butte to the heads of mountain valleys and over old passes, some of which can be crossed only in late summer after the winter's snowdrifts have melted. To me one of the most beautiful short drives in Colorado is from Crested Butte to Gothic, especially in late July when the wild flowers are at their peak and carpet the high pastures and slopes above and below the road. Whole meadows cut by the twisting East River are blue with columbines, and clumps of Indian Paintbrush vary from creamy white through fuchsia pink to blazing vermilion. These, intermingled with purple Pentstamen, yellow Cone-flowers, blue Flax and other species, make the drive a kaleidoscope of color.

Although little of Gothic remains, for even the old cabins are remodeled and used by those attending the Rocky Mountain Biological Laboratory, the weathered two-story store still sags beside the roadway that goes through the town.

C. A. Martin lived in Gothic in 1912 and reminisces as follows (Nov. 30, 1951):

Crested Butte, 1964. Elk Avenue and the Butte.

Gothic, 1960.

My wife and I went to Crested Butte about Aug. 20, on my vacation. We drove up to Gothic, and noted many changes from the time when I was a kid there in about 1912. A Mr. Sampsell used to brag to us that it was he who had made the greatest wire silver discovery ever at Gothic. Our old ranch, 4 miles south of Gothic Mountain, seemed a complete ruin only fit for grazing. The buildings were falling apart, but there was a nice trailer parked in the yard.

Two of the mines had quit years ago, and there was hardly a sign that there ever had been any there. A bit heartsick, I turned back to Gunnison for the night. Almont was built up pretty, but the fence of elkhorns was no more. Jockey Thorpe (the famous one who rode in Paris to win a great race and much wealth) was going to buy Almont, but never did. We knew him in Geneva, Neb., where I went to school with his nephew. He was wealthy then, but about 1917 I donated a dollar to help save him from a pauper's grave, after he died in a saloon owned by Jabez Cross, who let Thorpe do porter work around his place.

Letters from Frank O. Long, Jr. mention Garwood H. Judd (whose name I misspelled as "Gatwood" in *Stampede to Timberline*), who witnessed the rise and fall of Gothic during his fifty years of residence there and whose ashes were strewn over the town that he loved.

Nov. 26, 1969

I own several claims up near the Biological Laboratory. In fact I own Garwood H. Judd's original claim. My, how I would have liked to have known him. I also own one of Mr. Harris's claims called the Mountain Chief that overlooks Crystal.

Dec. 8, 1969

Garwood Judd's claim "Power" is on the right side of the road about two miles up Copper Creek from Gothic. I do not know exactly where the claim is since I have not found the corner posts. The *Crested Butte Chronicle* has it that Crested Butte may be one of the finest if not the finest Ski Resort in the U.S.

Matthew Oblock, Jr. in his letter also mentions Garwood Judd (July 12, 1951):

Several years ago a friend of mine now living in Glenwood Springs turned over his collection of old newspapers to me. This man was an Englishman and had been sent to Gothic, Irwin, etc. to look over some mines for an English syndicate. Among the papers, I have a copy of the *Silver Record* published at Gothic. The issue which I have is Vol. 2—No. 35 dated May 26, 1883. One of the ads in this paper is Patterson & Judd's Billiard Parlors. You mentioned Garwood Hall Judd in your book only you called him Gatwood. He was known as you pointed out, as the "Man who Stayed." When you were in Gothic did you notice a Forest Service sign where Copper Creek crosses the road? This sign reads—"Judd Falls ¼ mile." At the falls is a stone and concrete bench about four feet long and 15 inches wide with a back rest made of polished granite with the following inscription in letters about two inches high:

TO THE WAYFARER, FRIEND, OR STRANGER
THIS SEAT IS OFFERED IN THE NAME OF
THE MAN WHO STAYED
GARWOOD HALL JUDD
MARCH 14, 1852   MAY 15, 1930

I was a freshman at Western State when Judd died. A few years later several stories were circulated about Judd. One was that some lady in Wisconsin claimed that he had been her husband and that he had deserted her and their daughter and went West where he stayed. Some claimed that this was just someone who tried to get hold of his money of which he seemed to have quite a bit. Another story was that he was of noble English birth and that he had run away from home and had severed all connections with his home folks.

At the end of my Freshman year at Western, Horace Williams, whose father ran a country store at Marble and his uncle was postmaster there, and I went home by way of Gothic, Schofield and Crystal. We hitched a ride to Crested Butte and hiked to Gothic where we stayed overnight with the cattle range rider in a two story building which was still standing at that time. The next day we went through to Marble. Schofield was covered with snow. The trail down Crystal Canyon was narrow, frightening, awe inspiring, and still very beautiful. The punch bowl and the bridal veil are spectacular. Crystal is a welcome sight after coming through that canyon. We repeated the trip one year later. I also made four trips between Aspen and Crested Butte on foot by way of Gothic, Copper Creek, and East Maroon Pass all by myself during the years I attended Western State.

Beyond Gothic were two small, short-lived mining camps, Elko and Schofield. *Schofield*
Back in 1951, before the present jeep trips across Schofield Pass had been organized, a visitor from Alabama, E. G. Gates, wrote:

> There were no cabins standing in Schofield Park except some new ones that had been built, or perhaps an old one or two that had been completely reworked. In the lower end of the park there were signs of a house or two.
>
> The road from Crested Butte via Gothic, Emerald Lake and Schofield Pass was an ordinary mountain road, but not as bad as many that I have been on before. The road was closed for several years by a large snowbank just below Emerald Lake. Last year the snow melted and the road was opened up and cleaned out from there on over. It was in good shape to about three miles from the top of the pass toward Crystal. I believe that the entire distance down to Crystal could be covered with a jeep and expect to find out next summer.

I hope he returned in later years and made the trip.

On the north side of Slate Creek a road enters the highway from the west a *Pittsburg*
short distance beyond the Crested Butte cemetery and runs through the deserted coal camp of Anthracite (or Smith Hill) to Pittsburg, a once flourishing gold camp, now a mere site, part of which is staked out into lots offered by a real estate agency. According to M. J. Webber, a former forest supervisor of Gunnison National Forest: "No buildings are left at the site of Pittsburgh but one can drive to the old townsite."

A mine owner's wife, Mary E. (Mrs. Arthur) Starr, writes about a mine of her husband's high on Treasure Mountain, above Pittsburg (Aug. 21, 1950):

> I have lived my life on the Western Slope. Here we farmers were brought up on mining tales as we lived in the shadows of the San Juans. Our fathers were miners who took up farming to supply food to the "rest of 'em." My parents were Virginians who came to Cripple Creek over 50 years ago. Dad carried gold from the Butterfly to the train, he worked at the Tom Boy, Liberty

Bell (Telluride) and in Leadville. I lived as a ten-year-old at the Camp Bird (Ouray).

My husband is a partner in the Eureka mine above Yule Creek Pass up from Pittsburg. We have spent 3 years *building* a road, a camp and a tram to that mine. We also own the Augusta. Talk about thrills—the first day I went over that road! I cannot express my excitement adequately.

Incidentally there are few ghosts at our mine—just one sort of shack, 13,200 feet at the top of Treasure Mountain. I've never been up higher than that terminal. "They" say one can see every place up at the mine—the La Sal mountains [Utah] and the San Juans and even Grand Mesa's Land's End.

We are only 7 miles from Aspen, six from Marble, less from Crystal, and the Supt. of the Big Mine at the Buttes watched the building of the tram with his field glasses. It will always be a marvel to me "How did those old boys ever get those places." The roads are beyond comprehension.

If you are ever in this section just ten miles from Delta, toward Grand Mesa, we would be very happy to have you spend some time with us.

I suspect that the man from Crested Butte who wrote the following is describing the same mine:

If Dr. Sibell is something of a mountain goat there is a camp of just shacks practically at the top of Treasure Mt. (above Crystal) which is about 10 miles northwest of Crested Butte. This elevation, 13,200 feet is the highest I have ever reached in my hiking experience.

Anthracite
or
Smith Hill

A short distance down the creek from Pittsburg, a large mine known as Smith Hill Anthracite used to operate. According to M. J. Webber of the Forest Service, "Smith Hill is a recent ghoster! It had no stores, just homes and mine equipment."

Mrs. Elizabeth Osborne was my chief source of information about Smith Hill. While I visited with her in her Thornton, Colorado, home in 1960, she told me that her father, Ty Miller, was a carpenter in charge of building the miners' cabins on top of Smith Hill at the Anthracite coal mine. This was in the spring of 1919. For a while she and her parents lived behind the building used for a schoolroom. Later her father built the house down below at the Breaker, and they lived in it. He was the Breaker-man when the tram was built. The coal mine operated for about five years from 1919 to 1924. She was six years old at the time.

She remembers one day when her father, while working on the cabins, sent her home. She was mad, and told her mother, "I got even with him. I called him an S.O.B." "Did he hear you?" she asked. "No, but I called him that anyway." "What does it mean?" asked her mother. "I don't know, but that's what he calls his donkey when he's mad at it."

Other coal camps in a valley south and west of Crested Butte were served by a spur of the railroad which linked them with Gunnison. In one of his letters, written during the 1940's, M. J. Webber mentioned Castleton as a former coal mining town that had been entirely dismantled, Vidal's Spur and Teachout as only loading stations, and Baldwin as still in existence with a population of perhaps 200. Today Baldwin is a true ghost town, fenced to protect it from vandalism.

Floresta

About ten miles west of Crested Butte, just before reaching Kebler Pass on the highway to Paonia and Delta, a gravel road (much of which is the old railroad grade) leads south over Ohio Pass and back to Gunnison through stock-raising country. Visible from the highway as one turns onto this road is a sawmill; beyond

Smith Hill (Anthracite), 1946.

that the railroad spur to Floresta branched to the right. Though one can now reach the place with a four-wheel drive vehicle, my only trip there consisted of a three-mile hike over a poor trail, much of the last part uphill through a forest, until suddenly, at a break through the trees, I saw Floresta, deserted and dilapidated, on a flat below me.

Through the kindness of Miss Marian S. Van Dyne of Greeley, Colorado, I received the following note and a picture postcard with a photograph of Floresta. It shows the railroad track, which enabled the coal to be shipped, and the town, crowded on a flat meadow surrounded by timbered mountains. Close to the coal company's big buildings and huge tipple are the miners' cabins. Here is the message on the old card:

> Thank you so much for the catnip leaves. Just finished two cups of the tea. About a month ago I wrote a card asking for a new supply and today Mr. Barnes fished the remains of said card from his pocket, so thank your mother for her thoughtfulness in sending some by Chellis. Little Mary still grows in sweetness as well as flesh. Best regards and love,—Mary E. B.

Miss Van Dyne's note (Jan. 25, 1951) explains why she sent it to me:

> Having read your "Stampede" with much interest and noting your request for additional material, I thot this old card might be of value to you. The year is blurred, but the baby mentioned in the context was born in 1916, so it antedates your visit.

During the 1930's, E. V. Carter of Whittier, California, visited both Floresta and Irwin and describes what he found (April 12, 1962):

> As to my visits to Irwin and Ruby, they were in the late thirties, and your descriptions and drawings are more graphic than any of mine would be. The water system was still trickling, a few cabins and buildings were still standing, but most, even at the mines themselves, were on the verge of total collapse, or had already reached that state.
>
> Floresta, however, is a different story, indeed. My first visit there was prior to the collapse of the 11-story "Tipple," as we called it. My father was employed at the time by the Fire Mountain Canal and Reservoir Co., who purchased many of the timbers from the railroad trestles, as well as a considerable number of the steel coal chutes which came from the tipple itself. On my last trip in there, however, the tipple had collapsed into a heap of rubble.

Irwin   Between 1879 and 1884 Irwin emerged as an important and flourishing silver camp, ten miles west of Crested Butte and one mile north of the highway. There was little of it left, however, when I first saw it in 1941—just a few cabins partly hidden by shrubbery and water hydrants at intervals along the washed out gully that must have been the main street. These, however, reassured me that I had reached the townsite, and I scrambled up the dump to the Forest Queen mine to look down over the camp. From that height I could see traces of what must have been streets. Piles of weathered and broken lumber indicated buildings long since collapsed. Yet in a photographer's window in Gunnison I had seen a picture of Irwin when its population totaled 5,000, when streets were lined with two-story buildings, and when a steepled church occupied the center of the town.

Irwin, 1880's *Photo by a Gunnison photographer.*

During its peak years, Irwin was a show place, and many dignitaries and colorful characters visited it and enjoyed the hospitality of its exclusive Irwin Club. The mines caused the town to grow and prosper; but by 1884 when production in good properties like the Forest Queen, the Ruby Chief, and the Bullion King began to taper off, even the skill of the *Pilot*'s dynamic editor could not prolong its life.

William L. Martin of Carrollton, Texas, worked for the *Pilot* as a boy. He writes (Aug. 8, 1949):

> My father and other of his relatives were in Irwin and Gothic during the early 80's. . . . Gothic had a population of 5,000. I was a newspaper boy and worked on the old *Elk Mountain Pilot* in the later 80's while it was being edited by John E. Phillips, one of its founders in Crested Butte. I was a boy then and Mr. Phillips told me much about the starting of the paper.

The Schlichters of Los Angeles share a bit of personal history in their letter:

> We have just read your article "Ghost Town Explorations" in the July 1949 FORD TIMES . . . a really thrilling reading experience for Mr. Schlichter and me.
>
> Not long ago Mr. Schlichter was telling our doctor how he happened to be named Irwin and our friend brought out these little Ford Times to find their July number. You see Irwin's mother was born in Irwin, Colo. where her father owned a gold mine. This mine was sold to an eastern syndicate but was so much a part of the family history that my husband's mother named him IRWIN.
>
> In 1932 he and a mining engineer cousin visited the abandoned mine. They waded through several feet of snow and had a big thrill finding the old shaft.
>
> Mrs. Irwin W. (Martha) Schlichter

When Everett V. Carter mentioned Irwin and Floresta (in a letter already quoted, March 24, 1962), he spoke of places which he knew well, for, as he says:

> I am a native of Colorado, having been born in Paonia, in Delta County. As a kid I tramped over Irwin and Ruby, helped haul timber and ore chutes out of the old tipple at Floresta, and in 1944 I worked in both the Treasury Tunnel at Red Mountain and the Camp Bird at Ouray. . . .
>
> In checking the cemetery at Irwin, near the road to Floresta, did you perchance notice the coincidence of dates on two of the graves—widely separated, of course, one being in "boot hill"? The occupant of the grave in the boot hill section was hanged for killing a 15 year old girl, accidentally, during the course of a gun fight. The girl's grave, bearing a date just one day earlier, was marked as being killed as an innocent bystander to a gunfight. I believe it was the summer of 1931 that I saw the two graves. As I recall, the marker on the girl's grave was a stone, while the one in the boot hill section was a badly weathered board at that time; so no doubt it's completely obliterated after another 31 years.
>
> Also, another question I wanted to ask you was whether or not your research disclosed any connection between "Uncle Charlie and Mother Howe . . . from Vermont," paragraph 4, page 201 in Gothic, and Jack Howe, the founder of Jack's Cabin, bottom of page 197. My great aunt, Izotta Sims, married Harry Howe, Jack Howe's son, therefore my interest.

This last request I cannot answer, but perhaps some reader can.

Mrs. Clyde Jewell's letters are so vivid that it is easy to picture the people she mentions and the incidents in their lives which she describes (July 4, 1951):

You mention your visit to Irwin, and the fire hydrants and the piles of old lumber which remained of the collapsed buildings. I was there several times, the last time being the summer of 1927. Most of the buildings were standing then, and two of the tall old houses were occupied. One was the summer home of Mrs. Russ, who was to Irwin what Mr. Judd was to Gothic. *As I understand it,* Mrs. Russ and her husband came to Irwin in the boom days and built their tall house. On account of the snows, and the social requirements of their young lady daughter, they wintered in Gunnison. Major (I believe that was his rank) Russ died, and the daughter married and went to live in California. But Mrs. Russ continued to spend the summers in Irwin in spite of her age and the fact that she was usually there alone except for the forest ranger near by. I recall seeing her there the summer of 1926 and 1927 or so. She must have been close to ninety then, for she had married *before* the Civil War and her husband was a Union army officer. I understood . . . from others, for she never spoke of her experiences . . . that she had made her debut in New York City and had married there in style in those pre-Civil War days.

Whatever was true about that, her husband had had considerable mining interests, and she always seemed to have adequate means for her tastes. She said she had lived the happiest days of her life in Irwin with Mr. Russ, and that she wouldn't leave the region until she knew she was about to die. She said she had promised her daughter that when that time came, she would start to California in time to die at her daughter's house. I was told she did know, and she did precisely that.

I first saw Mrs. Russ when she came to board at the school cooperative boarding club when I was a second-year student at Western. She was the innocent by-stander who caught a shower of snowballs intended for some flirtatious girls. The snowballs disarranged her spectacles (the nose part) and knocked her hat awry. She was rather tall, almost slender, and extremely dignified, but she laughed in such a way you knew at once that she was *ageless.* Later she stayed a winter where I roomed, and I knew even more that she got on perfectly with college age people.

She was built for a long life, for she had her own teeth with only a filling here and there, hair only partly gray, and an ability and determination to walk miles each day over the route she laid out for herself on the Gunnison streets. Briskly too.

After a summer at the Irwin home, she would move to Crested Butte for a few weeks. Then, toward Thanksgiving, she would come down to Gunnison for the snowier months. After each storm, a man from Crested Butte would snowshoe to Irwin and clear her house of snow to keep it from being too heavily loaded. In spring, Mrs. Russ would go back to Crested Butte, and then to Irwin as the snow was gone. Irwin was a lovely place in the summer. Last time I was there, the old bank still stood, with a sign saying "CLOSED" on its window shade.

Western State College had its Fourth of July fish fries there in 1926 and 1927, and you could still go up there by chartered narrow gauge train.

Irwin, 1920. *Courtesy John G. Curtiss.*

Irwin, 1946.

The hazards common to all big mining properties that operate throughout the year are graphically described by Myrtle (Mrs. T. R.) Hussey of Austin, Colorado, who grew up in several of the camps in the Crested Butte area. Two items in this letter (Feb. 26, 1953) give evidence that the usual winter snows in this corner of the Rockies were so deep as to require the two-story outdoor toilets, which had so amazed me when I first saw them in Crested Butte:

> You see I lived in Crested Butte, Irwin, Pittsburg, also at a mine named Key Stone. It was above Smith Hill about two and a half miles above Crested Butte. It was there that Daddy spent all his savings thinking he would strike it rich. I thought you would be interested to know the house you sketched this side of the Catholic church was my old home in Crested Butte.
>
> I was born on the Muddy about 28 miles above Somerset on a homestead Dad took up in 1892. When I was a year or so old the folks moved up to Irwin. Dad packed all our things on burros, and that is how we moved. Dad mined a couple of years up there that time. There were lots of houses there then. Then we moved to the Butte and Dad bought that house.
>
> We went up to Irwin again when I was six. I went there my first and second year to School. We lived in a large two-story house those two years and the snow got so deep we had to go upstairs to see out. We went to school on snowshoes part of the time.
>
> My Dad worked on the Bullion King mine. The winter he and mother spent there, a big snowslide came down and took the whole boarding house with it. It killed several people including the boss's wife, the cook and a two year old child. My Dad helped dig the bodies out and helped take them to Crested Butte on toboggan sleds. After the snow had melted, mother found an old gravy boat that I now have as a souvenir. My Dad worked at the Sylvanite for several years and when they went up in the fall he usually stayed there till spring—it was too dangerous to do otherwise. But this once the weather had been so nice and clear and the boys were getting hungry for a few Christmas treats and the mail, so Daddy volunteered to make the trip. He made the trip to the Butte all right on snowshoes and got the turkey and other things to go with it and had started back when it got cloudy and about noon began to snow and when it does snow hard in the hills it snows till you can't see very far ahead of you. He plodded on and on but it soon grew dark and Dad lost all sense of direction. He said he knew if he quit going he would soon be covered with snow and would freeze. He said he had gotten to the crawling stage and he struck something. When he heard someone swear and say, "Boys, something has knocked that stove pipe over," Dad called to them to get him which they did. It was a camp two miles below Daddy's camp and the boys thawed him out and they had the Christmas treats. Dad was with them a week, he said. He never forgot what a close shave he had and how good it sounded to hear that cook swear.
>
> He worked in Telluride at the Tom Boy, the Smuggler and Sheridan mines; also had several claims of his own that kept him stripped. He was working at the Smuggler when he died—only 45 years old, he was like all miners he knew he would strike it rich some day. He was the best Daddy in the world. He always loved the hills and mountains and I guess I too have lived there long enough to love them also.

I live on a farm 9 miles north of Delta, Colo.; have lived here for over forty years. My son that is still single runs the ranch. If you should ever be in this part of the country drop in and see us. . . . I have a neighbor that came from Crested Butte. Her father and his two brothers ran the mule trains to the mining camps. They were the Runnol (?) boys and my father's name was Herman Schraft.

Mr. Will Womack you mentioned was killed on Grand Mesa and is buried in Eckert. Several other names I recognized also—the Brown brothers who helped make Lake City. The world is a small place after all.

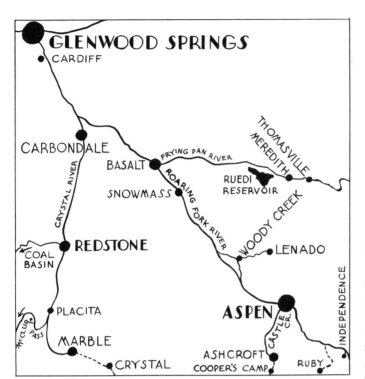

CHAPTER XI

Aspen and the Crystal River Country

In the summer of 1879 a few prospectors from Leadville struggled across the Continental Divide by way of Twin Lakes and worked their way down the western slope to a spot they called Ute Spring. Another party of men from Crested Butte entered the re-

Aspen gion at about the same time, and both groups staked claims on Aspen and Smuggler Mountains. One of the first to enter the area was Henry Tourtelotte, who struck out up Castle Creek and climbed a mountain on its east side until he came to a natural park near its summit. Here he found outcroppings of ore and located several mines. This area became known as Tourtelotte Park.

By the spring of 1880 so many men had pitched their tents near their claims that B. Clark Wheeler, Isaac Cooper, D. M. Hyman, C. A. Hallam, W. L. Hopkins, and others organized the Aspen Town and Land Co. and selected a wide meadow at the foot of the mountain for the townsite. Many lode mines were located and developed between 1880 and 1884, certain of which, like the Aspen, Durant, Smuggler, and Mollie Gibson, became heavy producers of silver. By 1888, when Aspen had a population of 8,000, two railroads reached the city—the Colorado Midland and the Denver & Rio Grande—and the silver camp was recognized as the leading commercial center between Salt Lake City and Denver.

Then in 1893, with the demonetization of silver, mining tapered off, but the owners of most of the big mines agreed to pay the miners $2.50 a day rather than lay them off. This kept the mines open, old properties were tunneled, and new development work was done.

Aspen was rather quiet throughout the first two decades of the twentieth century, but since then it has been rejuvenated, first by summer visitors who discovered its trails and fishing streams in the midst of impressive scenery, next by Walter Paul Paepcke of Chicago, who found in the quiet mountain village the perfect setting for his Cultural Institute. Through outstanding programs and workshops centered around music and the humanities, he brought thousands to the scenic high country where the festivals have been held.

Aspen was one of the first major ski areas to be developed in Colorado, and

Aspen, 1959. Durant mine and dumps in background.

largely through its attraction for lovers of winter sports the town itself has kept stride with the added year-round population by providing lodging and restaurants, "Swiss" atmosphere, and an ever increasing number of condominiums which creep up the lower slopes of Aspen Mountain.

Matthew Oblock, Jr., of Aspen, wrote a very helpful letter, which corrected mistakes I had made and added facts with which I was unacquainted (July 12, 1951):

Two years ago I purchased a copy of *Stampede to Timberline*. . . . I would like to make a few comments about some of the places you mentioned and perhaps add a few items that might be of interest to you.

Have you ever ridden on the chair lift up Aspen Mountain? The second section cuts across Tourtelotte Park. There was a great deal more mining there than what your book shows. J. E. Spurr in his Geology of the Aspen Mining District, Monograph 31 of the U.S. Geological Survey 1898 shows two excellent pictures of the Park opposite page 180 in his book. I have been told that at one time there was a store and a schoolhouse plus several dozen cabins. They even had a voting place for the elections. A tramway was built from there to Aspen, but some disappointed mule-skinner, who was put out of business of hauling ore down to Aspen, one night cut the cable with a hack saw and repairs were never made. There were a number of famous mines too, among which were: Best Friend, Mayflower, Sarah Jane, Justice, Last Dollar, Good Thunder, Little Lottie, etc. The Last Dollar had the deepest shaft.

On page 230 you quote a newspaper. The man's name was Joe Vonnah, not Vannah. We used to call him "Groundhog Joe" when we were kids. He lived under the hill in East Aspen between the Durant dump and the Ute cemetery.

On top of page 231 you list Aspen stores pictured there as burned. The Aspen stores which burned were across the street from the ones sketched. The two on the left—Berg's Confectionery and the building by it were razed. The others are still standing. One now houses Louie's Liquors and the other is Ed Tiedeman's place with the White Kitchen in between.

On page 232 you say that Aspen was once called "City of the Sulphurets." It's funny but I never heard it called that before. Charles Dailey who used to publish the Aspen Daily Times and The Aspen Democrat Times used to call Aspen, "THE CRYSTAL CITY OF THE ROCKIES."

On page 238 you give Jerome B. Wheeler credit for establishing and editing the Aspen Times. Here you have the wrong Wheeler. It was B. Clark Wheeler who edited the Times. Some other men, as the masthead of Vol. I issue 1 shows, established the Aspen Times. I have their names at the bottom of a trunk of books but haven't time to look them up now. B. Clark Wheeler was also responsible for changing the name of Ute City to Aspen.

"The rustic Pavilion," which you say, "is still standing in the middle of the square," which has the fire bell in it is not as historic as someone has led you to believe. That monstrosity was built by the WPA during the depression around 1934 or 1935. The bell was originally housed in a tall tower on the back of the building now called "Hillside Studio" on the corner of Mill and Durant Streets. The quotation at the start of this paragraph is on page 240.

The Catholic church at Glenwood is called St. Stephen's and not the Aspen church, as indicated on page 241. The Aspen church from the start was and still is St. Mary's.

P.S. You should get a copy of "Aspen on the Roaring Fork" by Frank L. Wentworth. Kendrick-Bellamy advertised it last Sunday in the *Denver Post*. It is very good and accurate. It is the best and most authentic article written on Aspen.

I don't recall whether you said that Telluride or Silverton used to celebrate Robert Burns' birthday. The miners in Leadville used to celebrate his birthday also. He is more or less the poet accepted by the miners as being theirs. Frank Vaughn in his "Spirit of Leadville" has several poems telling how the miners in Leadville used to commemorate his birthday. "The Spirit of Leadville" gives interesting sidelights on Leadville history.

Hope this letter may be of some help to you. It is away past my bedtime. I am very tired, so I haven't had time to make this letter free from errors of grammar. I hope that you can understand what I have written. I like your book very much. I also have *Cloud Cities,* I was born at Leadville.

In the mid-1940's I visited the widow of Elias Cohn who, in 1910, was manager of the Smuggler and Aspen mines at the time that flooding of the Free Silver and Smuggler shafts threatened to shut down both properties. Her husband went to Aspen in 1889 and within a few years was made general manager of the rich Aspen and Smuggler silver mines by D. M. Hyman. As she talked to me in her Denver apartment, surrounded by mementos of her Aspen days, she relived certain episodes of her long and busy life.

"They mined deep in those mines," she began, "and when the rich ore petered out and only low grade deposits remained, a portion of the mines was closed off. As soon as the pumps were stopped, water poured in and covered them. Gas accumulated and the shale began to smoulder. Since the company still wanted to work the lower levels, some means of dewatering the shafts had to be found. Mr. Cohn believed that deep-sea divers could reach the pumps and start them, and that, of course, would cause the water level to recede. Two divers from New York City were sent for and paid $100 a day from the time they left the east until their return.

"On December twenty-fourth the two descended through sixty feet of water to reach the pumps. Before noon they'd reached the bottom, started the pumps and the water level began to lower at once. As soon as this was known there was a call at our house from friends at the hardware store that they were sending a sleigh down to pick up Mr. Cohn and drive him to the main office of the mine. All the miners were in front of the office. His health was drunk in iced champagne in the back office. They gave him an ovation in town too, across from the Armory.

"The divers left on the afternoon train. One of them was to be married and he wanted to get back. Two days later the pumps stopped, and the divers were recalled. The mine company bought a diving suit, and the professional divers taught certain of the miners how to use it in case of further emergencies. The company also installed a new pump as a standby."

During the interview, Mrs. Cohn recalled other items of interest: "We had six solid weeks of snow one year. A dog was buried in a snowslide for thirty days, but he came out alive. When the late shift came off work at the Mollie Gibson, it was a beautiful sight at night to see the lights of their lanterns winding down the mountain.

"Silver profits from the big mines were immense. When D. M. Hyman, the owner of the Durant mine, celebrated his silver anniversary he sent east many sacks of its high grade ore to be made into a silver table service."

Cooper's Camp   Small mining camps sprang up fairly close to Aspen during the 1880's; one was Cooper's Camp. The road to Ashcroft skirts meandering Castle Creek all the way. I'd heard of a small camp beyond Ashcroft called Cooper's Camp and had tried to find it, but was never sure whether I had or not. On one of my trips to the district, Natalie A. Gignoux, who ran the Little Percent Jeep and Taxi service, drove me to the Montezuma mill, and during the ride I asked her about Cooper's Camp. She promised to find out what she could and let me know later. That fall I received the following informative letter (Nov. 28, 1950):

> This afternoon I had the opportunity to take Mr. Fitzpatrick, of whom I spoke to you, out to Ashcroft to his cabin. On the way I pumped him about Cooper's Camp, thinking of you and your search for it, and this is what he told me. Cooper's Camp was, as you say in your book, the location of a coal deposit which was used as a flux in the milling operations in Aspen. The teamsters hauling the coal always referred to the run from the mine to the Postoffice in Ashcroft as five miles. Mr. Fitz told me that the camp is located about three miles from the turnaround at the end of the Ashcroft road and about 1½ miles from the point where the Montezuma road takes off up the hill in the aspens. Just at that point in the aspens where the road takes off the flat for Montezuma there are, he says, the ruins of an old hotel and saloon and some other buildings which were known as Kellogg. I believe that this must be what you thought might be Cooper's Camp. He says that Cooper's Camp is about a mile and a half beyond this below the road and that there is no evidence left that there was ever a town there. (In other words the location would be south and east of Kellogg.)
>
> Fitzpatrick has been what he calls a "Mountain Robber" all his life. He came out here as a small boy to live with an uncle who had a Homestead at the Turnaround on the Ashcroft road and quite naturally took to the mountains and prospecting. I dare say that there is very little of this country around here that he has not explored looking for minerals. He is a man about 55 to 50 years old and a great deal of his knowledge is hearsay, but for what it is worth to you there it is.

Bowman   In talking to old-timers I had heard how Cottonwood Pass, west of Buena Vista, had been the main freighting road across the Divide into Taylor Park before Independence Pass was built. From the foot of Cottonwood, a road ran northwest, crossed Taylor Pass, dropped down on the other side into Ashcroft, skirted Castle Creek, and entered Aspen. Bowman, a stage stop and shipping point on the way to Aspen, was between Taylor Park and the summit of Taylor Pass. Naturally I wanted to see it. I therefore wrote to M. J. Webber, asking how to get there. He promptly replied (June 29, 1942):

> There is nothing left at Bowman except one frame shack which appears to have been built by a prospector since the boom days. The site can be reached by car from Taylor Reservoir by taking the road through Dorchester and continuing to Bowman.

M. A. Deering of Gunnison also wrote:

Bowman was the first town in Taylor Park other than Tin Cup. That was in the early eighties. All provisions to Aspen went up the Arkansas valley over to Taylor Park. Bowman was the shipping place. It had a stage stand, a log hotel, barn, camp cabins for freighters and mail routes.

William K. Martz of Aspen in his letter cautioned (Sept. 6, 1946):

The only way you could get from Ashcroft to Bowman would be on horseback as the road has been washed out and never fixed. Since the Enterprise mine shut down, they have never kept the road in shape.

When I finally did reach the place, it was not by way of Ashcroft but, as Mr. Webber had suggested, via Dorchester; and I found Bowman just as he had described it.

Whenever I crossed Independence Pass and left the ghost town of that name   *Ruby* miles behind me, I would look for the narrow road up Lincoln Gulch to a distant mountain barrier and wonder just where the camp of Ruby stood. From Mr. Martz's letter I learned how to reach the site (Sept. 6, 1946):

Your letter handed me by the Post Master at Aspen to answer. There is the Ruby Mine about 10 miles above where Lincoln Creek empties into the Roaring Fork River called the Junction. The Ruby Mine was last worked about 30 years ago. This was a silver mine. There is an old mill and boarding house still standing on the property, also a few cabins in the Gulch. I spent many years in this Lincoln Gulch district and have eight mining claims this side of the Ruby. The road to the Ruby is in fair shape and there are a number of cars go up to the Lakes to fish. If you care to go up this next couple of weeks I would be glad to show you around. I have the Martz Tourist Cabins at Aspen and you can contact me there.

At the time I could not accept his offer, but on my next trip I did get to Ruby and found the mill well preserved, the assay office with much of its equipment still intact, and a few cabins standing among the trees above the road.

After my book was published, Matthew Oblock, Jr. wrote to ask why I had   *Lenado* not mentioned Lenado (July 12, 1951):

Why so much emphasis on Ruby up Lincoln Gulch? You completely forgot or weren't told about Lenado, which is about 7 miles northeast of Aspen. J. E. Spurr has a picture of it opposite page 198. There are at least 15 cabins there. To quote Spurr on pages 199 and 200 of his text he says, "Northward from the Alta Argent Mine on Smuggler Mountain no ore has been found for several miles. Following the metalliferous zone, the first productive locality is in Lenado Canyon (up Woody Creek) where the mining village of Lenado is situated. Pl. XXVII gives a general view of this village in a rather dilapidated condition. This view is taken, looking across Lenado Canyon and up Silver Creek, from the hillside west of the creek, just above the Leadville mine." The Aspen Contact, Leadville, Bimetallic, and Tilly were the principal mines. There used to be a big mill there but I am told that it is quite dilapidated. The Flogus sawmill is about the only thing at Lenado at the present time. There was a forest fire there last week-end.

On one trip in the late 1940's I had driven to the settlement, where I found a large, empty mill beside some dumps, a few people living in cabins or frame houses, and a sawmill that appeared to be the only active property. It was several years later that I learned the camp's history.

The small settlement on Woody Creek dates from the 1880's, after A. J. Varney found a rich vein of lead and zinc ore and with the help of several associates developed his claim, calling it the Varney Tunnel Company. In time the population grew to 300, after the mining company built a sawmill, a boarding house, two saloons, and a store. A lead mill was constructed near the tunnel entrance. When I saw the place, the houses were surrounded by big trees and lush vegetation—only the deserted mill stood above the green townsite.

Crystal   Crystal, though hard to reach, is worth the effort. A stiff three-to-four-mile climb out of Marble levels off at Lizard Lake, beyond which the shelf road dips and twists the rest of the way. A short distance before crossing the Crystal River at the entrance to the town, the road passes deserted mines and a few foundations of buildings or mills, the most spectacular of which is the ruined but dramatic Sheep Mountain Tunnel mill that clings to a rocky cliff beside a waterfall. Crystal itself occupies a high mountain meadow completely surrounded by peaks—Mineral Point, Crystal Mt., Bear Mt., Sheep Mt., and Treasury Peak.

Since I have not seen Crystal since 1947, I cannot enumerate the present number of its houses, although in recent years it has had summer residents and is no longer the quiet, untouched settlement that blossomed in the 1880's, when silver, lead, and some gold were discovered in appreciable amounts in several mine properties.

My initial inquiries concerning road conditions were directed to the postmaster at Carbondale, who replied:

6/25/46

Dear Madam,

I drove to Marble last Sunday in an hour, 30 miles. The road is good. Wm. Knapp is located at Crystal, what work has been done on the road was done by Knapp, he is hauling ore from Crystal. I would guess there are 15 families at Marble and 7 or 8 at Crystal. Whether a Ford can make the grade to Crystal depends entirely on the shape it is in. No P. M. at Marble or Redstone. Mail goes Star Route from Carbondale.

Wm. J. Pings

From H. E. Tays of Price, Utah, came reminiscences of Crystal as he knew it as a boy (Oct. 31, 1955):

I first heard of your book from an old friend of mine, who at present, is a patient at the Veterans Hospital in Grand Junction. . . . As my father George H. Tays, was among the first groups of prospectors that came into Gunnison County and to the camps of Gothic, Schofield, Crystal, Irwin, etc., I naturally would like to read it.

I was born in the old town of Crystal in 1901 when it was quite a booming little city, but one would never know it to see it as it is today. . . .

I can give you some information about Crystal but it will be somewhat limited as I was about seven or eight years old when my folks moved away from there and went to Marble. I remember my first year of school at Crystal, . . . about the fall and winter of 1907 or 8 and then my folks moved to Marble and I started in the second grade that fall in Marble. But I do remember some

Crystal, 1947. Looking toward Schofield Pass.

of the business houses, etc. that were in Crystal then. There was the Bronson Hotel, The Crystal Club (a saloon), a barber shop, the Printing Office and a general store with a hotel on the top floor. The hotel and store were run by my uncle, Horace H. Williams and his two brothers John and Ambrose. My uncle is dead but the two brothers are still alive and live in Phoenix, Ariz. They are old men now, 85 and 90, I believe.

The roads into Crystal have never amounted to much but the road from Marble up was always the best of the two and the one traveled the most. The other one from Crested Butte to Schofield was kept in pretty fair shape, but from Schofield down through the narrow canyon to Crystal, was always bad and closed the biggest part of the year by snowslides. But that is the road the first prospectors came into Crystal by. The road from Marble to Crystal is practically closed now but I believe you can make it in a jeep. The last time I drove up there was in 1951 but that was before the twin bridges, about half-way between the towns, went out.

There is one incident I remember when I was a kid there and that is the time when about a twenty-team outfit pulled in there pulling a big ball mill for one of the mines there. They brought it up from Marble and how they ever made it I'll never know.

I can also remember long strings of burros, each with a sack of ore packed on each side going through town. And then on the return trip they would be loaded with supplies for the different mines.

Dr. D. Wilson McCarty of Longmont, who had nearly as much trouble in reaching Crystal as I did, wrote of his successful entry (Nov. 21, 1952):

This is a fan letter! During the brief vacations we have taken, *Stampede to Timberline* sits beside us on the front seat of the car as handy as a Bible, and Fran reads to me from it as I drive along.

We have especially enjoyed the vacations spent at Redstone Inn over the past five years, and have made numerous visits to Marble, before and after the recent great mud slides. And, as always, we have wanted to go on to Crystal, but were always discouraged because of the bad road. I was interested in reading in your book that our experience so much paralleled your own.

After reading of your several attempts, and then of the final rewarding trip in which you described Crystal as the most beautifully situated of all the old mining towns—and the old mill, with its background of peaks, looking like a stageset from some Wagnerian opera—well then we were determined that our next visit to Redstone would include Crystal—if we had to walk.

We had to do just that—cars can be driven to Lizard Lake—over a terrible road—a little beyond the bridge slides have covered some of the road. With Clinton Biggs and his wife from Grand Junction we delighted in the views from McClure Pass, and in the Aspen foliage which was so beautiful this year. We drove to the town of Marble and then to the lake. We hiked for what seemed an endless distance and were about ready to turn back in the late afternoon when we came upon the cemetery, and around the bend the late sun was full upon the old mill on the rock. We were amazed at the beauty of it and the village beyond.

Marble   The first prospectors who entered the Crystal River valley in the late 1870's came down from Schofield (near Gothic), and after finding silver ore, laid out the camps of Crystal City and Yule Creek, the latter named for George Yule, a pioneer who discovered the marble deposits in White House Mountain above the town. Thereafter the camp was called Marble. In 1890 a small smelter was erected on the flats by the stream and was blown in the following year.

The quarries were developed by a succession of owners and companies, and according to their output the town flourished or foundered. In 1910 the population was 1,500: in 1914 it was 4, and in 1916 it had risen to 1,000. The marble plant consisted of finishing mills, an electric power plant, a cable-and-electric tram, and a hundred cottages for the workers. A company-owned railroad connected with the D.&R.G. at Carbondale, and on it were carried the blocks of marble that were cut at the mill. The quarries closed down in 1942, and during World War II the railroad tracks were torn up and sold for scrap metal.

Disastrous floods, caused by heavy rains which loosened the coal shale and mud along the sides of Carbon Creek, more than once raged down the mountain and swept aside everything in their paths. Two such floods, the first in 1941 and the next a few years later, cut a wide swathe through the center of the town and left a thick layer of gray waste and debris over crushed and shattered buildings. Only the metal fire tower with its bell survived the mud flow and still stands in the midst of it.

Marble is situated in a grove with its buildings nearly hidden by foliage and shaded by the huge trees which arch over its streets. The firehouse, the movie theater, St. Paul's Episcopal Church, some private homes, and a few of the company houses, all of which are back from the banks of Carbon Creek, are what is left of the once-prosperous settlement. Even the cutting sheds have lost their roofs.

Marble, 1941. Marble Company's homes for employees.

Marble, 1941. Store built of white marble, ruined by 1941 flood.

The floods and their damage, the loss of the railroad, and the curtailed use of marble for building purposes have all contributed to Marble's decline; it is not deserted, however, and the remaining residents have a fierce pride in their community. But when I visit it today, no stores built of white marble greet me. On earlier visits I recall seeing marble sidewalks, marble slabs used to rip-rap the railroad right of way, and the marble foundations of a rather large building that had been a church. The Rev. W. O. Richards, former Vicar of St. Barnabas' Episcopal Church at Glenwood Springs, identified it as follows (Feb. 15, 1951):

> Former residents of Marble now living in Glenwood Springs tell me that the ruins you mention were the basement of the Roman Catholic church. Dur-

ing the winter of 1922-1923 heavy snows caused the roof to fall in. The town at that time was practically deserted. I am told that the furnishings of the church were salvaged by the few remaining residents of the town.

In the early 1950's I talked with R. H. Pearson of Denver who worked in the Marble shops in 1913, when the stone for the Lincoln Memorial was being cut. Most of the workmen were Austrians and Italians, and during World War I they liked to argue as to which were the best fighters overseas. Pearson also mentioned that in the marble cutting department special orders were taken for gravestones and that lambs were the most frequent subject selected.

When I spoke of the Catholic church in Marble, he told me that Col. Meek, the superintendent at the plant, may have been the one who contributed the marble blocks used in its construction. Meek was killed in an accident before it was completed. The tram from the quarry to the mill had a seven to eight percent grade. Col. Meek and others were coming down from the quarry on the tram when the car went out of control. Col. Meek jumped, as did three others, all of whom were killed. Those who stayed with the car were saved. His successor, named Manning, was a hard man, not interested in doing anything for the men; so Pearson felt sure he had not continued to donate marble for the building.

Many of the company houses were up at the quarry. Some of the men lived there, while others went down on the tram every night. The cars were four-wheel hand cars with brakes on each side, and the pilots prided themselves on throwing the brakes around corners.

The Colorado Yule Marble Company was succeeded, for a time, by the Knoxville Marble Company of Tennessee. A snowslide, about 1932, ran down the mountain across from the mill and flattened the buildings. It was not the mass of snow that did the damage, but the blast of air from the slide. This happened when Mr. McClusky of Tennessee was in charge of the plant.

In 1955 Mr. Pearson wrote me:

> As a high school boy, I had a summer job at the Colorado Yule Marble Co. at Marble and often hiked to Crystal. I have never been back there and was most interested in what you have written of that area.
>
> The Larkin Hotel in Marble was near the terminal of the railroad. Most of the train crew were Larkins. I lived at the Larkin House, where they killed a beef and served beef for the next ten days. Then they killed a pig and we had pork for the next ten days, then a sheep or lamb, and then around the circle again.

On my last visit to Marble, in 1974, I naturally went for another glance at the circular marble foundation of the locomotive turntable, near the cutting sheds, and found it only after scrambling through the bushes and weeds that had grown up and completely hidden it.

Matthew Oblock, Jr., whose accurate details about Aspen were given earlier in this chapter, writes of the old towns along the Crystal River between Marble and Glenwood Springs and of the railroad which served them (July 12, 1951):

> In speaking of the old railroad going from Carbondale to Marble you called it a spur of the D.&R.G. on page 218. Later in the chapter you corrected yourself and said that it belonged to the marble company which was operating the Yule quarries at Marble. The marble company owned the roll-

ing stock and the tracks from Redstone. From Redstone to Carbondale the track belonged to the C.F.&I. and had been built by J. C. Osgood. I earned part of my way through one year of college by working on the section of this railroad between Redstone and Carbondale during the summer vacation of 1930. It was called the Crystal River and San Juan Railroad Company. I was never able to figure out where they got the San Juan end of the name. A spur of railroad used to take out to the mountains of the west years ago up Thompson Creek, which flows into the Crystal River about 6 miles south of Carbondale.

There used to be extensive coal mines in all the hills west of the Crystal River from PLACITA down to GLENWOOD SPRINGS at one time. RED-STONE itself was not a coal mining town. It was rather the place where coke was made from the coal which was mined at COAL BASIN, now a ghost town some 12 miles west of REDSTONE. A spur of the C.F.&I. used to bring the coal down from Coal Basin. Osgood used to have his miners' band, which used to come up to the Fairs at Aspen and to the Strawberry Festivals at Glenwood Springs in the 1900's.

Placita, Redstone, and Coal Basin

Two letters from other informants mention Coal Basin. The first is from Jeff Dunn of Pueblo (Sept. 16, 1969):

Have you ever heard of Coal Basin? In the early 1940's a complete town remained. The history of Coal Basin is very obscure. I do know it lasted from the late 1890's to 1908 or '10 and was a C.F.&I. company town.

The second letter came from Mr. and Mrs. Carol F. Hunter of Glenwood Springs (Dec. 23, 1952):

There is one town which you didn't mention, while it was a coal mining town and you didn't write about the coal mines. It was Coal Basin located 8 miles up Coal Creek from Redstone. It was the town where they got the coal for the coke ovens at Redstone. The railroad used to run up there, and about 1945 the Oil Co. built bridges on the railroad bed and drilled an oil well about two miles from there but just capped it up.

If you ever care to visit the old town let us know and we will arrange to take you up. Perhaps it will be on horseback, as ever since the oil co. was in there the cloud bursts and beaver have taken lots of the road.

There is some talk of opening the mine again next summer although I doubt it, but it is a very good grade of cokeing coal. I understand there used to be from three to five hundred people there. There are not many of the old buildings left.

To return to Mr. Oblock's letter:

Janeway and Avalanche

Do you remember the place several miles below Redstone where Mt. Sopris seems to come right down to the Crystal River? Two small settlements in that vicinity were called Janeway and Avalanche. There is a little bit of galena ore up Avalanche Creek and some attempts have been made to mine it, but it never amounted to anything. When I worked on the track, a couple of men had a small mill about 12 by 20 feet over near the hot springs, but they couldn't make it pay.

Marion,
Cardiff,
and
Sunlight

A number of miles out of Carbondale to the west is the now ghost settlement of MARION where coal was once mined. Have you noticed the coke ovens across the road in going from Carbondale to Glenwood Springs? That is where the ghost town of Cardiff still stands. It is about 4 miles from Glenwood. The coal was brought down to the ovens from the ghost settlement of SUNLIGHT by a spur of the old Colorado Midland Railway which went up a steep mountain.

Frying Pan
Mining District

At one time there was some mining done up the Frying Pan River across the hill from FULFORD and from HOLY CROSS CITY. I don't believe that it amounted to a great deal except that there was a Frying Pan Mining District. NORRIE used to be mostly a sawmill city.

Mr. Oblock supplies details concerning other places in the Aspen-Snowmass-Crested Butte region, which for those who like myself want to know all they can about the minutiae of an area will prove most helpful:

I have a very interesting early day map. It is, "Bachtell's Topographic Map of part of the Elk Mountain Mining District, Gunnison County, Colorado—1881." The scale is 1 mile per inch. It was printed by the Strobridge Lithographic Co. of Cincinnati. It covers the region from a little south of Crested Butte over to Maroon and Snowmass Peaks to the north. Marble is not shown on the map, but Crystal City on Rock Creek is. There are several townsites indicated which I don't believe you mentioned. CLOUD CITY is shown to be in Oh-Be-Joyful Gulch near Crested Butte. SNOW MASS is on North Rock Creek near Little Snowmass Lake just a few miles north of Crystal City. In the townsite of Snow Mass Brownell's Cabin is shown. This Snow Mass is not to be confused with the present day settlement of that name on the Roaring Fork River in Pitkin County. Another townsite is that of LOST TRAIL MINING CAMP, north of Sheep Mountain near Marble. In it is Warrior Cabin.

CHAPTER XII

Off the Beaten Track

So much of Colorado is mountainous that certain mineral areas and their towns may be close to each other as the crow flies but miles apart by road or trail. Ouray and Telluride are about ten miles apart geographically but fifty-three miles by auto road. Many mines at high elevations are three to ten miles from a settlement, but access to them is still only by old pack trails or wagon roads which are no longer maintained and which require jeeps or other four-wheel drive vehicles to maneuver them. The places described in this chapter though accessible are not easy to reach.

Gold Park and Fulford are several miles south of Highway I-70, which in this area passes through the Eagle River valley. Gold Park is in the vicinity of Redcliff but is reached by a ten-mile drive up the Homestake Creek road. Before starting up Brush Creek on the way to Fulford, inquire locally at Eagle for directions. Dowd has disappeared, but its location and beginnings are described by a descendant of the founder.

Lulu and Gaskill (now a mere site) are at the foot of the Never Summer Range near the headwaters of the Colorado River, several miles north of Grand Lake. Just across the Never Summer Range in North Park are the remains of Teller City. It is on Jack Creek in the vicinity of Rand, which is a ranch center. Farther north in the Park, near the Wyoming border, is the site of Pearl, reached by driving some twenty miles west of Cowdrey.

In the Cache le Poudre River valley, a few miles above Rustic, is Manhattan, a short-lived gold camp dating from the 1880's.

Eagle River Country

During the summer of 1972, I revisited Gold Park, which I had not seen since the 1940's. I expected to find great changes, and such was the case. The Homestake road was wider but still very rough, which meant that our progress was slow on the ten-mile stretch to the deserted camp. We met numerous cars on the road, not because of interest in the few cabins left standing at Gold Park but because Homestake Creek attracted fishermen.

Gold Park

Mrs. Jean Kelleher of Pittsburgh, Pennsylvania, had written me (Feb. 16, 1955):

> Mr. Edward H. Wyatt, 910 Estes Ave., Lakewood, Colo. is my father, and one of the four present owners of the section of land on which Gold Park was built. Last September my husband and I visited Colorado for the first time, and fell in love with the country, much as you must have.

A letter from William H. Fricke described one building in Gold Park that differed from the others; he wondered about it. From his description I believe it was

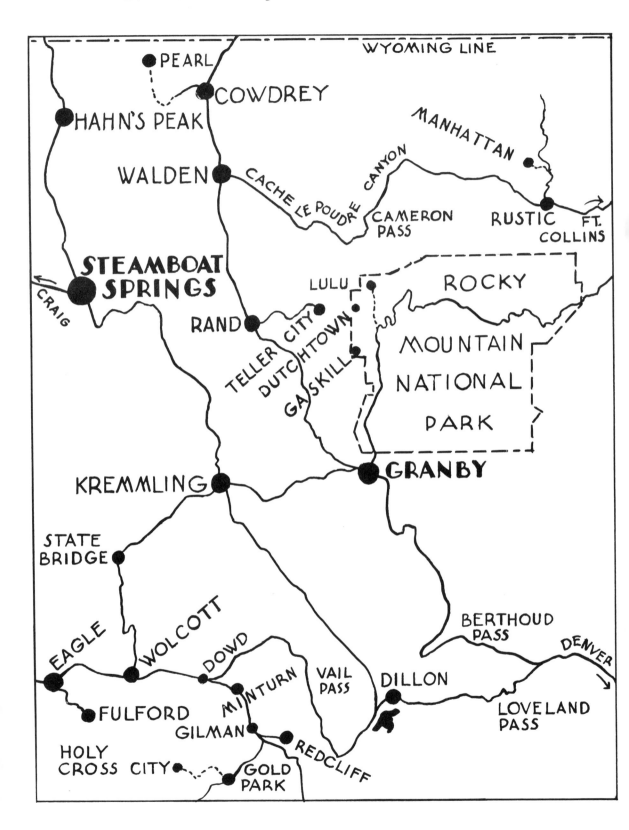

Gold Park, 1944. Community House.

the Community House, of which the postmaster at Redcliff had spoken on my first visit to the town. I was never inside the building, but Mr. Fricke had been. But by 1972 the big house had vanished. Mr. Fricke wrote:

> I have been to Gold Park and Holy Cross City a number of times, but know nothing about their history. There is one interesting thing about Gold Park, there is one frame house which was quite amazing for a mining town. It was recently torn down by the Army for a machine gun range. This house was in excellent shape and very well built, although obviously old. Somehow it reminded me of a summer house of some wealthy citizen more than the house of some miner. But comparing it with some of the houses in Leadville, it fits the period perfectly. Very ornate designs around the roof, large, spacious porches and welled stair case. There were about six rooms in the house. Possibly you are acquainted with the building I have described.

The history of Dowd, a town that has completely disappeared, is given by O. A. Lamoreux of Denver (July 30, 1956). It was situated northwest of Minturn, which is on U.S. Highway 24 and east of Avon.

Dowd

On our official state tourist map and on all petroleum company road maps as well as Rand, McNally maps of Colorado, the town of DOWD shows up at the intersection of highways 6 & 24. Just for fun, I wrote Jack Foster to see if he knew anything about this ghost town. He, or his researcher, could give me no information on it. Because the town originally was so small and has completely vanished, you likewise may have no information on it. As a matter of first hand record, and because you may care to incorporate it in any future revision of "Stampede," here are the facts:

My brother-in-law, Leonard E. Dowd, 1572 Leyden St., Denver, gives me this information. The discussion came about after he and my sister drove up to Kokomo last week for a final visit to his Father's and Mother's graves, (James & Charlotte Dowd). His (my bro.-in-law's) Dad, James, operated the Dowd Mercantile store in Kokomo in the eighties, closing it up around 1908. It was the usual general store of the period; miners' supplies, groceries, clothing, fuel, etc. There was a need for a sawmill for mining timbers and building of cabins, so at the intersection of present highways 6 & 24, James Dowd built a small sawmill surrounding it with quite a few cabins. Hence sprung into its short life, the "town" known as DOWD, Colo. My bro.-in-law, who also lived later in Redcliff, and in Denver as a baby in 1901-2-3 tells me that the old store in Kokomo really "gave up the ghost" and tumbled down. Three years ago, when I took a trip with them up there over the fourth, the old two-story building was then on its last legs.

Inasmuch as the Dowd's have sold their home on Leyden and are moving to California to live, Aug. 15th; I thought you might now be interested in getting this information, and anything further you might want that he could supply before he left. Some vandal even stole the sign DOWD, so there is now no identification.

Fulford   It was in 1944 that I set out for Fulford, not knowing what I would find. The camp is reached by turning off from Highway I-70 at Eagle and driving south up Brush Creek for about fifteen miles to the forks in the road and then taking the left fork for another three miles. At that time these last miles were difficult to drive, but now, I understand, a logging road gives easy access to the town. Fulford lay on a meadow surrounded by trees, its grass-grown street dotted with houses and a hotel, and next to it a store. Beyond the meadow toward the mountains was Upper Town, in which one cabin with an immense ridgepole had been the assay office.

If ever a place had a tragic history it was this gold camp on Nolan Creek, eighteen miles from Eagle, which sprang up in the nineties and became known as Fulford. Its hills and mines are both graves and storehouses of treasure; its early history is grim and consists of a succession of rich strikes, snowslides, deaths, and lost mines. The first prospectors, who arrived in the fall of 1849 enroute to California, found color in the streambed, and after further digging uncovered nuggets of gold near the surface of the ground in paying quantities. As fast as the ore was taken out, it was stored in a drift, until such time as it could be sent to a mill for processing. When their provisions ran low and a snowstorm isolated them, one of their number was chosen by lot to go 150 miles to the nearest camp for supplies. New storms delayed his return, but after six weeks had gone by he became so worried about his friends that he started back in search of them. When he reached the camp, nothing was as he had left it. During his absence, a snowslide had crashed down into the valley, burying both the cabin and the men and uncovering the rocky slate deposits that composed the mountain. What had been its surface was now a huge mound at its base, filling the gulch and sealing the drift where their gold was stored.

In 1881 James Fulford, a rancher, met a prospector who said he had found the abandoned tunnel in which were tools, human bones, and the piled up nuggets that had been stored more than three decades earlier. He wanted a partner to help him get out the cache. Although skeptical, Fulford was interested; however, the prospector was killed two weeks later in a drunken brawl, so Fulford sought out and found his cabin and in it notes describing the location of the gold. With these Fulford went to Aspen to consult a mining expert whom he hoped could help him locate the drift. He then started back to Brush Creek. That winter he was lost in a snowslide on New York Mountain, and again the treasure was unclaimed. Other attempts to find it were also unsuccessful.

More extensive prospecting of the region began about 1887, when William Nolan and a group of men scoured the hills for gold. In making camp one night, while Nolan was crossing the creek on a log, carrying his gun muzzle up, the trigger caught and the gun went off cutting out his tongue. His companions, unable to administer effective aid, watched him bleed to death. They cut his name and the date into the bark of the aspen tree at the foot of which they buried him.

During the summer of 1890 a rich strike was made by Dick Morgan on Nolan Creek, and many hurried to the place to prospect and to stake claims to the valuable deposits they uncovered. Dick Morgan and Art Fulford were two of the leading men in this new camp. Fulford, like his relative, also perished in a snowslide.

The town itself was a bustling place in the nineties—really two towns, the upper and the lower, both built at the same time one quarter of a mile apart. At the peak of the boom the population reached 2,000-3,000.

Letters from three people gave me additional information. From Charles H. Hill of Lakewood (May 9, 1947) I learned about "The Wheel," a huge arastra built in 1893 or 1894 not far from the townsite:

> The old mining town of Fulford, about 18 miles south of Eagle, Colo., is one containing a store, an old gold processing mill and about ten cabins, all of which are abandoned except one, which was built by my uncle in about 1870 and which was occupied (at least about ten years ago) by a William F. Colerick who is, or was, still mining in that area. Fulford is inaccessible by auto from Yeoman Park, unless the car is small and powerful; even then there is a bridge about one mile from the town which probably could not be crossed by car. Fulford is about 5 miles from Yeoman Park. Yeoman Park is accessible by car.
>
> There is also near Fulford a very interesting place called "The Wheel" which may be reached only by trail, about 2 or 3 miles up Brush Creek from Yeoman Park.
>
> "The Wheel" is the remains of a cabin and an old water powered gold processing mill. My last visit there was in 1928, and we stayed in the cabin for about a week while my uncle (George T. Richards of Colorado Springs) and myself were searching for a fabulous lost mine which had been his about 35 years previous. The story of the mine is in itself quite an interesting, even eerie story, which time and space will not permit at this writing.

The wheel is about 20 feet in diameter by about 4 feet in width and was still resting on its enormous wooden bearings at that time. It could easily be turned by walking on the inside (like a squirrel in a cage) even though it did squeak and groan as tho' it were being tortured.

Near Yeoman Park is also the old locally famous Fulford Cave which I explored by myself and am indeed fortunate to be here. . . .

It was Mrs. John R. Barry of Denver who provided the most information (May 13, 1947):

Was so very interested in an article in last Sunday's Rocky Mt. News, concerning your deep interest in ghost towns of Colo. . . . So many times I find reference to them by different "explorers" but have yet to find where any have visited the old mining camp of Fulford.

It too had a colorful history in the early 90's when my father first went there. It was a really bustling place, consisting of two towns, upper and lower town; and a population around 2000. Boasted a newspaper, two ore mills, saloons a plenty, livery barns and the other necessary mining establishments. And above New York park, below New York Mt., another little village which still has a few crumbling cabins left, called "New York Cabins." Also east of the upper town was what was known as "Adelaide Park" and a few cabins are still there.

The town originally was called "Nolan Creek Camp" after an early day settler who was accidentally shot and bled to death. I have many times seen the aspen tree carved with the date of death where he died. In fact I took a picture of it several years ago. Later, the camp was renamed for Art Fulford, a member of an early day family who ran the half-way house between Eagle and Fulford and was killed in a snowslide on New Year's day, I believe about 1892 or 3 while he was on his way to "jump" claims, as they called it in that day when assessment work had not been done before New Year dawned. He ate his last meal at my mother's hotel and though I was a very small child I still remember about it.

I am sure if you have never visited that particular area you would immensely enjoy it. Really beautiful scenery and in all my travels of Colo. I have yet to see mts. quite like the New York Mts. with their curved outline against the sky, so different to our pinnacled peaks of Colo.

If you are interested in more information I should be delighted to furnish whatever I might give you. You know I feel my old home camp has been slighted in this new interest in the glorious past of an era the world will never know again, and determined if I could bring it to the attention of a person like you who is trying to record it for future posterity I was certainly going to make the effort.

Later that summer I went to see Mrs. Barry to learn more about the place. She told me that Mr. Lamming, who was her father, went to Fulford about 1890 and lived in a tent until he could build his own home. The Upper Town had a log hotel which her father built, some smaller boarding houses, saloons, a livery barn, a store, and an assay office run by Ed Glenn. The post office was later moved to the Lower Town.

A mill was built in the Lower Town as well as additional assay offices and

Fulford, 1944. Hotel and store on left; upper town behind buildings.

Fulford, 1944. Store with post office.

more cabins. I mentioned the store that I had seen, still stocked with supplies, and the hotel next to it, whose inner walls were papered with Rotogravure pages from the *Denver Post,* and Mrs. Barry laughed and said, "My mother ran the Lamming Hotel in Lower Town for twenty years. The hotel became well known because of this unusual decoration."

The third letter about Fulford was from Mrs. Mabel S. Colerick of Eagle (June 4, 1947):

> A few weeks ago I read in the Denver paper of your interest in "Ghost Towns." I was so fascinated I was prompted to write you but being a procrastinator I failed to do so. This morning during a conversation with Mrs. Ella Barry she spoke of a recent call you made her and the fact you had visited my favorite haunt Camp Fulford, a few years ago. . . . If you come again this summer . . . I am asking you to be my guest while in Fulford. . . . I wonder if you found the spring on the road between Fulford and Upper Town. The barrel or tub has been in the spring for over fifty years and is quite moss covered. . . . Then too, there are the old mine ruins on New York mountain. I have not climbed the mountain for many years. I came to Fulford in 1907, after graduating from Chicago Art Institute to teach the little school. I have made my home in Calif. since 1911 but have visited Colorado every summer since 1922. . . . Hoping you will visit us this summer.

The Never Summer Range

Lulu   At the foot of the Never Summer Range of mountains in the valley below the old stage road to Thunder or Lulu Pass is the ghost of Lulu City, a camp whose span of life lasted from 1879 to 1883. The ore was low-grade silver not rich enough to ship out. Anyone traveling on Trail Ridge Road today crosses the Continental Divide at Milner Pass, and the car swings around the switchbacks which lead westward down to the valley of the Colorado River. Nearly four miles to the north on the valley floor is Lulu; nowadays a trail leads from the highway at the foot of the pass to the site. It was a letter from Frederick A. Sweet of Chicago that sparked my curiosity and caused me to go to Lulu on horseback (April 20, 1942):

> I have read your article in "Design" on Colorado Ghost Towns and believe you are the one person who can tell me the whereabouts of Lulu. I understand that you have to go in by mule back or walk but that it remains just as left—dishes on the tables, clothes in the closets, and all. . . . I never found anyone who knew where Lulu was. If you have written other articles on the subject let me know.

I found no dishes or tables, no clothes, and also no roofs on the several cabins that were scattered around the marshy meadow, only trees of various sizes growing inside the log walls.

Next, H. N. Wheeler, who had been forest supervisor at Fort Collins and Estes Park from 1907-1920, except from Nov. 1911 to April 1913, wrote me (July 17, 1946):

> Lulu was located at the very head of the main branch of the Colorado River, between Specimen Mt. and Saddle Back at the foot of Lulu Pass which

crosses the Divide between the Never Summer Mts. and Saddle Back. The Pass starts at the head of the Michigan River in North Park and at South end of the Medicine Bow Range of Mts. A road was constructed from near the defunct town of Elkhorn on the Elkhorn Creek Branch of the Cache le Poudre River in Larimer Co. This road passed between the two peaks of the Bald Mts. and kept along the Green Ridge up to Chambers Lake up Joe Wright Creek on to the head of the Michigan River and over the Pass. This road, or rather a branch, also dropped down the Michigan to the Owl Mts. and Teller City. It was built from Cheyenne, Wyo., came up past Livermore and on to the Elkhorn. I'm writing it a bit sketchily. Another road was built from Stony (?) Prairie, west across the South Poudre River and to the Long Draw region and on to Joe Wright Creek. If Mr. Ben Flowers is still living in Fort Collins he can tell you of this road since his father built it.

There are some foundations still in existence at Lulu but no buildings. When it was abandoned the Post Office still remained and mail was delivered there long after everybody was gone.

Irrigation Companies at Fort Collins can tell you about the Lulu Pass-Michigan River country. Probably old files of the Courier (now Express-Courier) of Fort Collins can tell you about Manhattan, Lulu and Teller City. Ansel Watrous was a pioneer, and editor for many years of the Courier. I'm not sure I've helped you but have tried to give some hints.

Southwest of Lulu, nearer to Grand Lake, was a still smaller mining camp called Gaskill. It, too, was at the base of the Never Summer Range. A number of years ago I was told that the Forest Service had burned the few remaining cabins as they were considered fire hazards to the area. L. Hanchett of Salt Lake City mentioned Gaskill in one of his letters (Aug. 6, 1949):   *Gaskill*

Capt. L. D. C. Gaskill [for whom the camp was named] ran the Wolverine mine. In 1886 the Wolverine Mine near Grand Lake was quite a producer of zinc-lead ores. Capt. Gaskill was the officer in charge of the capture of Jefferson Davis when Davis attempted to escape.

North Park Camps

Two ghost camps, Pearl and Teller City, lie east of the Continental Divide in northern Colorado close to the Wyoming border. Both Pearl and Teller City are reached from North Park, a high mountain plateau which extends south almost to the Colorado River.

Near the head of Jack Creek in North Park a camp called Teller City sprang up about 1879, the same time that Lulu did, because of silver deposits in the vicinity. Many mines were opened and at least one mill was built. The camp as laid out had several streets, nearly 300 log cabins, and a post office named for Senator Henry M. Teller. By 1883 it was almost deserted, and now the streets are covered with fallen trees and new growth.   *Teller City*

R. O. Throckmorton, county clerk of Grand County, Hot Sulphur Springs, wrote me (July 20, 1945):

I don't know just the year that the town of Teller was started. But I think about 1880 and finished about 1883. I went to Grand Lake in 1882 and the town of Teller was about at its peak at that time. There are a few old log buildings there yet. But the roofs have fallen in and the trees are growing inside them.

The road up Jack Creek is rough and has been cut through dense timber all the way to the camp. In 1946, when I visited it, there were many broken down cabins, others partly burned, and all were weathered and rotting. The untended cemetery contained several unmarked graves.

Teller City, 1945.

Teller City, 1945. Cabin foundation with tree growing inside it.

State Highway 125 bisects North Park. At the extreme northern end of the *Pearl* Park is Cowdrey, from where a gravel road runs west through rolling sagebrush-covered hills to Pearl, a distance of about twenty miles. The camp was laid out on a flat beside Copper Creek in 1878 or 1879 but was not thought of as a mining camp until the late 1890's, when copper, silver, lead, and some gold ore were found in the surrounding hills.

On neither of the occasions when I visited the place was there anyone to talk to. I learned about it from friends who had fished at Big Creek Lakes, two miles beyond the townsite, and their recollections of the place included many buildings and streets and made me anxious to see it. When I got there, scarcely half a dozen frame buildings or sheds, some weathered wooden sidewalks hidden by the sage-brush, and a deserted smelter on a hillside south of the town were all that remained.

Pearl, 1945. Wooden sidewalk remains.

Pearl, 1945. Smelter on hillside.

Therefore the following letters from persons acquainted with Pearl and its past were of unusual interest. The first is from Miss Marvis Richardson of East Lansing, Michigan (Sept. 12, 1952):

> After so much work it is unlikely that you will seek more information on North Park but, if you ever do, Mr. F. H. Hill of Fort Collins is a man of excellent memory if somewhat feeble legs. He homesteaded in North Park, rode the ranges, was interested in mining, and during his years in Pearl was mayor and owned the livery stables. In the later years, until he was 81, he was Larimer County road superintendent.
>
> Mr. Hill said you were misinformed about the origin of the name of Pearl. He went into North Park in 1886 and was one of the signers of the post office petition for a P.O. on the Wheeler ranch. They requested that it be named Pearl in honor of Mrs. Pearl Wheeler who was the first woman in that section of the Park.

This information is corroborated by Mrs. Pearl A. Wheeler herself, of Bay City, Oregon (July 6, 1951):

> I went to Colorado from Michigan as a bride in 1887. My husband owned a ranch where Pearl is now, he had cattle. Soon after that the settlers desired a postoffice. Names were sent in. Mr. Wheeler sent in the name of Angus as he had the black pole angus cattle. The Gov. chose Pearl, which is my name. I was the first postmistress. The settlers came from all around to get their mail. Before that we got our mail at Pinkhampton, 25 miles from Pearl, which meant in those days staying all night unless a cowboy rode over.
>
> As far as the mining, that took place after I left Colorado. Mr. Wheeler went back on mining trips and located what, I am sure would be a rich mine now. We still have some pitch blend he brought home. . . .
>
> My daughter was born in Fort Collins. She was in Pearl after the smelter closed but still intact. It was watched at that time. She told me the smelter was fine and the huge pipe line from the mining done by hydraulics was massive. A great deal of money was invested both in the smelter and the pipe line.
>
> On Easter Sunday 1889 my little niece died and we buried her on the other side from the smelter a little to the north. The snow was more than six feet deep on the level as that was a terrible winter, many cattle died.
>
> I am 85 years old and many of the early events have slipped my memory but if I can help you in any way I will be glad to do so.

Esther L. (Mrs. Hobart) Chambers of Clearwater, Kansas, provided valuable information (Feb. 26, 1957):

> I was interested to know what you had written of mining around North Park. I was sorry that you didn't find more at Pearl. I lived on a sheep ranch near Walden in the '30's. Our sheep range was on Independence Mt. not far from Pearl. We made many trips to Big Creek Lake above Pearl. About 1935 the gov't. built a good road up to the lake. Previously, we paid toll at a ranch just above Pearl. We learned quite a lot of the history.
>
> The most interesting ones we knew were the Blosser family. They are Kansas Chevrolet dealers. If my memory serves me—Blosser Motor Co., Concordia, Kansas. There were 2 or 3 families and they had cabins (summer

homes) below the lake, on the oiled road. They had gone to Pearl every summer since they were small children. They knew all the details. One house had the big mirror from the bar at Pearl and other items.

They said the mine was a swindle scheme of eastern investors. Everything came in by wagon from Laramie. When word was brought in by wagon that some of the easterners were coming to inspect the mines, it was decided to leave immediately. By morning everyone in camp had gone, leaving furniture, supplies, newspaper office, etc, etc. intact.

The Blossers used to go up by train every summer, traveling in a freight car with cow, supplies, friends, and rented the hotel which was left complete even to silverware and linens. The last time I was there, the big range was still in the hotel (between 1936 & '41). Cattle made themselves at home inside. If my directions are correct, it was to the north of the wooden sidewalks and was quite a building.

Near the smelter was the brick newspaper office. As late as 1936, Ernest Altick, who is director of Cheley Camps at Estes Park had several camp-as-you-go trips for boys and girls, from Wichita. They also camped at our ranch, and we would go to Pearl and Big Creek. The boys loved to explore the old mines and found old tools. They say the cabins at Pearl were looted of contents and finally torn down and hauled away for lumber.

Have you ever been interested in the fluorspar mines at north end of the Park? You pass them going from Walden or Cowdrey to Laramie—near North Gate. I have been there. Last spring when I was at Walden these mines were working. The former Mrs. Guy Allard was employed there in the office.

Seventeen years later, in 1974, Mrs. Chambers wrote again:

I saw what I described in the years I lived in Pearl during the 1930's. The information I gave you on the Pearl mine came to me from Walden residents who had spent their lives there.

Poudre River Country

Manhattan lies north of Cache le Poudre Canyon and some distance east of *Manhattan* Cameron Pass. It was a small, short-lived camp northwest of Fort Collins, reached via the Cache le Poudre Highway 14 by driving about forty miles to Rustic and then climbing two miles to the top of Pingree Hill. There were vestiges of a side road nearby, but I drove to the next fork to the west, where a marker pointed toward Red Feather Lakes, and within a mile I was at the site of Manhattan. This was in 1950. Today one had better stop at Rustic to ask directions.

Stanley W. Thornton of Grand Island, Nebraska gave me my first lead to the place (July 30, 1949):

We could refer you to a very interesting "gold miner" of the last century, now in his late 80's, whom we discovered a number of years back, and we heard from him as late as sixty days ago. He spends his summer months at his "mine" located a short way from Rustic, about 45 miles west of Fort Collins in the Poudre canyon. Steve Prendergast is his name, perhaps you have heard of him or met him. We have found him most interesting.

I wrote S. A. Prendergast in 1950 and received this answer (Sept. 15, 1950):

Your letter Rec'd last Sat. & I'll try and give you a little of the history of Manhattan mining Camp.

The prospecting for mineral in Manhattan started in 1886; but not much was accomplished until 1887 & 1888, when Fort Collins business men mobilized a few good mining prospectors to search for minerals in the Camp of Manhattan. It was gold ore and some of it assayed high in Gold, which caused an excitement & many claims were located there. Considerable work was done there for many years. The mines that did the most work are as follows—The Emily, Monti Cristo belonging to a Mo. Company, Gold King, Moonlight, Ida May, Elkhorn, Denver Lode, & lots of small mines.

Probably 200 or more miners and mine owners were there at its peak which lasted up to 1896, when the excitement died out. The ore decreased in values with depth. I went to the Camp in 1910 & have been there, with the exception of about two years ever since.

There will be mines there some day but not in my time. . . .

The name of my mine on which I have done the most development work is the Rockfield group. Had a mill on the property and I think nearly 1000 ton of ore was milled. Somehow it didn't turn out to be profitable, and so it was shut down.

No mining being done here during World War 2, nor since, exception of what little I've done. It will take money to find out whether the Camp will have pay producing mines or not. It's impossible to get anyone to put money in for development work. Fort Collins are our worst knockers of Manhattan. I wrote to the U.S. Geological Survey to see if they would send a practiced geologist up here and thoroughly geologize this Camp. They wrote me a very nice letter, but said it was impossible to get up here this summer as there was so much work ahead of them, I think in the Uranium District. Said they had Manhattan on their list.

You see if the geologist gave a favorable report, the people wouldn't take any notice of what the Fort Collins business men would say, and then we, no doubt, would be able to get moneyed men interested here.

I also wrote to Harry S. Thayer, mining engineer in Algonquin, Illinois. He mentioned mining properties in Colorado with which he was familiar and also described Manhattan and another small camp farther up the Poudre Canyon (Dec. 29, 1949):

In the course of my work I have had occasion to visit many of the old camps you mention. In 1903 I was with the Liberty Bell Company of Telluride; later a superintendent of the Colorado City mills of McNeil and Penrose for about eight years; and in 1918 or 1919 diamond drilled the old Bassick mine at Querida for eastern clients and erected a small test mill. Results were not satisfactory and it was abandoned.

You mention the Cache le Poudre River section. . . . My father bought a ranch on the upper Poudre in 1891 and we spent our summers there until 1902. In 1894 I carried the mail route from the old camp of Manhattan to the Postoffice called Home, which was a summer resort run by Zimmerman. The main road at the time from Fort Collins went through Livermore, Log Cabin and down Pingree Hill to reach the upper Poudre. The road to Manhattan branched off from this road at the top of Pingree Hill, the camp being about two miles west of this road. The only mine operating there at that time was the Forest or Forster, about two miles south from Manhattan in a small park. This mine operated intermittently for a number of years and a small amount of high grade gold ore was recovered but the mine was not profitable.

There were a few miners' cabins and some prospecting done near the old stage station, about five miles up the river from the Zimmerman resort, but nothing developed and no camp was established. This station was a change point for a stage line operating between Fort Collins and a camp on the Michigan River in North Park about 1880. I have always understood this camp was called Kelker with some hundred or more buildings and cabins but it might be the Teller City you mention. It was a real ghost town in 1891 with no inhabitants. Most of the cabins were left furnished and goods were left on the store shelves. My information on that came from a driver of the stage line who worked for my father for a number of years. . . . This is all quite fragmentary and probably of little value to you.

A more recent account of Manhattan is given by Robert K. Wattson, Jr. of Wichita, Kansas, who led a Boy Scout troop up there in 1960 and reports what they found (Nov. 14, 1962):

I have seen Manhattan—I think. Oddly enough it still shows on an occasional road map. In 1960 I went with my Boy Scout troop to the Ben Delatour Scout Ranch for the second one-week camp there. Our Committee Chairman had reconnoitred the area and knew of the existence of Manhattan; so one fine morning we piled the Troop into our cars and drove west from Log Cabin to the site.

No sign of any sort marks the town. Perhaps a quarter-mile east, up the hill north from the road, we stopped and viewed the tiny cemetery, perhaps as big as my living room, fenced in with chickenwire. Half-a-dozen wooden headboards and one wooden cross, as I recall, are within the enclosure, but only two boards have decipherable inscriptions: George Grill and one other man whose name escapes me died on the same day in 1893, and our Scouts had a fine time conjouring up a gunbattle in which poor George and his antagonist both bit the dust. Someone paints Mr. Grill's headboard occasionally. I wonder who.

The road rises gently to the west from the cemetery, and makes a sweeping turn to the north and up, while a pleasant little valley through which courses an intermittent stream continues the general east-west line of the approach road, rising and narrowing as it is closed by pines and aspen. The road in circling to the north curves around a solitary oak tree on a grassy mound. This is the Hanging Tree of Manhattan, and local characters apparently make much of it, nailing on steps and hanging ropes. Directly opposite this tree, across the road to the west, is the little valley, floored with tall grass.

Up this valley our Troop swept slowly, in twos and threes skirmish lines, for this was said to be the townsite, and we didn't want to miss anything. A Committeeman picked up a clear small bottle embossed "Pond's Extract, 1846." . . . Somebody found a lard pail—modern. Otherwise—nothing. The Chairman, my son and I reached the top of the valley and took a turn around two or three new prospect holes and a shallow lateral marked "Teeta Lode Mine." . . . Disappointed, we halooed the Troop together and turned to retrace our steps east down the valley. . . .

Three-quarters of the way down the north side of the valley, we seated ourselves on two or three rocks to absorb the scenery. At our backs a lone pine tree stood, perhaps twenty feet tall, and tickled our necks with its needles. As we sat, I glanced briefly at the rocks next to where I sat. They trailed into the weeds to my right and stopped. Or did they? I got up to look. The rocks went around to the west of the tree. . . . The pattern of the rocks was rectangular!

We had been sitting on a foundation whose presence we had not detected either on the way up the hill or on the way back. We inspected the foundation more closely. No wood remained to mark walls or door sills. We conjectured what the building had been. Then a Scout found a piece of bubbly, purple glass, and another, and soon we had quite a collection of bottoms of old bottles in our lard pail, all from one area just outside the foundation. This identified the building to our satisfaction—it had been a store or a saloon, and we were rummaging around in its trash pile. With our eyes more alert to what we should look for we again searched the slope. One of us went up to the top and looked back. In doing so he discovered what we had not before; the outlines of several old foundations could be seen from the top of the slope but not from the road. Being out of time, we left the townsite with its pine tree growing through the middle of the saloon and returned to our camp.

The same week I drove to Cheyenne to visit a friend, who told me the town had had a population of some several hundred and had once been involved in a county seat war, Kansas style, with newspapers discharging broadsides and bands of loyal citizens swooping down to swipe the county records. But I never found anyone who could tell me more. Can you?

CHAPTER XIII

The Wet Mountain Valley

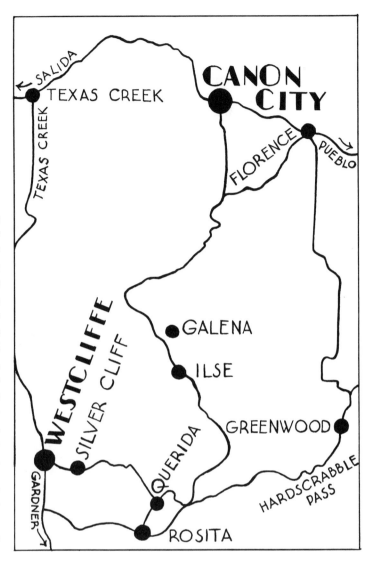

The most dramatic entry into the Wet Mountain Valley is by way of Hardscrabble Pass. Now that the pass is paved it is an easy climb from Wetmore to the summit and across high meadows and over gently rolling hills which descend to the vast valley. Ranch land stretches as far as the eye can see both north and south, while to the west looms the Sangre de Cristo Range with its continuous row of snowcapped peaks. On the valley floor are two towns—Silver Cliff and Westcliffe—and in the low eastern foothills are tucked a few smaller settlements, one of which is Rosita.

A narrow, rather poor road twists south from State Highway 96 in among the foothills of the Wet Mountains and passes through what is left of Querida and Rosita. According to Alfred Friedheim, who grew up in the district, "Rosita was created in 1872 by the discovery of the Humboldt Mine and was the county seat until Silver Cliff became its rival. Its peak years were from 1875 to 1877, during which the population reached 2,000." In 1878, when silver ore was uncovered at Silver Cliff, the town was virtually deserted as men left for the new camp, some moving their cabins to the new location. By 1882 business was at a standstill; so it is not surprising that about 1885, after a hot election, Silver Cliff became the county seat.

In 1972 I revisited Rosita, only to find that almost every old building is gone. Somewhere in the cemetery is the grave of Stephen Decatur, one of the commissioners for Colorado at the Centennial Exposition in Philadelphia—a colorful figure in the life of Rosita. He is spoken about in the third of the letters that follow.

Through a friend I heard about two brothers, Charles A. Barker of New Brunswick, New Jersey, and John H. Barker of Brooklyn, New York, who grew

Rosita

211

up in Rosita. I wrote to each and received the following two letters from Charles and the map which identifies many early sites within the town:

June 9, 1944

I lived as a small boy in Rosita, Custer County, Colo., when my father was superintendent of the Hayden Mine. I understand that Rosita is now a ghost town, but when my parents lived there (almost 60 years ago) it was an up-and-coming community. . . . Our home was at the base of Pringle Hill at the entrance of Hungry Gulch, and beside the road running from Querida to Silver Cliff and that was a long, long time ago.

July 31, 1944

I left Rosita when I was a small boy, but my brother at that time was in his late "teens" and I am beholden to him for the enclosed data on Rosita in the early "eighties."

We do not remember the house as sketched in your letter, but the following description of the map will give a fairly clear picture of what the town was like. Of course we have only hit the high spots.

"T.P." The town Pump.

"B." The Brewery (concrete bldg.)

(1) The Town Hall. Community dances and entertainments were held here; when the town declined, church services were held here also.

(2) School House. The Town Hall occupied one corner of the school house grounds.

(3) The cabin of "Commodore" Decatur. Veteran of the Mexican War and the Battle of Sand Creek, a character in one of the unwritten stories of the Old West.

(4) The Calaboose. Where the killers of Levi Kurtz were lynched in '84.

(5) The home of "Black" and "Red" Cooper

(6) Miller's Grocery Store (Stone bldg.)

(7) Payne's Grocery Store (stone)

(8) Slavik's Clothing Store (concrete)

(9) Maverick Mine

(10) The Barker Home

(11) Home of L. W. Smith. Ordained preacher and prospector, an authority on local ores.

(12) Blacksmith Shop

(13) Hotel Windsor

(14) Lem Kygor's Saloon

(15) Akin's Home. He was charged with being one of Quantrell's Guerillas and was convicted of murder, during the Civil War, many years after the war.

(16) The Smelter

(17) Snowy Range Hotel

(18) Humboldt Mine

(19) Pocahontas Mine

(20) East Leviathan Mine

(21) Virginia Mine

(22) West Leviathan Mine

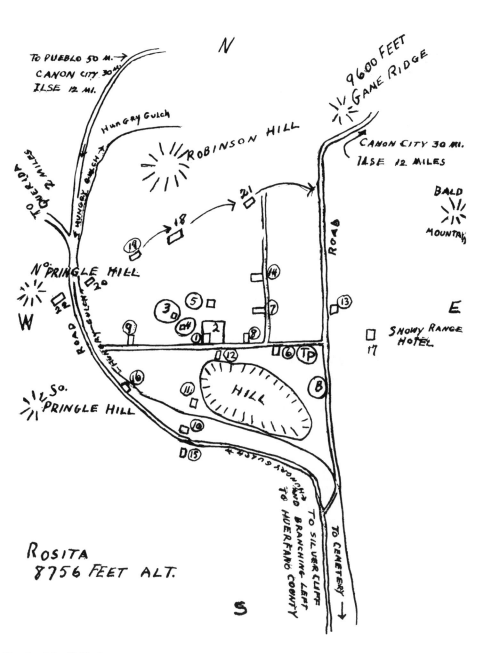

N

TO PUEBLO 50 M. →
CANON CITY 30 MI
ILSE 12 MI.

9600 FEET
GAME RIDGE

Hungry Gulch

ROBINSON HILL

CANON CITY 30 MI.
ILSE 12 MILES

TO QREIDA 2 MILES

Hungry Gulch

21

BALD
MOUNTAIN

18

19

ROAD

N° PRINGLE HILL
20

3   5

14

W

22

9

CH

5

7

13

E

SNOWY RANGE
HOTEL

1   2

8

17

12

6   TP

B

16

11

HILL

So.
PRINGLE HILL

10

D 15

HUNGRY GULCH

TO SILVER CLIFF AND BRANCHING LEFT
TO HUERFANO COUNTY

TO CEMETERY →

ROSITA
8756 FEET ALT.

S

Map by John H. Barker,
who lived in Rosita 1880-1888.

There are many stories about life in Rosita and some of them would, no doubt, be interesting today.

John H. Barker wrote this highly informative letter (Aug. 22, 1944):

... My brother Charles was visiting me and wrote you at my house. The map I sent was done in a hurry and ... there is one mistake in it. Number 21, the Virginia Mine, was located more to the right next to the right hand road and 19 and 18 should be moved over proportionately. To give you an idea of distance it is just about a mile from the road next to the Virginia (as properly located) to Number 22.

I moved to Rosita in 1880 in my thirteenth year and left in '88 in my twenty-first year. The census of 1880 showed a population of over 2000. When I left I don't think there were over four or five hundred.

The decrease in the price of silver contributed to the decline but principally the mines around Rosita, except the Bassick, just did not pay. The Bassick was the biggest enterprise in the neighborhood and a great many Rosita people worked there. The people who bought it from Old Man Bassick were a bunch of New York Stock jobbers. All they were interested in was trading in the stock. Anything which would effect the price of the stock was of interest to them.

We once had an exciting time over the alleged discovery of a plot of "Molly Maguires" who were going to do things to the Bassick and a dynamite bomb was exploded in the office yard. A group of vigilantes was promptly organized who arrested several men supposedly in the crowd and ran them out of town. My father was asked to join this committee and was willing to do so to guard the property but refused to participate in running the men out of town. The whole thing was a plant, engineered by the above mentioned New York bunch to effect the price of stock.

Old Man Bassick who found the mine was a pioneer. He made a fortune in California in '49 and lost it and another in Ballarat, Australia, about ten years later and ran through that. When he found the Bassick he was too poor to pay the assay and cut and split a load of wood to pay for it. But after that he could not spend the money as fast as it came from the mine. He sold it for $200,000.00 and one million in stock and returned to his home town of Bridgeport, Conn.

Querida was called Bassickville until the company took it over when it was re-named Querida. It was not a pleasant place to live as it sloped away from the sun and was colder than Rosita. The closing down of the Bassick was the final blow in both towns as there was nothing there to support them. The County Seat was moved from Rosita to Silver Cliff but was actually located between Silver Cliff and Westcliffe. This was a hard blow to Rosita.

One of the characters who was a prominent factor in the early history of Colorado was Commodore Decatur. This was not his real name which was reputed to be Stephen Decatur Bross. It was said he came from Mississippi or somewhere down that way and was married, and owned and ran a newspaper when something happened which caused him to abandon everything and go West. This was sometime before the Mexican War. He was associated with early pioneers such as Kit Carson and took part in the Mexican War. He was with Col. Doniphan's command in a famous march from California. He told me many stories about these things.

Comm. Decatur ran a newspaper in the northern part of the state, I think in Central City or Georgetown, and was one of the commissioners for the State at the Centennial Exposition in Philadelphia in 1876. The appropriation was exhausted about six weeks before the end of the Fair but Decatur would not close it and continued until the termination. The State never reimbursed him and he went broke. He came to Silver Cliff hoping to recoup his fortunes but met with an accident and lived in Rosita for several years "on the town." He lived about a quarter of a mile from our house and many a time I carried up from my mother a piece of pie or cake or something of that kind for the old man, and we were not the only ones who did it. I often called on him as I was much better read than the average and the old man liked to talk to me because I would discuss history, etc. with him. He told me many things about his experiences in the West but never a word about things preceding that and, of course, I never broached the subject.

Decatur had a wonderful head. He always reminded me of Michael Angelo's statue of Moses with a high brow and flowing hair and beard.

Several years ago I saw mention of Decatur in one of the papers here and the statement that his burial place was unknown. I can testify as to that because I helped dig his grave and helped cover him up. He was buried in Rosita cemetery. I have no exact data as to the time but from the season of the year and the time that I left Rosita, it must have been the summer of 1887. He had a great funeral considering the population. He was very much esteemed. He was one of at least three men I knew who took part in the battle of Sand Creek. All of them told stories of Sand Creek but that would have no place in this letter.

Silver Cliff was a boom town started, I think, in '79 or '78. It was supposed to be a new Leadville but soon "petered out" as the mines were failures. It got its name from a cliff projecting from the valley floor which had stains on it which assays showed to be silver; but the silver was of a form which could not be concentrated and the ore was too low grade to ship otherwise. The town site was laid out, fortunes made in lots, water works built, a fancy hotel was built, etc., all of which were failures.

The Rio Grande built a branch line up Grape Creek canon and following the usual piratical Rio Grande custom started a new town at Westcliffe to make money selling town lots. I understand Westcliffe is now the principal local town and that the railroad has been abandoned years ago and communication is via stage to Cotopaxi.

Ilse was twelve miles from Rosita on the road to Canon City. This branched off from the Pueblo road at Cold Springs about four miles out. Ilse was dependent upon the Terrible Mine and Mill. When this played out I assume Ilse faded off the map. The Terrible Mine gets its name from a local use of the word "terrible" as an adjective; i.e., instead of a very big vein they would say 'terrible big vein.' When the prospector found this enormous deposit it was really a terrible big deposit and they so named it. This ore body was crystallized lead and a very desireable form of ore to the smelters. It was finally bought by the Grant Smelter of Denver to get control of the ore.

Several miles further down Oak Creek was Yorkville at the junction of the road coming in from Silver Cliff. There were but a few houses there and I am quite sure they had a post office. About eight miles out the Pueblo road there was for a time a small settlement known as Silver Park. This soon faded away after local mining properties were failures.

I never heard how Hungry Gulch got its name. There was no local reason for it as if anything it was better watered than adjoining gulches. In Harper's magazine in about '81 or '82 there was an article about Colorado and a picture of Hungry Gulch. I never could identify it exactly but thought it was about opposite the Pocahontas. A back file of Harper's might help you.

Rosita was started, I think, in 1872 and it was known as a good town for a bad man to avoid. The early settlers were old timers who did not like bad men. Shortly after the camp started a prisoner escaped from the prison at Canon City and, hearing of the new mining camp, thought he would be safe there but, instead, the miners arrested him and sent word to the authorities who recaptured him. I think his name was Walker, and shortly after this his term expired and he organized a gang and went to Rosita and "jumped" the Pocahontas, driving away the owners and workmen. The following morning a group of local citizens captured him down the hill below the mine, shot him without ceremony, threw his body in a prospect hole and started up the hill for the Pocahontas. The rest of the gang, who had seen what had happened to Walker, escaped by the time the committee got there. That established a precedent for the conduct of affairs in Rosita.

The Williams-Gray lynching, I think in '84, was an example of what the old timers thought. They perpetrated a perfectly senseless murder just before noon on Sunday and were hung at two o'clock Monday morning. You can check the data of this by old files of the Denver papers which raved about it but I thought then and think now it was the right thing to do, and it certainly cleared the atmosphere. It was a lesson to a lot of boys with whom I was raised and who commonly carried guns.

I could go on at great length about Rosita but time forbids. I hope the above will be of some help to you. . . . I will be 77 next month.

I received no answer from Mrs. Helen A. Downer, attorney at law at Ouray, to whom I wrote requesting permission to use the following letter which she had written me. I did, however, hear from her brother, George R. Hurlburt, Jr., who informed me of her death but granted my request:

Aug. 13, 1952

Just ten years ago my husband, Roger Downer, closed the assay office in Goldfield, Nevada, which he and his brother had operated for thirty-five years. Their tales of Goldfield's boom days were so fascinating that I persuaded him to write his recollections of its history. We worked on it together, he dictating to me and had quite a manuscript typed before his illness which resulted in his death in 1947. Being a mining engineer he was interested in technical details and someone in Reno . . . suggested that I work it into a story of the Downer brothers and their part in Goldfield's life. . . .

To start out I told a little of his parents who came from England and settled first near Rosita, where Mrs. Downer died when Roger was only eleven. On receiving your book I turned first to "The Wet Mountain Valley" in the hopes of finding out something of the Bassick Mine, and something of the na-

Rosita, 1941.

ture of that locality. I cannot describe what I felt when I turned to page 282 and read "Elizabeth Deborah, the beloved Wife of Tho. Downer." That was my husband's mother, and it gave me something to calculate dates from; that chapter was of much help. . . .

I may mention that my father was one of the three partners in the Bachelor mine (p. 372)—referred to as a mail carrier, which is a fact because he carried mail by snowshoe out of Rico one or two years, often through blizzards. He was killed while out surveying at the age of 86—he gloried in his work and did a full day's work the last day of his life.

Sincerely,
Helen Downer Croft

Silver Cliff

Seven miles west of Rosita is Silver Cliff, once the third largest city in the state. This is hard to believe today, for many of its ten miles of streets can be located only by fireplugs and faint trails amidst the sagebrush. The discovery of silver ore on an abrupt cliff thirty feet above the prairie occurred in 1877. This brought a rush to the area, and the town of Silver Cliff was created. After pure horn silver was found all over the ground no deeper than the grass roots, the surface was thoroughly worked, and the Racine Boy, Geyser, Bull-Domingo, and other good mines were discovered and developed. Gradually the silver ore played out and the population of 10,000 dwindled. It was 50 in 1946.

In its prime Silver Cliff bonded itself for $140,000 to build the water system. As the town went down, it defaulted on these bonds and they were refinanced for a new bond issue of $40,000. This issue also defaulted, and, owing to the heavy tax levy, the merchants and home owners put their buildings on wheels and moved one mile west to the new railroad town of Westcliffe, thus evading the heavy city tax. The D.&R.G.R.R., which started this new town so as to make money by selling lots, began construction on its Grape Creek Extension from Canon City toward the Cliff in 1880, and reached Westcliffe the following year. Passengers for Silver Cliff traveled between the two towns by Concord coach.

Silver Cliff's rapid growth seemed to justify the removal of the county seat from Rosita to a more central location. In the next election Silver Cliff, with Westcliffe's support, won and prepared to move the courthouse to its new site. Rosita, however, refused to give it up. Only when a committee of vigilantes marched to Rosita, took the records, and brought them to the Cliff was the transaction completed. The papers were later deposited in a new courthouse which, in appreciation of Westcliffe's aid, was built halfway between the two towns. The last time I saw it, it was a private home. As Westcliffe prospered and Silver Cliff declined, the county seat was moved again—to its present location at Westcliffe.

Two letters written from Silver Cliff in 1881 are from an exuberant young man, F. S. Schenck, who came west for his health and was caught up in the mining excitement. Both letters were sent to me by F. H. Belden of Texas City, Texas, who found them among some old family papers and was good enough to share them with me. Now I share them with you. The map shows Silver Cliff at the beginning of the silver boom:

Aug. 11, 1949

Dear Mrs. Wolle,

I was very interested in your article in the July 1949 issue of "Ford Times." Shortly thereafter, by coincidence, I was going over some old family papers which had recently come into my possession and discovered numerous stock certificates in a company known as the Silver Prize Mining Company which were issued by my great grandfather back in 1880 and 1881. Along with the stock certificates were two letters written in pencil on tablet paper by a Mr. F. S. Schenck. Mr. Schenck, if I remember the story correctly, was a grandson of the above-mentioned great grandfather who lived in and around New Brunswick, New Jersey, and who made a trip to Colorado for his health. It also appears that his trip to Colorado was either the cause or result of his grandfather's participation in the fortunes of the Silver Prize Mining Company. He went to Silver Cliff, in Custer County, Colorado, and since the letters contain a certain amount of local color, I thought they might be of interest to you. I therefore had them copied and attach them for your pleasure. I rather doubt that Silver Cliff comes under the heading of a ghost town since I find that there is such a place on the map just a mile or so east of a larger town called West Cliffe. However, perhaps some of the data in the letters or in the little map which was included may tie into some of the information you have and be of interest. I was amused to find that the Mr. Dickinson mentioned in the letter as being "slightly interested" in the Silver Prize Mining Company signed the stock certificates as vice-president of the company. As far as I know, no money was made on this investment whatsoever, and the old stock is worthless now.

Very truly yours,
F. H. Belden

Silver Cliff, Custer County, Colorado
Wednesday, June 22, 1881

Dear Father and Mother:

Today I suppose you are feeling too big for your boots and your shoes. You are as proud as Cuffee. You think that the band is playing for you—that the big crowd has gathered to hear your wonderful boy speak his little piece— and that all New Brunswick and the rest of creation are looking upon you in rapt admiration. Now let me tell you that at this distance you are perfectly invisible. From the height at which I am you are too small to be seen. You need not feel so big—it is all in your feeling—we off here can't even hear the band play. I only hope you are having such a pleasant day as we have. These Colorado days are just splendid. You know what nice days you have when the wind blows from Orange County after a rain when the sky is clear as crystal and the fine Orange County air comes down upon you with the cool north-west wind. Well just such days we have all the time. The sun shines brightly and it is a little warm in it, but in the shade it is cool and there is almost the whole time a fine breeze blowing; at night it is delightful for sleeping. I have my window closed tight and go to sleep under a heavy comfortable and before morning I pull up a second comfortable to keep warm. We are about eight thousand feet above the sea, so if I could move on a level until I stood over your heads I would be a good deal over a mile above you way out of reach of the malarial fogs of your old swamp. All around us are the hills and mountains. The hills are the rocks bearing silver, they are of but little consequence to one loving fine scenery as I know you do, but the mountains are just grand. To the west of us running north and south is the magnificent range called the Sangre de Cristo, stretching far off in either direction as far as the eye can reach and just opposite us, seeming so near that one feels he could throw a stone to their feet, a multitude of cone-shaped peaks stand out distinctly against the clear blue sky. On their sides in ravines and hollows lie large masses of snow shining in the sunlight and dark forests climb half way up their rocky peaks. Some of these peaks are over six thousand feet above us—over fourteen thousand feet high. I suppose a further description of the scenery here and which we passed through in coming here would be more interesting to you than anything else I can write, for I have known your passion for the beauties of nature ever since you so rapturously admired the sunset when we were returning from our fishing excursion to Greenwood Lake. Still I would like to say a little concerning a certain mining property called the Silver Prize in which my companion, Mr. Dickinson, is slightly interested. By the way, I find Mr. Dickinson a first rate travelling companion. I could not have had a better. He is cheerful, always in good spirits, not at all exacting, on the contrary very accommodating, does not force himself upon me but is always ready for conversation and is very interesting. I am very much pleased with him and consider myself highly favored in being with him instead of alone in the Adirondacks. I think the Silver Prize property has been very moderately described by the prospectus and by Mr. Clark. Of course, I am not as good a judge as I hope to be within the next three weeks but it seems to me we have an immense quantity of ore as good as the Silver Cliff is turning out and they claim to be clearing about $2,000 per day above all expenses. Besides the town is growing toward the Railroad Depot beyond Silver Prize, and the two main streets run right through Silver Prize, so it bids fair to be the heart of the city before long. And Silver Cliff bids fair to double itself in the next three years. Mr. Dickinson has set four men to work getting out the ore. They have been working since Monday morning and are doing finely. A piece of ore by my side now, gotten out yesterday morning, Mr. Foss says is equal to any ever got out of the Racine Boy, and there is plenty of it. But the Duryea Furnace is not finished as we expected it would be and will not be in three or four weeks. Perhaps it is as well; we can examine it more carefully and other mills too. I hope to be able to tell you more fully next week what we are doing and propose to do. Much will depend upon this week's work in the mine. Much love to you all. . . . I am expecting letters often from you and will probably remain here a month or more.

Your son,
F. S. Schenck

Thursday, June 23rd, 1881

Dear Father and Mother and all at home:

I received your letter this morning and was real glad to hear from you poor mortals living so low down in the world and so far off from silver and gold and Colorado air. . . . I will send you another today. Especially as I have just drawn a picture. . . . Let me explain it for your delectation. You, of course, recognize the noble chain of mountains which I described in my last and will no doubt linger in rapt gaze upon their noble outline and it is hardly necessary to describe the rest. . . . The Silver Prize seen in the center of the picture is the property in which my friend, Mr. Dickinson, is slightly interested. You perceive the circle marked in it. This describes the bluff forty feet high. It seems to be solid trachite rock of a yellowish white color. Much of it stained with black oxide of manganese and red oxide of iron. The rocks grow deeper in their stains as you dig down. All this trachite rock is said to contain silver, the stained rock contains it in greater quantity and also gold in small traces. With cheap process of treating the whole would return handsome profits—with the present modes of treatment the stained rocks will yield rich returns. The rock now being got out of the cut marked "C" where we have four men working is pronounced by good judges as fine as any the Silver Cliff are using and they claim their mill is turning out nearly $2,000.00 per day above all expenses of milling and mining.

The whole ground beyond the circle of bluff is supposed to be underlaid with the same rock. To get at it we would have to have shafts. The stained rocks have been ascertained by mill runs, assays, etc. to contain silver amounting to $20 per ton. Sanguine persons expect much larger returns.

Main Street is the principal business street in the town. Ohio Street is the principal dwelling house street and runs direct to the depot—Broadway is also a fine residence street. Cliff Street stops with the bounds of our property and the bluff; Main and Ohio run straight through it giving lots for sale on both sides, and Broadway runs above it giving sale of lots on one side. The main part of the town is now about one Hotel—our Hotel is on the eastern side of it in fact, and I can walk from the Hotel to the Prize property in less than five minutes. The R.R. Co. own a large tract of land about their Depot and located the Depot there for the purpose of drawing the town that way and selling their land. It is probable the center of the town will be on Silver Prize

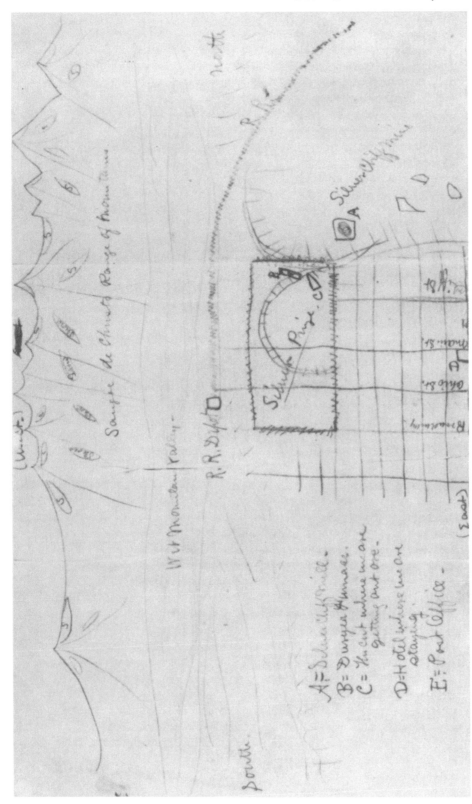

Silver Cliff. Map by F. S. Schenck, who lived in Silver Cliff 1881.

property before many years. Mr. D. thinks Cliff Street can also be laid across the property without interfering with the mining. The width of the property is 900 feet, so you see the building lots are of considerable value. The amount of rock in the bluff seems inexhaustible. The bluff is not nearly as high as that on which the Silver Cliff mine is located, but the stained rocks are on the surface in the Prize while they are low down and covered with the pure trachite on the Silver Cliff property—it is probable in these geological ages, the trachite has been washed off the Prize for our special benefit.

So Cantine and Marion are coming home. I am sorry I cannot be there to see them. . . . Take rides, my horse and carriage are at your disposal and have a good time. . . . I don't know how long it will be before I let loose of the Rocky Mountains. I will be here yet several weeks I suppose so write often. I have not had a headache or back-ache in over a week. Much love to you all. The second pair of glasses work well.

Your son,

F. S. Schenck

Theodore H. Proske of Denver, who worked in several of the state's mining communities, writes of the Geyser mine in Silver Cliff (Nov. 22, 1951):

. . . In 1894 I arrived in Denver . . . and not being able to secure employment of a permanent nature, in the Spring of 1895 I started for the mines; arrived in Breckenridge about the last of April. I had been told by a mine manager in Breckenridge if I would put in an appearance about the 10th of May he would give me a job, so I went on to Leadville, and there secured a job as Mucker in a mine. Later that year the West Creek district was formed, and I secured a job at a prospect driving a tunnel. Later still the boom at Puma City came on and I went there and opened a blacksmith shop. This like West Creek also soon busted. Then I went to Cripple Creek and worked as a blacksmith at several mines.

During these years I contacted several old miners who had followed booms for many years, and heard lots of experiences. Among these stories was one about the Geyser Mine at Silver Cliff in Custer County, whose shaft was said to be the deepest in the State, at that time being over 2500 feet deep. I also met men who claimed they had worked in what was then supposed to be a very rich mine, also in Custer County either at or near Rosita or Querida, and called the Bassick Mine, where it was reported that some very rich gold ore had been mined.

As the story went, the Geyser mine when located produced a lot of rich silver ore near the surface. This mine was reported to be owned by a company in Boston, composed of an organization known as "Spiritualists" and was what was known at that time as an assessible stock company. The manager would go back to Boston whenever his funds were low, and the stockholders would hold a seance, and the manager would report that the Spirits during these seances would have visions of immense bodies of Silver Ore, and on this report they would levy another assessment, and so the work went on. The report that was current at that time said they had the finest and best mining plant in the State, but like all good things even that came to an end after a while, and the next that I heard, they had shut down, and the surface plant was sold and carted away. . . .

Silver Cliff, 1949. Geyser mine waterworks and dumps.

Those who knew Silver Cliff in later years have different anecdotes and incidents to tell. One such person is J. M. Diven of St. Louis (March 21, 1958):

. . . During my first few years among the mining fields in Colorado I often listened "bug-eyed" to some old prospector tell about what a wonderful "property" he had—he was. "tunneling" and thot he would "cross-cut" the "vein" shortly. I would say to the local hardware dealer (my customer) "That old boy sounds like he really has something." "Pooh," would be the answer— "That old Codger has been about to cross-cut the vein for years. He is pretty near through the hill and if he keeps at it he will have to begin *tresseling* soon." To me there was something rather pathetic about these old boys—working outside to accumulate a few dollars with which to reach a little further towards the end of the "Rainbow." The fires of *hope* never died until they died.

Going back to St. Elmo for a moment. Once, I came back across the range from Tin Cup badly in need of a shave. Oh yes—St. Elmo had a barber. A fusty old chap (looking like he was some baths short) in a fustier shop. After the "scrape" I could not live with myself until I went down to the mountain stream and washed my face, drying it on my handkerchief.

I have mentioned that the Wet Mountain Valley is my favorite spot in Colorado. I have many happy memories of both Silver Cliff and Westcliffe. During the first few years of the Century, Silver Cliff was still fairly active and had the *only hotel,* The Powell House, later bodily moved down to Westcliffe. It was operated by Emmi and Fred Schible who deserve a chapter to themselves. They would have been "butter and honey" on YOUR BREAD.

I might explain that much of the joy of visiting the various "towns" assigned to me was in the individuals I met and with whom I became acquainted. The evening train arrived in Westcliffe about seven o'clock. With the hotel a mile up the hill from the station (before the hotel was moved down) transportation was supplied by an old character named Wadleigh, with an old bus (windows all gone) and team of old bony horses which could not move faster than a slow walk. "Waddie" was something to see and remember. I am a confessed Tight-wad but compared to "Waddie" I am a grand and glorious spendthrift. "Waddie" wore a full beard and was too saving to even wash the accumulated contents out. One winter the Authorities found "Waddie's" horses with nothing to eat except the bark off the fence posts and made him feed them. As the town faded out and the residences were vacated and could be bought cheap "Waddie" bought them in until he owned a good portion of the "Cliff." Believe it or not—any bit of rusty iron, old toothbrush, bit of leather or any other scrap found in the street was gathered up and deposited in his treasury.

I had heard of his accumulation and once when I had a bit of business with him and he invited me into his "office"—I saw it. Truly, words fail me when I try to describe it. This particular cottage had two rooms. Starting just inside the front door and piled up toward the far corner—reaching almost to the ceiling was this accumulation—wagon loads of the most conglomerate mass of small odds and ends imagination can conceive. The second room— with barely room to move in—was stuffed with files of old newspapers hanging on the walls. "Waddie" has, of course, long since passed away and I have often wondered what became of his possessions (he had no relatives that I ever heard of) and especially the old bus through which those Valley winter breezes used to whistle—while we rode the "longest mile" on record.

Oh yes,—"Waddie" was known to buy anything offered to him (at the right price) and the Silver Cliff kids used to steal "Waddie's" chickens and sell them back to him—when in need of spending money. But so much for "Waddie" and this load of "poles."

In the late 1960's I revisited Silver Cliff and found that the Fireman's Hall had been made into a museum, with the old fire equipment downstairs and mementos of local interest filling the second story. The work of sorting and arranging the variety of donated objects was not completed, but a couple of dedicated women were making remarkable progress and many old photographs, news items, pieces of furniture, weapons, household appliances, clothing, etc. were well displayed. When Priestley Toulmin, 3rd, of Arlington, Virginia, visited the Hall in 1961, the upstairs collections had not been organized. In his letter to me (Jan. 12, 1961) was a photostat of one of the old programs that was there then:

> In your book you mention a two-story Fireman's Hall as having been built in Silver Cliff in 1882. (Incidentally, the date 1879 is painted over the center door of the building). . . . I was there in the summer of 1954, when I was doing geological field work in that area. At that time there were still two old fire wagons in the building, one of which (bearing the painted name "W. J. Robinson") was in excellent condition and surprisingly complete. Lanterns, fire hose, leathern buckets, and sundry fire-fighting implements all were intact.

The second floor was in considerable disarray—papers and trash of all eras were strewn about in such profusion that we couldn't begin to sort through all of it. I did find, however, a few items of some interest: some supper tickets stamped by the Canon City & Rosita Stage Line at Silver Cliff (or perhaps it is the Silver Cliff, Canon City & Rosita Stage Line?) on November 27, 1879, and the minutes of a meeting of the 4th-of-July Celebration Committee of Silver Cliff, held on June 22, 1884.

Here is a copy of the program:

Office of THOS. FLYNN

MANUFACTURER of

HARNESS, SADDLES AND WHIPS

COLLARS, ETC.

Silver Cliff, Colo. June 22nd 1884

Committy met at Flynn's Store. Tom Flynn Elected Chairman, Geo. R. Partridge Secretary.

The following program adopted:

Mulligan Guards at	9.00 A.M.
Fireman Parade at	10.00 A.M.

Reading the Declaration of Independence address by Hon.—Grant Warner

	Prizes
Pony Race 300 yds. First Best	10.00
Pony Race 300 yds. Second Best	5.00
Burro Race 200 yds. first there	5.00
Foot-Race 100 yds. first Best	10.00
Sack Race 25 yds. first Best	2.50
Wheel Barrow Race 50 feet	2.50
Hub & Hub Race between the "Kids" & Zeigler's: 400 feet—to lay 200 feet Hose, break coupling and put on pipe.	
First pipe up to win the Prize	10.00
Last and best, the fastest dog down the track with or without a can	1.00

CHAPTER XIV

The San Luis Valley

In the early 1940's I reached several of the older mining camps at the northern end of the San Luis Valley. For anyone entering the huge mountain park from the summit of Poncha Pass, on U.S. Highway 285, the first town on the road is Villa Grove, which was established in the 1870's as a supply point for the ranchers and for the mining camps of Orient, Oriental, Bonanza, Claytonia, and others in the same vicinity. To reach each of these meant driving several miles east or west from the highway.

Orient   On my first visit to Orient in 1939, Victoria Siegfried was my companion. A five-mile drive on State Highway 17 south from Villa Grove brought us to Mineral Hot Springs (now Valley View) where we left the main road and drove east six or seven miles up onto the first bench of land at the base of the mountains. Here we found Orient, the deserted Colorado Fuel and Iron Company town. Rows of small, frame homes overlooked the valley; behind them loomed huge dumps below the mine portals.

In 1950, after revisiting the empty camp, Victoria wrote me:

> There are only cement foundations left. I shouldn't wonder if a couple of those tiny frame buildings that were there on our earlier trip hadn't been moved down to Valley View. The panorama of the valley, however, is as exciting as ever.

Oriental   From Mrs. Stanley M. Nevin of Baton Rouge, Louisiana, I learned that Orient and Oriental were different places, although not far apart (Dec. 18, 1952):

> In your book you mention a town called Oriental but state that you can find no trace of its existence.
>
> In a recent visit to Colorado I visted my uncle, Mr. Claude S. Rogers of Canon City, who told me about it. His mind went back to May, 1881, when, as a young boy, he moved to the town of Oriental with his family. He says it was situated at the west entrance of Hayden Pass, about six miles north of the town of Orient. Further to orient the location, he said that it lay five miles east of Villagrove in the San Luis Valley, in Saguache County. He remembers that the mountains were the Sangre de Cristos.
>
> The Rogers family lived there less than a year; that is, from May 1881 to March 1882. My uncle said that my grandfather was prospecting and that my grandmother ran a small restaurant, which was supplied from Pueblo, the distributing point for that area. He further said that there was a small grocery store, a school for sixteen children presided over by two teachers named Alice Moore and Charley Taylor.

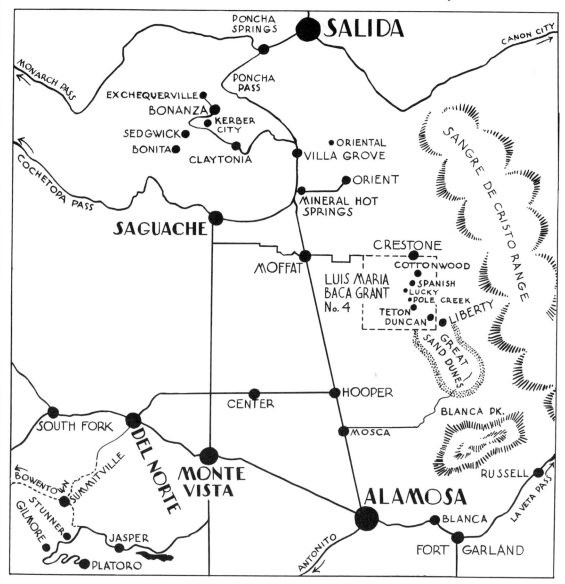

At my grandmother's restaurant, he said, such city fathers as the barber, storekeeper, and blacksmith boarded with her while he, as a lad of eleven, worked in the Vanderbilt Mine. I suppose the town completely passed away because the ore deposits of lead and silver in small veins were not large enough to make big mines. Such mines as the Andover, The Mountain Lion, The Vanderbilt, and another whose name I have forgotten shipped only a few tons of ore apiece.

He recalls that the Andover was operated by a man named Bushnell and that the teacher, Charley Taylor, and a Jack Kennedy, who was later found dead at the bottom of the shaft, supposedly of gas fumes, evidently financed Bushnell. My uncle did not make this last statement; it is only a supposition on my part.

Bonanza   On the highway south of Poncha Pass a sign at the northern outskirts of Villa Grove points to a side road west of the highway and indicates that Bonanza is some twelve to fourteen miles in that direction. Much of the drive to Bonanza follows Kerber Creek up a narrow valley between sage-covered hills and small groves of cottonwoods. The road passes an occasional ranch house before swinging around a corner onto the main street of the once-active camp.

Bonanza and its satellite camps date from 1880, for after the Leadville silver strike prospectors swarmed in and discovered both gold and silver veins up Kerber Creek. There were actually four towns—Sedgwick, Kerber City directly across the creek from it, Bonanza City one mile farther upstream, and Exchequerville two miles above Bonanza.

As Bonanza grew, it absorbed the other towns, and so promising was the mining outlook by 1881 that it was spoken of as the "New Leadville." Of the many lodes discovered, the Rawley was the best prospect. Although mills and smelters were built to process the ore, by the end of 1882 the initial boom was over. Some mining has been done since then, but only a few people still live there. On my last visit to Bonanza, I hardly recognized the place, for so many of the buildings as well as the mills were gone.

The letter from Clifford W. Kingsley of Los Angeles tells much about the town, although he could not have been there before the mid-1890's, having arrived from Victor, Colorado, which did not exist until 1893.

Oct. 16, 1950

When the Rawley Tunnel was being drilled at Bonanza, I was sent there from Victor. At that time there were perhaps 25 families in camp, and among business houses, 2 saloons and a two-story hotel.

I notice that you do not mention, that the most of those buildings were 2x6 studding, then up and down boards with muslin and wallpaper on the plain boards inside, then they filled in the space between with sawdust, making the places more or less comfortable in winter.

They told me that when silver dropped, they had stopes full of broken ore ready to ship. They stopped the pumps and they were never worked again.

During the boom, which I believe was at the time of the roller rink craze, they had built a rink, laying down logs and 1 inch oak flooring; as near as I can remember it was 75 foot square. We had a Christmas dance in it, and mostly stag, on account of so many extra men working at the tunnel.

There were a few other properties working at that time on lease. Also a property up on the pass, which I believe was about 5 miles from our tunnel heading, was working summers.

We had a mountain lion scare during that time, and they sent to Routt County and got the famous Roosevelt bear dogs. In one of the saloons was a mounted grizzly, a female about 3 years old that had been shot by a Dutchman with a 22 long rifle. After 16 or so shots he killed her. He then came into town, at the south end where his cabin was, and got his burro and went back to get the bear, tying the burro to a quakin asp he mosied over the hill, hoisted Miss bar on his back and as soon as the burro saw him coming through the brush he tore out for home Quaikee and all. The Dutchman was in bed for a couple of days after he got into town, lugging Miss bar. Had she had pups, he wouldn't even have been tired.

Bonanza, 1942. Empty store.

Orient, 1943. C.F.&I. company town.

I had not been in camp long before I became acquainted with another young haromskarom from Leadville. He was in charge of the orneryest mule on the tunnel job. One of your San Luis valley mule-skinners stories reminds me of one on Dutch.

Dutch Doyle was a typical Irishman. He had asked our boarding boss for a coffee can, and with the aid of advice from a bottle of Old Crow and his knife had made a wonderful reflector for his carbide lamp. That night, going up the gulch, he had remarked how far it shown. Well, this mule I speak of was a piece-worker if you ever saw one. Get him started and you couldn't get him stopped, but try to get him started! They were pulling five cars from the breast and to get him to pull a car with dull extra steel, you had to have some one push the car so he wouldn't know it.

Dutch had gotten into the heading perhaps a hundred feet, with the new light hanging on the mule's hames, when the mule decided he didn't need the light, so he just scratched it off on a squareset. So Dutch proceeded into the switch and borrowed a light. When he came out with it, the train had stopped at the blacksmith shop and unloaded some steel he had on the last car. I was in charge of the Compressors and also Powder Monkey.

When he tried to start the mule, nothing doing. I heard him and when I looked across he picked up a starter and says—"No wonder they chrusafied Christ, if he rode a mule like you son of a bitch," and hit the mule a smack between the eyes. Down went the mule, but when he came to he was ready to go, off the end of an 80-foot dump. The mule hit first, and then the cars. They had to get a fellow on horseback to go up in the pass in 4 feet of snow to bring him back—a few places on his hocks were skinned. Three of the 5 cars went to the blacksmith shop.

The Rawley mine was at the time owned by the St. Louis Missouri Traction Co., I believe. At the time I went up there, the narrow gauge in the Valley was supposed to have the longest straight stretch of track in the U.S.

Clifford W. Kingsley

Nov. 13, 1950

In my last letter I find I've neglected telling you how those towns looked when I saw them. The Roller Rink in Bonanza I mentioned was a log building, built low, two windows to each side and I'm not sure if the roof was shakes or shingles. I'm inclined to think it was shingles, a nice double door entrance and no supports in the center of the floor.

A couple of jolly school marms from Saguache were imported to liven up our Xmas dance and they were lovely girls, full of fun. The mountain lion tree'd by the Routt County Roosevelt bear dogs measured nine foot tip to tip.

In January the sun arrived in the canyon between 9 and 10 and left around 2 or 3 P.M. The main street was fairly wide. I believe we only had two streets, running with the canyon, and across the creek on the side of the mountain a few cabins with a swinging bridge; so we did not waste time geting to the Saloons that side of the street. As I remember there were 15 or so buildings.

A young engineer by the name of Russell was Supt. of the Rawley tunnel. He had just graduated from Palo Alto, Calif. I became acquainted with him and would often stay for supper and we'd play cards. He lived just back of the Boiler house. The other properties on up the sister canyon were worked by leasers. The tunnel was in 1200 feet when I left there, and they had discovered quite a few veins, one over four feet of Peacock ore. So there is no doubt the venture paid off.

The Kingsleys—Laura and Clifford W.

Farther south in the San Luis Valley, up against the western base of the Sangre de Cristos, there were other camps.

One summer as I drove east from Moffat, looking for Crestone, which nestles at the foot of the mountains, I passed the gate of a ranch whose wrought iron sign read "Luis Maria Baca Grant #4." What did that mean? In Crestone I inquired and learned that it was an old Spanish land grant. When King Ferdinand VII of Spain in 1823 rewarded one of his subjects with the title of Don Luis Maria Cabeza de Vaca and granted him a huge tract of land in the new world, it included the present site of Las Vegas, New Mexico. As Mexico recognized the rights of the Vaca family, it authorized them to select four new ranch sites totaling 400,000 acres in return for giving up its claim to the Las Vegas property. One of the tracts they selected is in what is now Saguache County and is the 100,000-acre Grant #4. *Crestone*

After the Mexican War, when the United States came into possession by treaty of Spanish and Mexican land grants, this twelve-mile-square piece of property, with the summit of the Sangre de Cristo Range as its eastern boundary, was recognized by the surveyor general of Colorado in the name of the heirs of Luis Maria Baca, by authority of an act of Congress, June 21, 1860.

Crestone lies pressed against the foothills of the Sangre de Cristo Range, well shaded by the many large cottonwood trees that line its quiet streets. Except for the crossroads center of town, with its two general stores and post office, it is largely a community of homes half hidden by vines and bushes. It is hard to picture it as the business center of the twenty-five-mile long mineral belt that attracted prospectors to it in 1879. Its second boom was in 1890, when the main portion of the town was built. Mail, passenger, and freight service connected it with Moffat, its nearest neighbor in the valley. In 1900 when it was incorporated, Crestone had a population of 2,000; real estate boomed, and a railroad was completed as far as the town. The following year the big strike was made in the Independent mine, and prospectors scurried to discover other promising deposits. When gold, silver, copper, lead, and iron ores were found on the Grant at the base of the mountains, the camps of Cottonwood, Spanish, Duncan, Pole Creek, and Lucky sprang up south of Crestone. While the mines were productive, Crestone was a busy place, but when the ore was gone—it lay in "pockets and kidneys"—most of the population left.

While at Crestone, I found several people who had lived there during the mining boom and who gladly talked about their experiences in the short-lived camps which sprang up toward the end of the nineteenth century.

From Mrs. George (Beatrice) Niebuhr of Walsenburg I heard how close to town wild life came. Even her husband didn't like to fish by himself in lonely places because of bears. One woman she knew was going home when she saw a mountain lion as she neared her place and in fright she jumped over the front fence.

Cater-corner from each other, at the intersection of the two main arteries, stand corner stores. The second floor of one was used as the Masonic Hall. A small, false-fronted building several lots west of the main corner, on the north side of the street, was formerly an assay office. When Mrs. Niebuhr went there as post-

Crestone, 1901. *Photographer unknown.*

mistress, it was still cluttered with all the paraphernalia used by assayers, including a furnace. The front of the building had also been a barber shop.

Although Crestone never had a major fire, floods were just as devastating. Upon occasion they changed the course of the creek, nearly wiped out the town, and washed houses downstream to be piled up and splintered amongst other debris. High winds blew down the old log church; and the present small building which replaces it, The Little Shepherd of the Hills (Epis.), was originally built to be a private home. One Christmas the Community Church burned while gifts that were to be distributed were still in it.

Cottonwood and Lucky

The largest of the mining camps south of Crestone was Cottonwood. The Colorado State Business Directory of 1883 lists it as a "new mining camp 30 miles east of Saguache and 7 miles south of Crestone." As it was not on a creek, people had to buy water at ten cents a bucket. One woman recalls using one bucket of water all day—for cooking, saving some for dishes, and reusing that water to wash the floor.

Mr. Marshall, who had lived in Cottonwood, told me that he came to Crestone with his parents in 1879 when he was seven years old. He knew Cottonwood when it was a good-sized place with many one-story log houses and a few frame buildings. While still a boy, he worked for the Independent mine by leading sixty to eighty jacks up to the mine, waiting for them to be loaded, and then leading them down the hill to the big one-hundred stamp mill where the gold was processed. As the mine was developed, the ore changed to copper, zinc, and lead found in schist and quartz. The mine closed in 1898.

Mr. Handy, another of my informants, had also lived in Cottonwood. He remembered riding a handcar on the railroad from Cottonwood to Crestone. A bunch of mountain sheep were caught on the long trestle, and he had to stop and get them off before he could continue.

Mr. Noah Mayer spoke of Lucky on Spanish Creek as a rough camp, built up and down the creek, with a crooked street that was alongside the railroad. The railroad was completed to that point by 1900.

Just as the mines were beginning to pay off, in 1900, the San Luis Valley Land   *Duncan*
and Mining Co., a Philadelphia syndicate and owner of the Baca Grant at that time, took over the mines and began to develop the mineral resources.

In the southeast corner of the Grant was a mining camp called Duncan, near the mouth of Short Creek. The new survey included it, and when its miners were run off, they in anger and frustration laid out and built a new camp they called Liberty, just outside the Grant fence, and began working at the new diggings they uncovered. From the Niebuhrs I learned that Duncan had both a hotel and a saloon, in addition to the miners' cabins. After the men had been ousted from the property, having to leave furniture and all other possessions behind, one of them went back at night to the saloon. It was a one-story building on corner supports. He crawled under it and with an auger drilled through the floor and through the barrels of liquor, letting them drain on the ground.

I found all the above information valuable and interesting, but I still hadn't   *Liberty*
seen Liberty. So I wrote to Mr. Collins, the president and manager of the San Luis Valley Land and Cattle Co., which was working the ranch on the Grant, asking permission to cross the Grant and see what was left of Liberty and the other deserted mining communities. He sent the permission and added that he would secure a guide and provide horses for the trip. Some weeks later, when I reached the old adobe ranch house which served as the company office, I saw no horses—only a lanky cowboy leaning against an old and somewhat weatherbeaten car. Mr. Collins explained that in summer the hot sand was hard on horses' feet and that Sam, my guide, would take me in his car instead.

As we angled across the miles of pasture land, I saw wooden watertanks from which the stock drank. When we drew close to the mountains, the ground turned to deep sand. This slowed us down, but Sam's expert driving, which avoided the extensive patches of quicksand, brought us well before noon to the barbed wire fence at the southern edge of the Grant. Sam stopped the car, he unlocked the gate, and we began our trek up a sandy trail. By the time we reached the first few deserted cabins we were in the shade of the trees which lined the edges of Short Creek and covered the foothills east of us.

So as not to miss anything, I picked my way higher up the overgrown path. The underbrush on either side was so dense that it was hard to discover the few remaining sheds and cabins, so well camouflaged by vines and other vegetation. "Here's an old saddle," called Sam, pushing aside some deep grass with his foot. "Farther up the trail there's a wagon falling apart that you'd pass by without seeing unless you looked close," he added. On the way back to the car Sam told me that the men who were forced to leave Duncan mined from ten to fifteen years at Liberty and found enough gold and silver to warrant installing a five-stamp mill.

A letter from P. M. Maddrell of Freeport, Illinois (Dec. 18, 1949), was of special interest to me, for the camp in the 1930's must have looked much as it did in the 1940's when I was there.

In the fall of 1934 or 1935 I drove through the Baca Grant from Crestone (what a drive) and spent the night in the bunkhouse at Liberty—I believe it was the lower, left hand building of your sketch on page 307 [*Stampede to Timberline*].

Liberty, 1942. Cabin with square-hewn logs.

I remember the old blacksmith shop to the south about 100 feet, and across the creek the shop that had a "U.S. Post Office" sign over the large front door. There were also 3 or 4 scattered cabins south of Liberty proper, quite near the Sand Dunes.

From an interview with Mr. Cochran of Del Norte in 1947 I learned that gold from Milwaukee Hill at Liberty was sent to a smelter in Pueblo to be refined. It was arastra gold, which meant lumps of free gold. Prospectors had searched for years for its source but had never found the lode. At that time there was a small mill there, and some work was being done. Cochran seemed to know the entire mining belt, for he told me that "if you want to get the greatest collection of kinds of rocks, hunt twenty-four hours around Crestone."

Music City   "What do you know about Music City?" I asked Sam, as we drove back across the Baca Grant. "Wasn't there a pass of that name too?"

"Sand Creek is the first creek south of Short Creek, where we were," replied Sam. "Both are this side of the Great Sand Dunes. Music Pass is on Sand Creek higher up on the range. I saw a bunch of cabins one time when I was packing up that way. Maybe they were Music City."

The following letter from Edgar G. Dicus of Taos, New Mexico (Feb. 22, 1950), may offer a clue to the location of the elusive place, as well as provide information about the valley.

My father went into San Luis Valley and still owns a lot of land there. He has watched the valley change for the last 60 some years and is familiar with a number of the things about which you wrote. In 1910 father, mother, my brother Dick and I went out to the valley for the summer. While we were there we, in company with certain other families, left Mosca in two covered wagons with 16 horses to each wagon and went high into the Sangre de Cristos. We stayed at an abandoned gold mine way up on Sand Creek. I read your chapter on the valley and I am sure that the place to which we went is the one to which you wanted to go when you were down there. I have not been up there since. Incidentally, when my father settled in the valley I believe that the only towns up the valley were Monte Vista and Hooper. He started another town and named it Streator after our home town in Illinois. The railroad built thru later and changed the name to something, but I don't recall what. Then the post office department changed the name to Mosca, which it is today. As father's trips (usually two each year) have always taken him thru Poncha Pass, he saw all the activity around there.

U.S. Highway 160, down the west side of La Veta Pass, follows the curves of *Russell* Sangre de Cristo Creek all the way to Fort Garland, at the foot of Mt. Blanca. About two-thirds of the way down from the summit of the pass on the north side of the road is all that is left of Russell, a small, placer-mining camp that dates from the 1870's. A signpost, a few houses scattered over the townsite, and a frame schoolhouse mark its location. When the Rio Grande narrow-gauge railroad ran from Walsenburg into the San Luis Valley, there was even a turntable at Russell. Today nothing but a hole in the ground indicates where it stood.

Through a friend, Aileen Fisher of Boulder, who arranged for me to meet a former miner and resident of Russell, I learned something of its history. One Sunday morning in April, 1956, Aileen drove me up Sunshine Canyon to the Snowbound mine, where the seventy-year-old man lived. Though the mine has not been operated for many years, its buildings still cling to the edge of the canyon, and in a small house just above the road we were welcomed by Charles Guhse. His home was cluttered but warm and cozy and, to our surprise, was still decorated with his Christmas tree and pine cones suspended from the ceiling. He mentioned that the Snowbound mine was one of the earliest patents "on the hill" and that the Horsfal mine at Gold Hill was No. 59. He was eager to talk, and I scribbled rapidly to put down all he told me. Settling himself in an armchair, he began:

"Russell, which was on the Trinchera Grant in Huerfano County, was called Placer at first, because of its gold-bearing gravels. Later on the name was changed to Russell, for William Green Russell. You remember, he was the one who found gold in a gulch not far from the Gregory District in 1859. His camp there was called Russell Gulch.

"Placer mining at Russell began in the early 1870's and continued to about 1900. There was also mining up Grayback Gulch and Buckskin Gulch. The mouth of Grayback was hydraulicked and there was even some dredging done in the early 1900's. At one time the Colorado Fuel and Iron Co. had an iron mine up Grayback.

"When I went to Russell in 1909 I had a cabin six miles from the camp at the junction of Grayback and Buckskin Gulches. William V. Casey, superintendent of the Boulder schools, went in with me, and he built the cabin. Bill Thompson built another one in Grayback too. He left his six-shooter to me. See, here it is, with cartridges still in the belt. In 1909 or 1910 the Colorado School of Mines made a survey. The bench mark at 10,000 feet was on my cabin. While I was working my mine, my wife stayed at the cabin with me. That meant she'd go for eight months without seeing a woman.

"Russell, when I was there, had one main street and a few back streets. There were two general stores, a post office, a saloon, and a schoolhouse. Mrs. Margaret Sutton, who was born in Gardner, and her sister ran the hotel. Her husband was a roaming leaser known as Shortie. He'd come to Russell very early. While I was there she was living alone. Everyone called her the Angel of the Sangre de Cristos. She got presents from people whom she knew or who stopped to see her. I was there one year at Easter time. She received a big box at the post office. In it was an Easter lily. I got a friend to borrow a house dress from her and send it to me at the Snowbound. I bought yard goods with a printed design of the Royal Gorge scenery on it and had a friend make the dress and then mailed it and the old one to her. She wore it that summer while I was there on a visit.

"The cemetery was uphill from the post office. A few sluice-box robbers are planted up there. A colonel from Fort Garland was buried there too in the seventies. Another grave is that of a four-year old girl who drowned in the creek. Gil Palmer, a hunchbacked prospector who had mines at Russell, agreed to speak at the grave, since the camp had no preacher. He was drunk as usual, but I remember what he said: 'We're all common people. With the sighing of the winds and the whispering of the pines in the shadow of these everlasting hills that she loved, we're laying Little Gladys to rest.' Then he prayed. I can't remember the prayer. I couldn't bear to go to the post office for my mail for two months after that, because Gladys had always run out to meet me.

"In 1910 a gray-whiskered man in his eighties arrived from a soldiers' home back east. He had with him an anonymous letter from an old fellow who had included a hand-drawn map which showed where the old soldier's brother was buried. At that time two brothers—the Berry boys—lived in Russell, and the old man got them to dig for the body of his brother where the map showed it was. Near the railroad whistling post for Russell they found a skeleton and a six-shooter which the old soldier recognized as belonging to his father and which his brother had taken west with him. Also a buckskin belt which crumbled when exposed to the air. In it were sacks of gold nuggets. The old man gave the gold to the Berry boys, saying he couldn't take dead men's gold."

Guhse paused for a moment, and then added: "I wouldn't take a dead man's gold either. I might take a dead man's silver or paper money, but not gold. And that makes me think of Bob Hollenbeck whom I knew and of a lost mine which he tried to find. Up on Mt. Blanca there was a small camp called Columba—I'm not sure of the spelling. North of it and only reached by a burro trail was the White Pigeon mine. It was on a portion of Blanca known as Snowslide Mt. The mine had been covered by a slide, and many soldiers from Ft. Garland and prospectors scoured the area trying to find it. This goes back to the time when Ft. Garland was activated. Prospectors used to drift into the fort in the fall and hole up for the winter. In the spring the government would grubstake them and they'd streak off to the hills. One such prospector was an Italian. From time to time he would return with rich gold. Lots of men tried to track him back to his mine but none succeeded. Some soldiers who looked for the Italian never returned. Years later my friend, Bob Hollenbeck, puttered and hunted around that area while prospecting. Finally he found a tunnel, an old shack, some playing cards, G.A.R. buttons, and two skeletons."

Platoro   The San Luis Valley stretches south to the New Mexico border and beyond. On the west side of the valley twelve miles south of Monte Vista, a secondary road angles west through the foothills to the Conejos Mining District. Only a few houses remained in Jasper and none in Stunner—both small, short-lived mining camps through which I passed on the way to Platoro in 1942. At the foot of the mountains, just beyond the site of Stunner, the road crosses the Alamosa River and climbs a ledge to a saddle near a peak south of the stream.

On the way up the ledge road, I looked down to the valley to see where I had been, and there, north of the river, were the remains of Stunner. Rows of hollows and fragments of foundations where buildings had stood marked the location of several streets, all of which had been hidden from sight at close range by tall grass and weeds.

After crossing the saddle beyond the shelf road that I was traveling, the country opened up into a new valley, and on the mountain meadow below lay Platoro. I could see its streets and a number of cabins, which appeared to be deserted. Cattle grazed on the grassy flat.

Platoro, 1942.

Platoro was laid out in the 1880's after prospectors discovered promising ore in the immediate area. Although it was nearly inaccessible and high—its elevation is 9,700 feet—the population was 300 at its peak in 1890 when the Mammoth and other mines were producing.

Both it and Summitville, another mountain-top camp a few miles to the northwest, needed wagon roads to ship out their ore. The one built in 1888 from Del Norte to Summitville served that camp; an extension to the southeast was constructed to Platoro. In 1901 and 1902 large veins and pockets of gold and silver tellurium were opened in the District, and Platoro and Stunner became important mining centers.

Although Platoro was almost deserted when I saw it in the 1940's, beyond the town there was much activity. Across a gap between two hillsides a high dam was under construction, and trucks, machinery, and supplies littered the valley floor. Workmen swarmed over the concrete barrier, and bulldozers reshaped the forested slopes into access roads.

Charles O. Axell, civil and mining engineer from Chicago, writes of Platoro both past and present (May 31, 1950):

My father entered Platoro in 1882 and secured two or three properties at that time. Charles S. Barnes, Henry Reynolds, John Roush and Samuel P. Mix came about the same time. The above named men became prospectors. Reynolds and Roush did most of the prospecting, opening up many good veins, such as the Paroles, Merrimac, Illinois, Queen Bee, and many others. Barnes became United States deputy Mineral Surveyor for the district and acted in that capacity for a good many years.

The camp now has three lodges, the Green Roof Cabins, under the direction of Wm. F. Hawkins, the El Rancho and the Sky Line Lodge. There are many private cabins on the townsite. It is interesting to know that all townsite lots have been sold. There is not a single lot left to purchase. I have tried to get one.

The dam and reservoir now nearing completion will be an added attraction. Antonito and the surrounding farm country will benefit by irrigation from the storage water in the reservoir. It will also prevent losses to farmers in dry spells. Farms are now limited in area to 160 acres.

If you have any information you want, please write me. I have tried to think of everything.

A. C. Saunders of Castle Dale, Utah, knew Platoro in its boom days (Jan. 24, 1951):

I have been interested in the old mining camp of Platoro since the first boom. At that time, 1913, I joined the crowd, and we issued a weekly newspaper, "The Platoro Miner." Our first run of the paper was sold out. This paper was published each week until late in the fall, when most of the miners and prospectors were pulling out.

Well, the boom failed to materialize, and the following spring there were very few that came back to work their claims. I went back up the next spring, but there were very few that had made investments. The rest had pulled out their equipment, and the "camp" became a ghost town. However, I still have a couple of good cabins and a number of lots on the townsite. These cabins are rented during the summer to outing parties.

We were up there last summer, but on account of the blasting by the contractors, fishing was not as good as usual. I have all my cabins rented to parties working on the construction of the dam. This dam will be quite a boon to fishermen as soon as it is finished and stocked, as well as for vacationers.

Summitville   On my first trip to Summitville, my driver and I approached the 11,000-foot camp from the south. The last grade before entering the townsite was a series of short, sharp, steep switchbacks on which, I was told, even the ore trucks had to back up before completing each turn. Since then this road has been closed. According to R. L. Tikka, district ranger at Del Norte (June 25, 1968):

Summitville can be reached most easily by motor car via the Park Creek Road, which leaves Highway 160 about 12 miles below Wolf Creek Pass. The Park Creek road is currently being maintained to facilitate the mining activities in the Summitville area.

Summitville is an old mining camp which has seen several peaks and declines. South Mountain, near the summit of which the camp is situated, has been aptly de-

Summitville, 1942. Switchbacks on old road to Alamosa River valley.

scribed as one big gold mine, and ever since the first discovery in Wightman's Gulch in 1870, miners have searched and found gold within it.

By the mid-1870's the owners of several paying mines built mills and cabins near their properties and thus were able to spend the year-round on the mountain. So productive were the mines that by 1883 the District was the largest gold producer in the state. Yet by 1893 the area was almost completely deserted. In 1934, when the Summitville Consolidated Mines Company began to rework the old properties on a big scale, Summitville revived. Miners' homes were built on a new townsite, a schoolhouse and post office were added, and roads to the camp were improved. During World War II Summitville was converted to a strategic-metal copper camp, but when the gold ban was lifted it reverted to gold mining. Since then it has been owned and leased by different individuals and companies.

On one of my mining trips I met Mr. Frank Panghorn (or Pangbour) and talked with him about Summitville. As a younger man he must have lived in the town, for he knew about the early mines, such as the Annie in the upper part of town and the Old Adams at the top of the grade. The first road to Summitville, he assured me, was up San Francisco Creek and over Baldy. He spoke of a table with marquetry inlay that he had made from Stamp Stem guides from the Annie mill, and he recalled that Summitville at its peak had a population of over 1,500. It also had hotels, the Bowen House and the Riddle House, as well as fourteen saloons. His mother was the only woman in camp during the nineties.

In 1950 Inez (Mrs. J. R.) Richardson, writing from Granby, Colorado, described Summitville as she had known it in 1948. The old log buildings with sod roofs that she mentions are what is left of the original settlement of the late 1870's and early 1880's:

> I am the wife of a mining engineer. . . . Summitville lies like a ghost now. We rehabilitated parts of it in 1948; the small mill, a few houses and the staff house. I had the fifteen year old son of a dear friend of mine up from Florida

and he and I explored. We found an old, old camp, logs with sod roofs, sagging to their not-too-far end, and have wondered about them. They quite evidently predate any other buildings there. Did anyone think to tell you of the very first prospectors there? It seems they formed bars of their metal and buried them. A Bunch of Indians killed them, the usual one man escaped, carried the story and died, also as usual, leaving no maps. *But one bar was found!*

My life is most interesting, and although sometimes I lack city water, what is better than Mountain spring water; I may be forty miles from nowhere, over impossible roads, but think of the noise and confusion I miss.

Her next letter tells more about her year in the high mountain camp:

June 27, 1950

I would have loved to have seen Summitville as you did, you really should go back now and see its ghost. This is the story of Summitville as I learned it while living there nearly a year.

The very first miners in there were, as I have told you, killed by Indians. Then I can learn nothing more until a number of years have passed and a group of Frenchmen opened the old "French" diggings. From that mine group comes a form of copper ore called covelite much sought after by "rock hounds." I believe that particular form is found only in three places in the known world. Anyway, there is a story current locally that the Frenchmen high-graded both themselves and the owners out of the place.

Then in a few years another group, or rather groups this time, came in. There were dozens of small claims and all up and down the creek were stamp-type mills. One of the larger owners of some of the better claims was a man by the name of Reynolds, whose daughter inherited and who is now Mrs. George H. Garrey, a socially prominent Denverite. As nearly as I can gather Mr. Garrey and Mr. Poxson, now one of Gov. Johnson's advisors, bought and consolidated the properties under the name Summitville Consolidated Mining Co., Inc. and then leased it to a Mr. A. O. Smith. He was a munitions manufacturer from World War I and very wealthy. He built the Summitville you know. At that time it was second only to Homestake in the Dakota Black Hills. . . . You must have been there the last of its good days, for Mr. Smith was called to Washington; when he came back the entire town was closed down and moved out within twenty-four hours. Had Mr. Smith not died during the late war, I think Summitville would have reopened, but die he did.

Mr. Poxson and Mr. Garrey continued to operate as a copper, lead and zinc camp, but in a much smaller way.

In July 1948, Jones-Nylene Mining Co. of Leadville leased the camp. I went up that Hair-raising road in mid-July, with my husband and his pardner, Mr. E. D. Dickerman of Denver, who were to rehabilitate certain portions of the mine. The small flotation mill was caved in, due to the weight of snow, to the ground. The roof was off the big mill, the Reynolds tunnel caved in, rails and timbering rotted out. The staff-house with three rather nice apartments was inhabited by pack-rats and chipmunks. The offices were damp, littered with trash, torn records, rat nests, broken windows; the drafting-room and small (but complete) hospital and doctor's office was a sickening mess. The plumbing in what remained of the houses was simply minus. Water lines had to be dug out, and thawed a foot at a time. We were without water for three weeks. It was one of the ghastliest messes I ever saw, and was only made worse by the fact that we could see what it had been.

Jones-Nylene have poured thousands into Summitville, but it needs millions. The ore is low grade, and at present prices just doesn't pay. The labor situation is bad as there is nothing to do but work, no recreation of any sort. From October to May the roads are really terrible. They need one snow-plow on the road to Monte Vista twenty-four hours per day and another in camp. We had a bad time getting a teacher to come in at all. It takes a huge sum of money to open, rebuild and operate such a property. And once shut down, it is worse than going into a completely new property. But there it lies, a ghost town now. I often wished while I was there that walls could talk.

The last time I saw Summitville was in 1968 when I received permission from the present owners, the Reynolds Mining Co. of Denver, to make sketches of the property. The Park Creek Forest Service road from the Wolf Creek Highway was like a boulevard compared to that over which I had been years before, along the Alamosa River and finally up the switchbacks.

Summitville, 1968.

Times have changed. Although one of the mining properties was being further developed for copper and its new mill was running, there were no signs of life among the company houses, sheds, and mills in the once-busy camp. A talk with the mine superintendent explained why. With a good road up Park Creek and modern cars and trucks, all the employees live in South Fork or Del Norte and commute to work each day, even during the winter months.

Bowentown   Even before I went to Summitville, I'd heard of Judge Bowen in connection with mining properties and of a small camp called Bowentown, thirteen miles to the west, and twenty-five miles northeast of Pagosa Springs. To connect it with Summitville a trail of sorts crossed the Continental Divide at Ellwood Pass, while another trail from the camp followed the East Fork of the San Juan River down to Pagosa Springs.

In 1950 and 1951 Mrs. May Wallace Ross, formerly of Animas City, Colorado, sent me several letters full of historical data and minute descriptions as to how to reach Bowentown and also Logtown, another high town northeast of Durango. As I began to correlate the material provided by Mrs. Ross' letters, I gained a clearer picture of the isolation of the early camps, especially those that were located at or above timberline in the days before wagon roads had been gouged out of the forests and rocky slopes that led up to them. Let Mrs. Ross give her account of Bowentown:

Jan. 24, 1951

Now in regard to Bowentown. . . . Judge Thomas M. Bowen was a territorial pioneer at Del Norte. He became judge of the Fourth District Court and held that position until 1880. There was a mining property near Summitville which was not paying well, so Judge Bowen quietly bought enough shares to give him a controlling interest. He developed the property and became quite wealthy. He was a man of education and judgment, and he never was given to ostentations as H. A. W. Tabor and Tom Walsh were. He spent his wealth with dignity, and therefore was not remembered as the show-offs were. I will verify this and write you later. I am sure of the circumstances but have forgotten the name of the mining property, which I can get.

The road you will follow to get to Bowentown will be easier than the one to Logtown. . . . Maps show the Wolf Creek Highway and the trail along the east fork of the San Juan River. It would be a ride of about 12 miles from the East Fork Camp Ground. It is a trail now but it was once passable for wagons. Since you have been to Summitville, you will remember that Summitville lies at 11,300 feet elevation, at the head of Wightman's Fork of Alamosa Creek, but its supply route was a 25-mile road along Pinos Creek from Del Norte. That one road to Del Norte was Summitville's only connection with the outside world. Down the east fork of the San Juan River, and then along the main stream for a few miles, lay the hot springs which the Ute Indians had called the big Pagosa, the big medicine. The Indians had always believed that these springs had great curative value. The agreement of 1873, signed by the Ute Indians, relinquished the springs at Pagosa. The miners who worked at Summitville suffered from rheumatism and kindred ailments which made it seem desirable to make a way to the hot springs, which lay only about 30 miles distant. In 1875 some men took a wagon down to the springs. It required moving rocks and felling trees, and the progress was at an average rate of a mile and a half a day. This route was used between Summitville and the springs but was not improved very much. Late in the summer of 1878 a garrison was brought to the new log fort which the Army had built near the springs, and travel over the east fork road was increased, since soldiers had only two places to go on leave; to Summitville, where there was a saloon for each 60 persons, about 30 miles uphill; and Animas City (the one since annexed to Durango), 62 miles distant over the old Spanish trail, where saloons were abundant also.

The road to Summitville went through Ellwood Pass after passing Bowentown. I notice on the map that there is a Mountain States cabin near the old site of Bowentown. I understand that those cabins are well maintained. It is a comfort to know you could spend the night there if you needed to.

March 1, 1951

I have found an account written at the time, published 1885, which covers the mining venture of Judge Thomas M. Bowen in the San Juans near Summitville. In Ernest Ingersoll's *Crest of the Continent,* on page 167, you will find a story substantially as follows:

In 1876, when there was a great rush to the Summitville region, a corporation called San Juan Consolidated Mining Company was formed, with the Ida mine as its principal holding. Judge Thomas M. Bowen, later Senator, was the "most intelligent" stockholder of this company. Although he was busy with his duties as Judge, he became certain that the company was being managed inefficiently.

In 1880-81 the stock of the company became almost worthless. Shares of stock were played at poker in Del Norte. Two drinks were given for one share of stock at the bars of Del Norte. Judge Bowen, with an outlay of $75.00 gathered up stock with a face value of $300,000.00. Two or three other men bought smaller blocks of stock. Judge Bowen persuaded friends in Denver to erect a mill on the property, which became known as "Bowen's bonanza." Bowen was not spoiled by his riches.

Six miles across the mountain from Summitville there was a settlement of prospectors who had a lode of silver galena, the Cornwall. Farther on, among the springs at the head of the Rio San Juan were other lodes, the principal one being the Perry.

(The rough road from Summitville through Ellwood Pass and down the East Fork of the San Juan River was first used in 1878 and was the only trail toward Pagosa Springs at that time.)

In Frank Hall's *History of the State of Colorado,* Vol. IV, Page 296 is a statement that at the time of writing (1890 was the publication date of Vol. IV) there were three paying mines in the Summitville District, among them the Golconda, formerly owned by Judge Bowen. In the same volume at page 298 there is a short biography of Judge Bowen, probably not important for your purpose.

Since Mineral County was not formed until 1893, wherever those mines were located in the vicinity of Summitville and on the other side of the pass, they were in Rio Grande County after it was formed in the spring of 1874. Whatever records there are of Judge Bowen's mining activities must be in that county, if they kept records then. There must be tax records, assessment records, etc.

I haven't been able yet to find anyone who has been along that East Fork trail where the ruins should be. There is much time yet before the trail would be passable. It might be that the forest ranger at Pagosa Springs would be the best source. They have to ride all of their trails once in a while. It is unfortunate that there is no ranger station between South Fork and Pagosa Springs on the Wolf Creek Pass highway. The ranger at South Fork may have ridden the trail along the East Fork of the San Juan River. The rangers move around to some extent.

I am willing for you to make use of anything I know about the San Juan Mountains. I do not feel that any credit must necessarily be assigned to me. The history of the San Juan Mountains is a hobby of mine.

Cordially yours,
May Wallace Ross

Logtown   I had never heard of Logtown until Mrs. Ross brought it to my attention through her letters. Since she grew up in Animas City, her knowledge of these two places is orientated from the Durango area. In her first letter she describes ways to reach the deserted place.

Dec. 1, 1950

I am sending a section of the map published in 1942 by the Forest Service of the U.S. Dept. of Agriculture. They have undoubtedly published a more recent one. . . . The map comes in a bulletin entitled *San Juan National Forest*. This map shows Logtown at a point close to the boundary of the primitive area. The map shows three forest ranger trails leading to within a mile of it. I have compared this map with a much larger one published in 1932 by the U.S. Geological Survey. There are three or four maps accompanying *Geological Survey Professional Paper 166,* which bears the name *Physiography and Quaternary Geology of the San Juan Mountains, Colorado*. The map entitled Plate II shows elevations. Logtown is at approximately 12,000 feet, on a fork at the source of the Florida River and also very close to a fork of Vallecito Creek.

There are three main trails which cross the 12,000-foot contour level and come close to Logtown. The shortest, only about half a dozen miles, is also the highest one. It starts from Needleton on the Animas River (Las Animas Perdidas) and goes up Arcade Creek to the 12,000-foot level after a hard climb, and stays up there the remainder of the distance. Needleton cannot be reached by highway—it is a station on the R.R. between Durango and Silverton.

The next shortest would be the trail from Tacoma. Leave the Durango-Silverton highway near Electra Lake. Take the short secondary road to Tacoma on the bank of the Animas River. From there take the 10-mile Ranger Trail to the Logtown vicinity. This is a higher trail. The elevation is 10,500 feet a short distance from Tacoma. The trail reaches 12,000 feet at the end of about the first 5 miles. It proceeds at this elevation the rest of the way to the junction point with the other trail mentioned above.

There is another trail which leaves Wolf Creek Highway at Bayfield and skirts Los Pinos River for a few miles. Follow secondary road to the U.S. Reclamation Camp near the end of Vallecito Reservoir (about 10 miles from Bayfield). Then take forest trail for about 16 miles in a northerly direction to its junction with the Florida River Ranger Trail. This trail lies between the Florida River valley and the valley of Vallecito Creek, with elevations comparable to those on the Florida River trail.

There is also a route to Logtown along the Florida River. Leave the Wolf

Creek Highway at a point about two miles east of Falfa, where the Florida River is crossed. Follow a secondary road up the river (northwest part of the time, northeast part of the time, as the river winds) to a point about 17 miles upriver, where the road ends at about the 10,500-foot level. Take the ranger trail from that point to within about half a mile of Logtown. The approximate trail distance is 10 miles.

> Elevations: 10,500 at lower end of trail
> 11,000 at end of about 2 miles
> 11,500 at end of about 2 miles more
> 12,000 at end of about next 3 miles
> Last 3 miles at or above 12,000

I have noted the progressive elevations along these trails. For you they would hold no terror. For me they would mean disaster from nosebleed and heart difficulties. The trails lead in there, undoubtedly, because they furnish a way into the Primitive Area. Why Logtown is shown on the maps I don't know. It must have been a mining town because there would have been no way to haul lumber out.

> May Wallace Ross

To find out what caused Logtown to spring up in so remote an area, May Ross wrote to "a longtime lumberman on the Florida River," and received a prompt response:

> Jan. 4, 1951

Dear May.

Gene forwarded your card asking for information about Log Town at the head of the Florida river so I will write down what I know about it. I visited the place in 1919 with my Father and he told me that the place was started by a group of Army officers from Ft. Lewis when that was an Army Post. Along in the late 1880's.

The town is on the shore of a small lake which is the head of the Florida river. It consisted of about a dozen log houses of various sizes—all were in ruins when I was there. The lake is at about 11,000 feet elevation and the prospect holes were higher on the mountain. When silver went out of fashion all the prospects and mines in that country were abandoned, as it was a 20-mile pack to the nearest wagon road.

The only living man who might know more about it is Chas. E. Herr, who owned the Pittsburg Mine on Silver Mesa in the same area. I think he is still alive and lives in San Diego, Calif.

> With best personal regards,
> John R. Nelson

Shortly after receiving his letter, Mrs. Ross sent me a copy of it. Since he mentioned Charles E. Herr as a possible source of more information, I wrote to him and received this courteous reply from Mr. Herr (Feb. 16th, 1951):

Replying to your letter of February 11th last be advised that I no longer own the Pittsburg mining property, but I recall that lots of silver ore was taken out of the shaft 140 feet deep, and it took me six weeks to pump the water out of it. Then it was a twelve mile stretch to Needleton, the R.R. point, and that was more than the mine could stand.

This was all back in 1896 and Logtown was below the south line of the mesa, not much of it left. I was told that some of the officers at Fort Lewis, Colorado, worked the mine and their help lived there.

The ore was very rich in silver and there was a small vein of gold. I only shipped a few tons and after smelter and freight charges there was not much left. The porous quartz contained shot-size silver.

May Ross' next letter reads:

I am glad to begin the new year with the loose ends about Logtown well in hand. My nephew, Joseph Wallace, was at home from Gunnison for the vacation, and I asked him to get information about the place. He was able to reach a retired forest ranger, named Tom Price, who had been in the Logtown area. Mr. Price stated that Logtown was given that name because the construction of the camp was from logs throughout, instead of sawed lumber. Even the doors were made of split poles (Puncheons) instead of sawed boards.

The location of the settlement was at the "edge of timberline," according to Mr. Price. On the adjoining Silver Mesa the Pittsburg Mine was located. There was no placer mining there. Mr. Price had obtained information of the early days there from a man named Said, who was connected with the Pittsburg mine.

Mr. Price does not think any of the original buildings would be standing now, but I think your guess would be better than his, since he retired some years ago. You have been at Beartown, which you reached from the Rio Grande trail. As the crow flies, it would be about eleven miles from Beartown to Logtown, across the summit. About the same conditions would be found at both places, as regards weathering. The Geological Survey map published in 1932 showed 5 of those square dots for Beartown and three for Logtown. Those dots are supposed to indicate buildings which are standing. The Forest service map published in 1942 shows only 3 dots for Beartown and one for Logtown.

In connection with the trails to Logtown, my nephew's maternal uncle, Glenn Dalton, who is also a retired forest ranger, sends a warning that the climb is too steep from Tacoma on the Animas River, and that if you should go to Logtown, the approach should be made from the Florida River valley or the Pinos River valley. There is a forest ranger at Bayfield on the Wolf Creek Highway who could advise you. The Forest Service map shows a trail across the summit from Beartown, to a point near Logtown. Perhaps the guide from the Rio Grande Reservoir resort could tell you whether that is a safe trail.

With the experience you have had with mountain trails you can judge for yourself. You will find, in the San Juan Mountains, that because you are a woman, men accustomed to riding the trails will discount your chances for success. There are few women riders there in comparison with California mountain areas.

Mr. Tom Price expressed a willingness to be of help in any way he could, and I do not believe there is a more reliable source of information than he would be. I hope that you will not think that I am disloyal to the area in which I grew up when I say that the Forest Service demands of its rangers more in the way of intelligence and education than is commonly found among the population. All of my family know Mr. Price. . . . I believe he would be a good source for you. He can be reached at Durango, Colorado.

Mrs. Ross' final epistle (Jan 24, 1951) includes a suggestion that I ride a horse up to Logtown, and I should have liked to have done so, but I could never fit the trip into my busy schedule.

In regard to Mr. Nelson's suggestion that Logtown was built by a group of army officers from Ft. Lewis, I somehow doubt that. Fort Lewis was originally built as a log fort near the present town of Pagosa Springs, in the summer of 1878, and garrisoned for the first time that fall. In 1880 the fort was moved to a site near the La Plata River, and was finally degarrisoned in 1891. You undoubtedly know where it was, through your visit to old La Plata City. After 1891 it was an Indian school, and is now a branch of the state agricultural college. Silver mining was never very profitable until after the passage of the Sherman silver purchase act in 1890. It was not profitable at all after the repeal of the Sherman act in 1893. Army officers were far distant from Logtown when stationed at Pagosa Springs or on the La Plata river. We will look into that.

Mr. John R. Nelson will be operating a sawmill near Wagon Wheel Gap this coming summer. . . . It would possibly be worth while to talk to him about the best way of getting to Logtown.

I passed on to you a warning from a retired ranger, Glenn Dalton, that the Tacoma trail is too steep, but I have since had a letter from Mr. Dalton's sister . . . saying that she had ridden that trail from Tacoma to Lime Mesa, which is near Silver Mesa, and that she had no difficulty. She is an expert mountain rider, but I have an idea you are too.

CHAPTER XV

The Lake City and the Creede Areas

Lake City The first time I saw Lake City I was impressed by its wide streets shaded by huge cottonwood trees and bordered by attractive frame houses, many of which dated back to the 1880's. Two-story brick and stone business buildings stood on Silver Street and on Gunnison Avenue, interspersed with log cabins and false-fronted wooden stores. But where was the lake? The person I put this question to replied rather indignantly, "Lake San Cristobal is only four miles from here. You'd better drive up and see it."

Precious metal was believed to have been found in the area in 1848 by members of Fremont's party, but their discovery attracted no attention. In 1871 J. K. Mullen and Henry Henson located the Ute-Ulay veins, and in 1874 Enos Hotchkiss uncovered rich ore near the lower end of Lake San Cristobal and called his mine the Golden Fleece. That same year a treaty was signed with the Ute Indians by which a strip of territory in the San Juan Mountains was opened for settlement, and prospectors poured in and set up small camps where the mineral deposits were most promising. Lake City, which was one of these, was laid out in 1875 and became the supply center for the entire area.

That same year Hotchkiss and Otto Mears built a toll road from Saguache in the San Luis Valley to Lake City, and this further opened up the region. The next summer Barlow and Sanderson's stages rolled over it several times a week.

As mining increased, concentrators and smelters were built close to the city to process the local ores, but a railroad was needed to ship the concentrates. In 1889 the Denver & Rio Grande constructed a branch from Sapinero, up the Lake Fork, thus avoiding steep grades all the way. In the early 1890's two passenger trains ran daily over this route, but after the silver crash of 1893 most mining stopped, and by 1937 the rails were pulled up. Since then the high trestles have disappeared, but the roadbed can still be traced in places—sometimes close to the stream and at other spots high above it on the side of a cliff.

The road to Lake City from Gunnison is now paved and winds across sagebrush hills, past Powderhorn, and through rolling open country to drop down to the valley of the Lake Fork of the Gunnison River a little north of Gateview. From there it goes directly south to the town.

Lake City has never been a ghost town, for at certain seasons it attracts many fishermen and hunters, as well as the Texans who hurry north to avoid their state's torrid heat. Its streets are still shaded, and many of its older buildings remain, although a few have been modernized past recognition. New stores, restaurants, motels, and resorts cater to the summer visitors, but in spite of these additions the quality of the place, fed by mountain streams and surrounded by wooded slopes and craggy peaks, is unchanged.

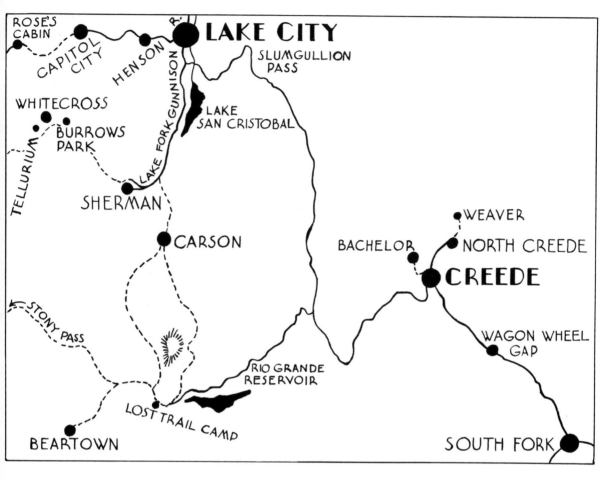

Lake City has several old churches, and to get accurate facts and dates related to their history I wrote in 1955 to Miss Jean Vickers, county clerk and recorder of Hinsdale County, whose office was in the Lake City court house. I had stayed more than once at the Vickers Ranch, and it was Purvis Vickers who finally got me up to Carson and across the Divide that I might sketch the eastern side of the camp with its numerous weathered shafthouses. She replied (Feb. 11, 1955):

> The church as pictured [*Stampede to Timberline,* p. 347] is the Baptist Church, and always was. I believe the date 1891 is correct. The rectory of the Baptist Church has been torn down for many years, and the Matterhorn Motel is located on that site. As I remember the building, your sketch is very accurate.
>
> You are correct in stating that the white church with a steeple almost opposite the small Episcopal Church is the Presbyterian Church [*Stampede to Timberline,* p. 345]. It is the oldest church on the Western Slope of Colorado.

A month later she wrote again and sent a plate with a picture of the facade of the Presbyterian Church on the front and a short history on the back (March 14, 1955):

I don't have time to get films reproduced and developed now, so am sending the next best thing, which will also give a little history of the church on the back of it that may help you. . . . I do hope you will like the plate. You are correct in stating that the manse and Church are on Fifth Street and Gunnison Ave. The manse is still standing, and is in very fine condition. The minister who was here used it, but now since he is gone, the Ladies' Aid uses it for a meeting place.

The plate is not a family heirloom, but a souvenir that the members of the Church had made to sell to tourists. It served to make money for the Church and also gave the tourists something for a souvenir. I want you to keep the plate; . . . I was glad to have something like that to send you.

I am sort of a fiend for history of Hinsdale County and Lake City, and if I can find articles about either, I keep them for my scrap book. That is the reason I like the County Clerk's job so much—I do a lot of abstracting, and when I have an abstract to make, it gives me a chance to look up old records. Needless to say, it takes me a little longer to make an abstract, as I take time to read all the deeds, powers-of-attorney, or anything pertaining to the property. My mother was born in Lake City, and has told me of different places and people, and when I find deeds, etc. with names that are familiar, I feel that I knew the people.

Anytime I can help you, please let me know, and I will be looking forward to seeing you this summer.

The story of Alferd Packer, the "man-eater," whose victims are now buried in a plot on Cannibal Plateau outside of Lake City at the foot of Slumgullion Pass, is familiar to many people. All do not know, however, that a fish fry followed this interment and was well attended. Paul H. Gantt, a former Denver attorney, authenticated the event through the following letter (April 16, 1951):

I am an attorney in the U.S. Bureau of Reclamation in Denver and in my spare time engaged in historical research. I just finished a book on Packer, the man-eater, and noticed that you report of a rumor that the dedication of the Packer-plaque was celebrated "by a fish-fry." In my research I ran across an invitation to the dedication ceremony in 1928, to which all comers were invited and as a special premium the serving of a fish-fry is held out.

The invitation to the dedication exercises of the Packer Pioneer Memorial Tablet reads, in part, as follows:

FISH FRY
A magnificent fish fry with all trimmings
will be served on the grounds for all in
attendance, by Committees appointed by the
Ladies' Union Aid Society.

Said Packer, either in heaven or in h . . . : "Bon apetite."

The Occidental Hotel in Lake City, a two-story, comfortable hostelry which I was fortunate to stay in during the late 1930's, served good meals and was filled with massive Victorian furniture. In 1945, when I planned to return to Lake City to sketch, I had hoped to stay at the Occidental and wrote to reserve a room. The proprietor replied:

Lake City, 1941. Occidental Hotel.

June 1, 1945

Dear Customer:

This is to advise you that the Occidental Cabins will be open June 1st, for our summer tourist season. They will be under the control of Mr. Clarence E. Wright and I suggest that you immediately write Mr. Wright at Lake City making your reservations early as there have been an unusually large number of reservations already made.

The Occidental burned down in December 1944 so will be unable to have hotel accommodations, however there will be a Cafe one block from the cabins for people who desire to eat their meals out. The cabins are completely furnished for light housekeeping. Cabins range from $3.00 per day to $5.00 per day with weekly rates given.

Sincerely yours,
H. B. "Jimmie" Grant

The cabins are still there, but I'm wondering what the present rates are!

My first visit to Lake City was in 1935, and one of the buildings I recall seeing was the small, false-fronted, frame office of the *Silver World,* the local newspaper, whose editor was Henry F. Lake. Ten years later, when I needed to find

Lake City, 1941. *Silver World* newspaper office.

out more about the city's past, I hoped to read portions of the newspaper file, but I had waited too long. The office was closed. Therefore, I was especially interested in a letter from Mrs. Helen L. Hubbard of Craig, Colorado, whose background of newspaper writing made her as curious about editors and their newspapers as I was about mines and mining. She wrote (Nov. 3, 1949):

Mr. Hubbard and myself had the privilege of working five years for the *News-Champion* at Gunnison while Henry F. Lake was still at the reins. He died almost two years ago.

Another topic that should be put into book form is the newspaper. My interest has always been great concerning the weekly, particularly in Colorado. On July 4th, 1947, we were on the Lake Fork on a camping and fishing trip, driving around Lake City a number of times out of curiosity. In passing the *Silver World* office we noticed the door was open a crack. I just had to see who was in the old building and made Mr. Hubbard drive me up in front. Billy Blair (now deceased) was there, looking over old files and day dreaming. Mr. Blair was the last owner and editor of the *Silver World,* and we heard much concerning the paper from Mr. Lake. He and Mr. Blair did not care for each other—reason stemming from a personal feud over Mr. Lake wanting to purchase the files. Mr. Blair would never part with them.

Imagine our surprise on seeing the forms of the last edition of *Silver World* still on the stones—with not ONE line having been thrown in! That last issue was published in 1938—I believe the date was in October. Everything was hand-set, and the Washington hand press is still in good condition. The plant is complete and a year ago had not been sold. Have heard nothing yet as to what Mrs. Blair's plans are. Incidentally, Billy Blair survived Henry Lake by one month. Blair was a clerk of some kind at the state house in Denver, following suspension of the publication of "the oldest paper on the Western Slope." . . .

The Elk Mountain Pilot was published until this spring by the *News-Champion.* It was the paper established by Phillips at Irwin about 1880-82, later being moved to Crested Butte. The plant was absorbed by *News-Champion* a good many years ago, and Mr. Lake would not drop the head-right. The Lakes sold the *News-Champion* a year ago to a stranger to the state, and many changes have been made. Believe Mr. Lake was still living at the times you visited Gunnison and had access to the files. About that time we were working on the *Daily Courier* at Alamosa.

Two letters from people who had lived in or near Lake City in its earlier days describe many incidents that bring the place alive. The first is from Anna R. (Mrs. Clyde) Jewell of Grand Junction (Jan. 31, 1950), whose parents and grandparents were in Lake City in the 1870's and 1880's. Later she and her parents lived at Gateview during the years that the railroad ran between Lake City and Sapinero.

I wish I had started writing down the stories my father used to tell about the 80's around Lake City, so that I might have had them definite in my mind.

The moment I got your book in hand, I turned to the part about Lake City, and I saw there the story of the man who prayed that the Lord would help him get his oxen across the pole bridge. I was delighted. Sixteen years ago

"Uncle Al" Nunn, quite a character himself, and Montezuma County's last surviving Civil War veteran, told me that story with huge enjoyment. He was freighting from the Gunnison Country into Lake City sometime in the late '70's. On the way home from his funeral in 1942, I told the two Dolores ministers who had conducted it that story, as "Uncle Al" had told it to me: "O Lord, if you will help me to get these blankety-blank oxen across this blankety-blank bridge, I'll never call on Thy Name again." The old minister, whose funeral sermon had consisted of telling tale after tale about Uncle Al, said he wished he'd heard that one in time to include it, for it was so like Uncle Al. (I don't mean to imply Uncle Al was the man with the balky oxen. He just told the story.)

Henson, Sherman, Capitol, Rose's Cabin, White Cross, Animas Forks, American Flats, all sound so natural. Dad's parents brought him to Lake City as a high-school-age boy in June of 1881. His older brother had come out to relatives in Lake City in '79, and he was driving stage for Barlow and Sanderson. My grandparents ran the Occidental Hotel for a short time, then went to truck-gardening at Elk Creek to supply the camps.

Grandfather had gone overland to California in '49 when he was 21, not to prospect so much as to grow vegetables to feed prospectors, and he rather fancied that recipe for getting on. It had given him a liking for new places, so he and his sons made the run into Oklahoma in '89, but Dad came back in '96 or somewhere thereabouts when things boomed again in Lake City. Later Dad went back to Oklahoma to bring a bride back to Lake City where they lived the first year of their marriage. My mother was fascinated with Lake City, and after they moved down to the "Gateview Ranch" (where the road from Powderhorn leaves Indian Creek and heads up the Lake Fork), she always spent a week in Lake City once or twice a year at least. My memories of Lake City are confined to those visits since the folks left "Lake" before I was born. However, "Lake" was "town," so far as we were concerned, and very handy since the train made a regular stop in our field and left supplies, passengers, and picked up mail. It was as convenient as the most obliging bus, and the river residents simply "hopped on" and went to town for errands at will.

While living at Gateview, we "put up stoppers" officially. Gateview had been the stage station before the railroad was completed, and it was the only house along there large enough for "putting up stoppers" as a business. Whoever lived there was simply elected. People headed for Cebolla Springs to "take the baths" and people headed for Powderhorn or away from either got on the train at the Gateview stop in our field. And usually they had a meal at our house in the process. People going through other than on the train usually stayed over the night at our house. I found it very fascinating. The "stoppers" were from simply everywhere. I recall two English and two Scotch gentlemen with mining interests in Lake City who stopped twice on the way to and from Cebolla Springs to "take a *bawth* in a mudhole!" Also a former Lake City man who spent some years in the Klondyke. A repeated "stopper" had been in the Orient a lot with the French Navy. A German came often who had spent much time in the Near East. Many other people were full of reminiscences of the mining camps of Gunnison and Hinsdale Counties.

When the railroad to Lake City had started operating, a nearly new Concord stagecoach had been left at the Gateview stage station, and by the time

I was large enough to take an interest, it was much the worse for the weather. I played "train" and "house" in it though, and enjoyed it to the hilt. Gateview also had an *orchard* planted on it, believe it or not. A cherry tree was left when we lived there, but up in the "loft" over the diningroom was stored the cider press brought to the country by the planter of the orchard. I do not know whether it was the Mendenhalls who planted the orchard, or a predecessor. The place had been known as Allen and Barnum before being called Gateview, and it had had a postoffice, the window and boxes still remaining as I recall it. There had also been a school there until the local children grew up.

You mentioned a Davison who had a sort of hotel at American Flats, and I have a suspicion he was my teacher's father or uncle. And concerning American Flats, my father always said it was one of the most beautiful places he ever saw, . . . or so he believed. When he came to Colorado, he was too young for a man's job except in a pinch. But, as the baby of the family, he had been taught to cook and keep house very efficiently, and he got plenty of work along those lines until older. As cook's assistant, he worked at American Flats one summer when all the flowers were out. The Cornish miners wanted a "Jenny" instead of an American man cook and were so cantakerous that the cook quelled 'em with a cleaver. But it would have been unsafe there for him and Dad after that. So they tidied up the cook shack, packed their roll, and walked all night in bright moonlight headed for Lake City. Dad said it was supremely beautiful, and he never forgot that night walk.

Kenneth N. Ogle of Rochester, Minnesota, was born in Lake City and although he writes that his "rambling" will be of no interest, I disagree and share his letter with you (Nov. 28, 1952):

I was born in Lake City, Thanksgiving Day, 1902. The Dr. Cummings you mentioned brought me into the world. I heard he is still living in Gunnison.

My father was postmaster in Lake City until 1909, at which time we moved to Colorado Springs. . . . I have a brother buried in the Odd Fellows' cemetery above Lake City. The Baptist Church in your sketch was where I started Sunday School.

This early summer we visited the place again for the first time in seventeen years. The little white house in front of which are the huge elms on the main street to the lake is my birthplace and is still there and in excellent shape. Father planted those trees! The woman who owns the house now, Mrs. Mabel Rawsom (the County Clerk) bought it from my father when we left Lake City. She is a wonderful person, and at one time lived in Capitol City. Her daughter Celia (Mrs. Joel Swank) took us on a jeep trip beyond White Cross and up to where the snows blocked our way. Many times I have heard my father speak of Sherman and White Cross, but never before had I seen them. The scenery is magnificent and I want to go back over the range to Engineer Mountain and back via Henson Creek. The Ute-Ulay mine is busy, and there are many miners about the place. Two I met on Main Street in Lake City who were working an old mine in Carson. Of course, the Texans are taking over the place and there are modern hotels there now. Electricity and modern plumbing. The old brick school where I attended first grade was condemned and has been torn down—a cement block school has been put in its place.

Although I was small when I lived there I remember many of the things that would impress a small boy: the celebration on Silver Street on Washington's birthday—the greased pole and greased pig—miners everywhere—the narrow gauge railroad over Marshall Pass and from Sapinero to Lake City—the big pot-bellied stove in the station at Sapinero—my father's taking me to the station to see the train come in—the time coming over Marshall Pass when the snow stopped us and the coaches were uncoupled on top while the engine went ahead to force a path through the snow. My mother worried for fear the coach would start rolling down from the pass. The men trying to keep the coach warm by the little stove at the end of the car—the curfew and fire bell which was near our house. The ringing of that bell at night struck terror in the hearts of all. The first movies (held in the hall just in back of the bank building in your sketch)—staying in the Occidental Hotel the night before we left Lake City for good—etc.

I never could understand why my father went to Lake City (from Iowa where he was well established in business)—gold fever, I guess. More than that I cannot understand how he persuaded my mother to go to that rough and wild place, for she was an anxious person and a great one to worry. It was only the accidental death of my sixteen year old brother that turned the tide and my mother won out.

Well, this rambling is of no interest to you, but your interesting book has brought it all back to me.

Henson Creek flows into the Lake Fork at Lake City, having first cut its way through the silver-gray cliffs at its mouth. A road which flanks the stream climbs west for twelve miles to the ghost of Capitol City, passing on its way the old camp of Henson where the Hidden Treasure, Ute-Ulay, and Ocean Wave mining properties cover much of the mountainside.

Henson, 1942. Flume to Ute-Ulay mine. Schoolhouse on lake shore. Road to Capitol City on right.

Capitol City

Capitol City lies on a swampy, willow-grown flat surrounded by high peaks. It was a new camp in 1877, and its 200-acre townsite included a sawmill, a post office, a school, and about thirty cabins. When silver was found close to the surface of the ground, Capitol's future seemed secure and the camp overbuilt itself.

George T. Lee erected a smelting works a mile below the town on Henson Creek and a planing mill within the city limits. He also built his large home there using bricks, which were the most expensive material he could buy, since they had to be packed across the mountains on burros. The place was always spoken of as the Lee Mansion.

The small log schoolhouse was soon replaced with a large two-room frame building painted white, which cost $1,511.00. During the nineties, the population reached nearly 700, and when high-grade gold and copper were found in 1900 a second boom brought a rush to the area.

On several trips through the deserted city, I saw the schoolhouse and the Lee home and watched their steady deterioration.

Mary K. Mott of Long Beach, California, wrote down some of her memories of Lake City when she lived there as a child, and especially reminiscences about Capitol City:

Oct. 28, 1950

I was a little girl of 11 when I lived in Lake City and some way my heart and memory will live over the past, I think on account of my pioneer family who now sleep among the rocks and singing pines.

When I want to see a *ghost* I look at the picture of the white schoolhouse in Capitol City you drew. I was the first teacher who occupied it and we all felt so grand and important. About ten or twelve pupils but Capitol was quite thriving at that time—miners some 200, working near by—a saloon or two, grocery store and P.O.

I love Henson Creek and the days of romancing—the wonderful horseback parties and in my childhood the trips to the Ute and Ule mines when mining activities were carried on in a big way—John J. Crooke & Bros. were the owners at that time. But like "Mr. Phiney's turnip" I grew and grew and one day I became a teacher in the Lake City school—afterwards Montrose, then Mr. Mott took me off the scene.

We lived in Telluride 20 years and saw it in its zenith. If I were able I would love to explore with you and live it over, but everyone tells me I would be disappointed. Anyway the clouds, the lakes, the mts. are there. . . . Lake City has an airport now and they tell me Texas has taken over. . . . I expect a big mining boom there one of these days.

Nov. 14, 1950

I feel like chatting about the days that used to be. The Spirit is willing but the flesh is somewhat weak especially as to eyesight. . . .

I will start with *Capital City* which I think was the post mark and the way we spelled it. I know I have an autographed album of '83 and the children and others wrote *Capital* city. I cannot imagine the place ever being thought of for the Capitol of the State away up there most to timberline and with a wide valley at Lake City.

I personally do not believe the Lee Mansion was ever built for a Capitol building. It was simply one of the many ventures of mining promoters with

Capitol City, 1942. Schoolhouse.

eastern money. Those were the gay days in Capital, about two or three years before my time—One of my older girl friends was a teacher there however and she surely had a new experience of high life.

You asked me if there were any children in the Lee family. One girl, Rosa. She was a chum of one of my pupils. They had ponies and rode a lot. The Lees had left before I taught there but the beautiful home was there with its conservatory and porches and walks. It afterwards became a disgrace to the county with its obscene words and sentences and in time was practically stolen away by small lots.

I taught in a little log schoolhouse not far from there. I had about 10 pupils. Afterwards the big *white* schoolhouse was built. It was built for two rooms but not separated so rather resounding with so few pupils. We used to catch butterflies for sport. I never saw such large gorgeous beauties. I pressed so many and mounted them. So cruel! We had a visitation of millermoths one summer and my brother said they were spirits of the murdered butterflies haunting me—may be so.

There were some very nice homes there (a few only) but deserted when the town went down. We used to dance in the boarding house, someone playing violin or accordion or both for special occasions. I had the time of my life for I was not allowed to dance at home—my father thought it was the last step to hades.

The storekeeper then was a bachelor and so full of fun even though he had had a serious love affair with one of the young ladies. I thought he was the finest ever. He kept me in a constant giggle—I used to say I laugh because you are such a fool. He wrote in my album in '84:

> "If sometime in the sweet hereafter,
> You should chance me to forget,
> Turn to this page with joy for laughter
> And wonder if the old fool is living yet."

Well, he is not, and his life had a sad ending.

I must tell you of the luscious wild red raspberries that grew among the burned timber—folks put up gallons of them. Across from the Lee House on the hillside was a *wonderful patch* I shall always remember. The Indians and bears must have had many a feast. "Poor people have poor ways" so in lack of containers there was a neat trick of putting a heated ring over a beer bottle (of which there were many) then a gentle tap would break the neck off smooth and leave a glass. The jam was put in and white paper sealed over the top.

I'm told all those houses have been carried away even the "white" schoolhouse. They evidently are rebuilding again. New blood, youth, vision and money can do wonders but this will *never* be the Capital City that I know.

P.S. If you could get a copy of the *Sat. Eve. Post,* 1948, June perhaps, an article written by Jim Marshall called *Mountain Mutton,* it is a most graphic description of the Capital-Henson Creek, Lake City country. I'm sure the library would have it.

About 6 miles from Capital was a Post Office General Store. This was at the top of the range, owned by a man named Rose. It later became the property of one Charley Schafer. It was known as Rose's Cabin. Schafer used to freight from Lake City up beautiful Henson Creek. His outfit consisted of oxen mostly. After acquiring the store he disposed of the freight outfit.

When I with my 2nd Grade certificate taught school in Capital, I was so *green* I do not know why the cows did not eat me, they would perhaps but there was only one cow in the burg—and she was something on account of the milk she supplied. She was owned by a nice little lady, Mrs. Gregory by name. Her husband was a miner and Mrs. G. took the cow to the nice green pasture each morning and went for her in the evening.

One evening she came back and said she couldn't get the cow—that she seemed to be with a *buffalo.* This was a joke but she didn't think so. The next day she brought the cow home and a few evenings later she said the buffalo was there and struck off into the timber—more laughs, etc.—some of the men folks about town looked around a little and saw nothing—but in a few days there on the side of the hill was the cow and buffalo feeding—word was sent to Rose's Cabin and Charley Schafer and E. L. Schafer came down and they decided to shoot the beast. Mr. S. stood behind a large rock and made a weird noise, the animal looked up in perfect position for the bullet which felled him. I never knew a cow could make as much noise as that one did. She pawed the earth, bellowed and ran around crazed. It was found that the buffalo was an ox who in unheard of manner weathered several winters. Evidently the hair had been frozen from his back and left his hide a thick blue-black skin but around his horns tufts of wooly hair had grown and he did indeed look like a buffalo. He was as fat as butter. They butchered him where he fell and gave generous portions to neighbors and friends.

Like all love affairs, the cow soon quieted down and wended her way morning and evening as usual much to the delight of the customers. I felt sorry for the lonely cow. *Every feller needs a friend.*

Mary K. Mott, Pioneer

When Vernon K. Hurd of Chicago visited the camp in 1952, he found few buildings left, and the brick home that he mentions, the Lee Mansion, showed the ravages resulting from both neglect and vandalism (May 24, 1952):

Last summer I was in the Lake City area climbing in the mountains in that vicinity and passed through Capitol City. You may be interested to learn that the brick building at the east end of town is the only structure left and it is extremely dilapidated and the walls are crumbling. There is, at the northwest edge of town, the remains of a small mill or perhaps a store building of some sort, with some rusting machinery still inside, and a structure that is little more than a leanto nearby. Other than that, however, even less of the community remains than when you were there.

In the 1960's I was driven through Capitol City on a jeep trip, and when I failed to see the white schoolhouse which had been one of the largest buildings, we stopped and searched for its site. I did find a broken down rectangle of rock that had been its foundations. As we continued down the road, I was on the lookout to see what the present state of the Lee Mansion might be. The last time I'd seen it, nearly half of it had been torn down for the bricks that were in it. On this trip not

Capitol City. Lee Mansion, 1941 and 1959.

a trace of the house remained; and yet our driver, who had taken another group of tourists over the same route just four days earlier, had seen a section of it still standing and piles of bricks ready to be carted off. Today the main activity in the immediate Capitol City area is logging.

Rose's Cabin   Rose's Cabin was a haven for those crossing the range above Capitol City. The Cabin was a one-story cabin built of square-hewn logs by Corydon Rose, an early pioneer in the San Juan Mountains. As men began to struggle across the rugged terrain, his cabin, which stood at 11,200 feet above sea level, became a stopping place. After a stage road was built across the mountains and passed in front of his place, he enlarged it into a hotel. Neil Foreman wrote from the Gunnison area:

> I have not been to Rose's Cabin since 1919, but we had it fixed up and were living in it at that time. There are several good mines in that district that have good quantities of minerals and produce sometime.
>
> Our home was in Capital City in 1914; we lived in a log cabin before you cross the creek into Capital City. There was a store and post office and about four families living there at that time. Clint Buskirk had a stable of fine large horses there and hauled ore for most of the mines around Capital City. He lived there for many years.

Carson   Carson is the only mining camp I know that straddles the Continental Divide. It is twelve miles from Lake City, out beyond Lake San Cristobal. The last four miles are a steady climb up Wager Gulch nearly to timberline, and there the Pacific slope portion of the camp is situated. A further climb of a mile to the summit of the 11,500-foot pass leads to the head of East Lost Trail Creek on the Atlantic slope, where the best mining properties were located.

Christopher J. Carson was the first person to discover traces of minerals which, when assayed, ran high in gold and silver. He staked the Bonanza King and other claims. This led to the organization of a mining district in 1881, which included the head of Wager Gulch on the north and of Lost Trail on the south. Named for its discoverer, it became an active camp throughout the eighties and nineties. The first road, built in 1887 up Lost Trail Gulch, provided access to the camp, which boomed up until the silver crash of 1893. After 1896, however, and into the early 1900's, gold properties were developed, with most of the mining being done on the Lake City side of the Divide. Never a large camp, its population rose to only 500. Of its mines, the Maid of Carson and the St. Jacob's were the best known, with the latter producing the best ore.

The Wager Gulch side of Carson still contains quite a group of buildings, while the Lost Trail side has large but ruined shafthouses widely spread out over the steep slope that descends to the Rio Grande valley.

W. H. Cochran, the mining engineer whom I interviewed in Del Norte after I had been to Carson, wrote me later as follows (Sept. 23, 1949):

> Lost Trail and Carson City are one and the same. Your trip from Beartown to Carson must have been a most interesting one. I have never been over that Pole Mt. high trail.
>
> Lost Trail is a deserted camp south of Lake City. Eleven big properties are in one gulch, all with mills, smelting plants, etc, boarding houses (there 4-5 years ago). Have to pack in—above timberline. Also reached out of Creede. Lead ore. Lots of mills, wild-catted to death.

Rose's Cabin, late 1930's.

Carson, 1950. St. Jacob's mine. The camp straddles the Continental Divide above 11,000 feet.

A more recent report on Wager Creek road to Carson came from Kenneth A. Phillips of Amarillo, Texas (Aug. 18, 1964):

> I made a journey to Carson last May. . . . The old bridge you spoke about across Wager Creek was still there, although the road had been extended up beyond the bridge about twenty feet where it forded the stream. The same boulders remained in the roadway just where you described them. The ledge just above the bridge is still there, only narrower. The log section is still there but the beaver dams have disappeared.
>
> Most of the buildings in Carson are still there, only leaning a bit for the worse. We were unable to go up to the mines as they were under twenty feet of snow and ice.

Sherman   On one visit to Lake City I talked with a Mr. Ramsay who knew about both Carson and Sherman. At Carson he had built a smelter for the St. Jacob's mine and, as he put it, "I got the smelter up and then the lease ran out, so it was brought down to Lake City. The St. Jacob's mine produced both gold and silver ore—it was the best gold property up there."

Mr. Ramsay had also worked at Sherman, the camp at the junction of Cottonwood Creek and the Lake Fork; it was flooded out by a cloudburst. In 1906 he was commissioned to build a 147-foot high dam above Sherman. By the time he had raised it to the height of sixty-nine feet, the company went broke. Another outfit agreed to complete the dam. Just after it was finished, a cloudburst took it out and cut a new channel through the town, which left cabins on "islands" in the streambed and obliterated streets and many buildings.

Sherman, 1941. Houses in streambed.

He spoke of other mining camps and individual properties above Sherman, such as the George Washington on Cataract Gulch which produced high grade gold and silver ore; of Sterling, seven miles up Cottonwood Gulch; of Gardiner City, three to four miles above Sherman where work was going on as late as 1920-1921; and of some older mines up Cuba Gulch that were active in the 1880's.

In an endeavor to pinpoint these forgotten places, I wrote in 1948 to the county clerk and recorder of Hinsdale County and received the following information:

Sept. 10, 1948

Dear Mrs. Wolle:

Tellurium was about ¼ mi. above White Cross and below the Tobasco Mill. White Cross is above Sherman. There was a camp called Garden City (maybe Gardiner City), and Sterling was above White Cross. There were about 34 houses there at one time. Colorado Historical Society in Denver has a good history of this section, which was sent in by C. E. Wright of Lake City. Mr. Wright lived at White Cross when he was a boy and seems to know more of the location than any one else around here. Gardiner City was up Cottonwood or Cataract Gulch and not toward W. C. but above Sherman, south of the White Cross road. Hope I have made this clear.

Yours truly,
Mabel B. Rawson

The fifty miles from Lake City to Creede are through scenic but lonely country. *San Juan City* It is most spectacular when the aspens have turned and the sun shines through the shimmer of trembling leaves. State Highway 149 climbs steadily from the time it leaves Lake City's southern edge and the valley of the Lake Fork to the top of Slumgullion Pass, where a lookout offers a fine panorama of the San Juan Range and of Lake San Cristobal below. The road continues through miles of forest land and high meadows until it crosses the Continental Divide at Spring Creek Pass. About fifteen miles south of the Pass the highway meets the Rio Grande River and winds through open ranch country past fishing camps and pioneer ranches all the way to Creede. If you turn right from the point where the Lake City-Creede road joins the river, you will pass the Rio Grande Reservoir and continue to the river's source; and also cross Lost Trail Creek.

San Juan City, an early settlement in this area which I had been unable to find, was located for me through a letter from E. L. Bennett, formerly of Creede and author of *Boom Town Boy,* who mentions Carson and Beartown also (June 16, 1969):

I wonder if you have found the answer to your question, "Where is San Juan City?" Surely, someone wrote to tell you you went through its remains when you went to Carroll Wetherill's camp on Lost Trail Creek as a starting point for your trip to Beartown and Carson. On page 119 of their COLORADO'S CENTURY OF CITIES, the Griswolds have a mention of San Juan City and there is a picture of one of its cabins on page 103. When I first knew the place, in the neighborhood of 1900, there were several cabins similar to that one mentioned above and it is possible that a photograph of the early San Juan Ranch would show some of them. The place is now a combination cow ranch and dude-cabin area.

Am sending you a picture of the two cabins that I believe were part of San Juan City. I am sure the date, where it is 195, should read 1895 as it was in that year that my uncle, Dr. H. D. Newton, D.D.S., from Salida, took a lot of pictures in the Creede area and I am sure that it was he who shows against the wagon cover and it is his son, Clem, behind the wagon. In BOOM TOWN

San Juan City, 1895. Ranch in Antelope Park. *Courtesy Edwin L. Bennett.*

BOY there is a picture on page 105, of which we have an original print and it is labeled in the same handwriting as the San Juan and it is dated 1895. He also took the picture called Bob Ford's Bar on page 106 and the white patch on the side of the bar is one of his ads. [Fred] Mazzulla called my attention to that.

The picture in COLORADO'S CENTURY OF CITIES, of the San Juan City cabin was taken a little way northwest of these two cabins. When my brother and I were hauling hay, about 1900, this long building was still being used as a bunk house and grain storage space. I believe the Bents lived in the other house.

No one who reads paragraph three on page 336 [*Stampede to Timberline*] can doubt that you went to Beartown. The stock grazing on Grassy Hill could change from blackface sheep to whiteface cattle and the once populated towns of Junction City and Beartown could settle into the ground but the horseflies and deer flies would carry on so industriously they would win a place in every history of the upper Rio Grande.

Between the time I left the Pyramid District of the Rio Grande Forest in 1922 and the time you were there, Beartown changed a lot. If there was a Sylvanite Mine when I was there I never heard of it—practically nothing had been done since the 1890's. I heard when I went to Creede in the 1960's that there had been a renewal of prospecting but with small reward. In the proper season that mountain west of Beartown used to be covered with Colorado Columbines. I estimated over 100 acres of them when I saw them.

If there was a fire tool box on the little hill slope just below the junction of the two creeks, it was from there I saw my wife catch a big fish in lower Bear Creek, one I had hooked and lost twice. Was that frustration!!

In Beartown, I knew the little German and his English partner who made and drank the most worst alcohol drink I ever met up with and played sweet music on a little tin whistle he made, himself. His drink finally killed him, I believe.

Although I rode in the vicinity of Carson, first as a Forest Guard and later as a Ranger, for five summers between 1915 and 1922, I did not know that Carson had ever shipped any ore until I read about it in your book. On a trip to the part of town on the Gunnison slope I found several places where electric detonators had been strewn around and I got out of there pronto. Maybe they were duds but I didn't want to find out by way of a three-legged horse.

I am at a loss to know why the road was 500 feet higher than Stony Pass. I did not notice that when I was riding that country and I was on the pass several times. Made one trip down to Silverton around 1916 or possibly 1919. Whatever you know about Carson and other places there is that much more than I do. I just took them for places that once were and were no more and let it go at that. My interest in the high country was sheep and their herders.

In your drawing, CREEDE CLIFFS, on page 327 of STAMPEDE TO TIMBERLINE, the sign on the side of the Hord store ends in "***GHTING" where it should be "***GHTNER", the end of "Hord and Lightner."

There is only one woman who has reliable information about that country and she refuses to write it. She still lives near the ranch where I believe she was born.

Creede   The approach to Creede is dramatic. After miles of open country, a wall of high, reddish cliffs, dotted with trees, makes a formidable background to the town which sprawls on the flat in front of them. Willow Creek has cut a gap in the cliffs which guard the canyon entrance and rushes through the town to join the Rio

Grande River beyond it.

Nicholas C. Creede, for whom the camp is named, and his partner, George L. Smith, wandered in 1889 through the narrow rocky gateway to prospect upstream. They worked their way to the junction of East and West Willow Creeks and soon found float, which Creede traced to its source at the head of West Willow. After staking their location and working it all summer, they left the mine, which Creede had named the Holy Moses, and went out of the area for the winter.

North Creede, c. 1890. Main Street. *Courtesy Denver Public Library, Western History Department.*

When his discovery was known, prospectors waited impatiently for the spring of 1890 to let them stampede to the new diggings. In a very short time tents, shacks, houses, and stores lined the banks of the Willows, and as one man put it, "for six miles up and down the Creek there's not a foot that has not been staked." In the meantime Creede had uncovered another mine, the Amethyst, which paid from grassroots and from which a million and a half was realized.

The banks of Willow Creek could not hold all the buildings needed to accommodate the constantly growing population, so new camps sprang up both above and below the original townsite on East Willow. The second town lay on the flat (where the present community is) and was called at different times Jimtown, Creedmoor, and Amethyst. The third was Upper Creede or North Creede and was a continuation up the gulch on East Willow beyond the original location.

By October 1891 a branch of the Denver & Rio Grande railroad was completed to Creede, and to relieve hotel shortages the pullman sleeping cars were used as housing for the passengers.

The men in Creede were not all miners. Bad men and their gangs frequently dominated the pulsing city that was described by Cy Warman, editor of the *Creede Chronicle,* in this way:

> Here's a land where all are equal
> Of high or lowly birth—
> A land where men make millions
> Dug from the dreary earth;
> Here meek and mild-eyed burros
> On mineral mountains feed—
> It's day all day in the day-time
> And there is no night in Creede.

"Soapy Smith" and his followers were among those who ruled the camp. Bob Ford, who shot Jesse James, was also well known in Creede. Matthew Oblock, Jr. of Aspen writes of Ford's death in June 1892:

> I believe that you stated that Bobby Ford was killed at Creede by Ed O'Kelly in a drunken saloon brawl. According to Collier and Westrate in their book, "The Reign of Soapy Smith," O'Kelly was part of the Jesse James gang and had followed Ford to Creede for the express purpose of killing him for revenge. Soapy Smith guarded O'Kelly and kept him from being lynched.

Clarence Reckmeyer of Black Hawk, that avid seeker into Colorado's mining history, sent me the following letter (Oct. 5, 1953) and a photostat which raises a question concerning N. C. Creede's real name. Mr. Reckmeyer has since died, and I feel sure he received no further clarification of this mystery.

> I enclose a photostat of a letter from the late Luther H. North of Company A of the Pawnee Indian Scouts, of which his brother was commander. The Pawnee Scouts were peacable Indians who were enlisted in the U.S. Army to fight the hostile Indians. Until I moved out here in 1942 my home was in Fremont, Nebraska, and Luther North's home was in Columbus, while the Pawnee Reservation was what is now Nance County, Nebraska, until 1876, when they were moved to Indian Territory, now Oklahoma. From about 1925 until the time of his death we were very intimate and traveled the western trails up into Wyoming, but never came down into Colorado further than the site of the Battle of Summit Springs, 12 miles south and 5 miles east of Sterling. Luther passed away in 1935. He was a VERY CONSCIENTIOUS old gentleman and never made any statements that were not true. My home was at his home whenever I was in Columbus. The enclosed photostat of a letter was written to me when I was in Manitou, Colorado, one summer. Knowing Luther, so well, his letter to me is the gospel truth.

North Creede, 1959. Commodore mine.

Columbus, Nebraska
Feb. 24, 1931

Dear Mr. Reckmeyer,

Your letter just came, and I will answer it at once. The Officer I spoke of was known to us for many years as Billy Harvey. He came here in the early sixties and was with the Scouts in 1867, 1868-1869 if my memory serves me right. I think his people live in Iowa and his name was Nick Creede but for some reason when he came out to Columbus he called himself Harvey. He was a fine fellow and made a good Officer of Scouts. After he was with us he was employed by my brother, J. E. North to take charge of a trading Post and was with him until about 1873 when he went to Colorado and prospected for several years and finally struck it rich at what was afterward Creede, Colorado. Before he found the mine at Creede he wrote letters to Frank in which he said his name was Creede. It was said that he sold his mining interests to David Moffat for one and quarter million dollars. He went to California and sometime in perhaps 1896 he wrote to George Lehman (also an Officer of the Pawnee Scouts) that he was going to Washington to see the McKinley Inauguration and that he was coming through here and would stop off. I think it was not long after that he was found dead in his Rose Garden. He was never married and I had always supposed that he was quite wealthy. I never saw him after he left here but J. E. North was in Creede sometime in the nineties and saw him.

When we were in Omaha a couple of years ago you remember I rode in the car with Poker Alice. She started to tell me something about Creede but never finished. She knew him pretty well I guess.

Sincerely,
L. H. North

I cannot say whether Creede's right name was Harvey, but four years ago I secured the address of Charles S. Goodaker of Ouray, Colorado, who lived in Creede in the early days and wrote asking him if he knew if Creede's right name was Harvey and here is what he wrote me:

Ouray, Colorado, April 26, '50

Dear sir:

In answer to your letter of April 18, 1950, will say that at the time I was in Creede I was rather young, 17. Was working in a livery barn at the time.

And Creede kept two horses there that he and a man named Harvey used to be with most of the time and Harvey used to come more often to get the horses than Creede did. Nick Creede is the only name I ever heard him called.

Bill Harvey used to pay the bill as often as Creede did. They looked very much alike and I hardly ever saw one of them without the other.

Otherwise it is about all that I knew about them but they looked very much alike about the same size and dressed alike.

Truely yours,
Chas. Goodaker

CHAPTER XVI

The Red Mountain Mines, Silverton, and Eureka

The Red Mountain Mines

The snow-capped Uncompah-gres form a rugged barrier between Ouray and the rest of the San Juan country. Driving toward them from Montrose, their jagged peaks suggest no simple progress through them. A red-walled canyon that leads directly into Ouray seems to end against the rocky face of Mt. Abrams, miles to the south, but a gap in the mountains, cut deep by the Uncompahgre River, provides the needed break.

Otto Mears, the road builder of the San Juan, refused to be stopped by a rocky im-

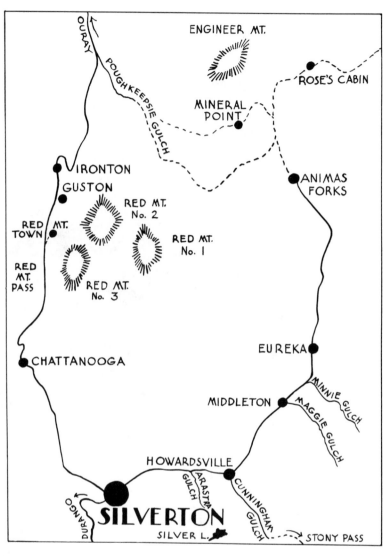

passe and literally hewed the first toll road from the solid rock walls hundreds of feet above the river. His vision and achievement made possible the present paved Million Dollar Highway, which in large part is built upon the original roadbed.

The climb to the top of Red Mountain Pass is steady and the scenery is stupendous. After several miles of ledge road, a series of switchbacks leads up to a wide, marshy meadow known as Ironton Park, through which runs an orange-stained stream; but the breath-taking sight is the Red Mountains, whose crests range from orange to crimson and whose sides are honey-combed with mines.

Ore deposits were found as early as 1881 when John Robinson's and others' discoveries created the Guston mine, a silver-lead property which turned out to be one of the best in the region. Within two years three camps sprang up—Ironton, Guston, and Red Mountain Town, each only a mile or two from each other. Before the present highway was built through Ironton Park, the main street of Ironton *was* the highway, passing between empty houses and a few stores before climbing to the

Ironton, Guston, and Red Mountain Town

271

top of the Pass. Narrow dirt roads angled away from the highway, skirting mine dumps before reaching Guston with its homes and mines. Red Mountain Town is not visible from the highway, and for years I failed to locate it. Finally I obtained explicit directions in Silverton and tried again. Just before reaching the top of Red Mountain Pass from the Ouray side, a narrow, rather indistinct side trail on the left disappears around a rocky knoll. Within a quarter of a mile this trail rounds a curve and there, in a protected hollow, lies what is left of Red Mountain Town. Its most conspicuous feature is a big dump crowned with the shafthouse of the National Bell mine. Almost all the buildings in the townsite have disappeared or fallen down. The dirt trail that leads to this spot is the roadbed of Otto Mears' 20-mile narrow-gauge Silverton Railroad, which reached the Red Mountain mines in 1888 and ran through the town on down to Ironton.

Red Mountain Town, 1947.
National Bell shafthouse on dump.

Ironton, 1942.
Coffee mill
abandoned in store.

Guston mine, 1941.

When I needed to ask more questions about these mountain camps, I wrote to N. C. Maxwell of Silverton who had been so helpful on previous visits in supplying me with accurate and valuable information. He replied promptly (Sept. 14, 1948):

As far as I am informed there is one Red Mountain town. The National Bell mine is adjacent; this is the stage stop and eating place. I have known of no other place in the 43 years of active contact. There are three mountains called Red Mountain, Nos. 1, 2, 3. This fact may give rise to the other idea of three towns. There were people living around the mines and in an almost continuous string from Red Mountain to Ironton.

Congress Hill has the Congress, San Antonio and other mines. There was never more than accommodations for men working the mines near or on Congress Hill.

The Genesee-Vanderbilt was not associated with the Guston, a separately operated undertaking and yet being prospected.

Ironton was incorporated, so I have been told, and an issue of bonds for water was sold. I imagine these are yet in the hands of heirs of the original buyers.

Another informant, Marshall D. Smith of East Orange, New Jersey, wrote (Sept. 16, 1944):

As nearly as I can recall, Red Mountain was booming during the later 'eighties. It was about that time that Otto Mears built the Silverton and Red Mountain R.R., from Silverton to Red Mountain.

There was one fabulously rich mine operating at that time, and it seems to me it was partly in response to its great need for transportation facilities that Mr. Mears built the road. Early shipments of ore—to the Argo Smelter at Denver—brought as high as $25.00 a ton. And it seems to me that armed guards accompanied the ore trains. The Denver and Rio Grande R.R. at that time was a narrow gauge line all the way to Denver.

Red Mountain Town, like almost all the mining camps, suffered from disastrous fires. Its worst blaze was in 1892. Julia Benton of Denver lived in Red Mountain Town as well as in Guston and writes her recollections of both places:

Nov. 4, 1954

I have just been reading your book *Stampede to Timberline,* and since I was living in the town of Red Mountain when it burned down in 1892, I thought I would write you. I was six years old at that time. The school was a one room one teacher affair, and the teacher a Miss Bergwinn.

My father, Louis Napoleon Ressouches and his brother Theodore, who were born in Paris, France, lived there with their families and worked, I think, in "The Yankee Girl Mine." Their mother, Mrs. Laurent Ressouches ran a boarding house at the White Cloud mine. My mother did dressmaking in Red Mountain.

The day before the fire I came home from the grocery and told my mother that I had seen two men pouring coal oil on the floors of a vacant hotel, I had passed on my way home. Since I was very young no one paid any attention to what I said, but the fire did start in that hotel, and it was thought later that the owners started the fire in order to collect the insurance.

My mother said that when she woke we children and told us to get dressed, I was very self-possessed. Dressed hurriedly, and went out to find our cat.

Our next door neighbors a Mr. and Mrs. Quigley quarreled violently during the fire, trying to decide whether to carry out the stove or the bed first. Meanwhile the house burned down. Another friend of ours, a young bride, Mrs. Haines, roomed with her husband at one of the hotels, and threw her set of Haviland china out of an upstairs window in her excitement.

One of the saloons was operated by a Mr. and Mrs. Fay, who had a son Charley. After the fire we moved to Guston for a short time, and from there to Silverton.

Nov. 11, 1954

My memory of Guston was that there was just a small cluster of houses, when we lived there. Supplies were brought in by burro trains. Our next door neighbors were a Mr. and Mrs. Leslie, with two grown sons, who took we children riding on their snow shoes.

We attended Sunday School in the little church where the attendance was so small that one teacher, an older man, was in sole charge.

We moved there in the summer, and the tops of some of the trees had been cut off at snow line, several feet above the ground. My mother who came to Colorado from Indiana asked about the trees and when it was explained to her, thought she was being teased, because she was a tenderfoot.

The houses and most of the business buildings were made of logs. A few were frame. I remember a sign on a doctor's window read: Dr. Pigg from Creede, which was considered very funny by the children of the town. The first Chinese I ever saw ran a laundry in Red Mountain.

Sincerely yours,

Julia Benton

Guston, c. 1895. Congregational Church at extreme left. *Photo from Marvin Gregory Collection, Ouray.*

In 1941 while sketching in the Red Mountain area I noticed a small frame church with a cupola, surrounded by aspens on a low hill not far from the Million Dollar Highway. Hiking over to it I found its weathered door ajar and went inside. Glass and debris littered the floor, long rafters broken loose from the roof hung down into the building, and the only furnishings were a few pews and a bench. Four years later I sketched it again, although by then the entire structure was leaning far to one side and continuing to sag. In 1947 it was gone. What was its story?

Guston, 1945. Congregational Church.

N. C. Maxwell offered one clue (Sept. 14, 1948):

> The church in question was near the Guston mine, and probably supported by people of Ironton and some from Red Mountain.
>
> Never in my time have I heard of more than one church in that part of the district. I have heard the story that some of the miners of the neighborhood thought a bell necessary; others added a whistle would be just the proper added aid to a call to worship. Have been told many men of Cornish ancestry were in that country and were the chief support of the church.

Walter S. Hopkins of Denver, who was compiling the history of Congregational churches in Colorado, also wrote me (July 4, 1960):

> You will be interested in what I have been able to find on Guston and Ironton. I cannot find in our church annals any evidence that a church organization ever was formed at Guston. One was formed at Ironton, and as nearly as I can learn the once attractive little edifice at Guston served both camps. As nearly as I can piece the story together a cabin was built at Ironton, which served as a parsonage. My findings seem to coincide with your account.
>
> Sometime when I am in Boulder I should be glad if I might have a brief conference with you with reference to some of the mountain churches, such as Ward, where Nathan Thompson preached occasionally about 1870.

It was Mrs. Annie D. Rogers of Los Angeles, California, however, who wrote the story I'd been waiting for (Nov. 9, 1957):

> Some time ago, it was my great pleasure to borrow a copy of your wonderful book *Stampede to Timberline* from the Public Library here in Los Angeles.
>
> It was most interesting and reminiscent of bygone days. Riding home on the bus, I thumbed the pages, for I had spent many years in Colorado, and many names and places were familiar to me, but on page 447 pictures of a little church instantly caught my eye even without reading anything for a moment or two. Then I forgot myself, and where I was. The bus was full of people but I exclaimed aloud, "I knew that church. It is the little Guston Church that my father built in 1892."
>
> Turning the page, I saw his name—Rev. William Davis—on the page's first line. Then I was really excited. The seat would hardly hold me, as I exclaimed again, "That is my father's name—Rev. William Davis. He built that Guston church in 1892, in that little mining town of Guston." A little lady sitting next to me smiled sweetly. I didn't know her, but I quieted down a bit and explained things to her.
>
> But now I saw the poor little church had withstood the severe elements of Winter in that very high altitude on the South Western slope of the Colo. Rockies. Heavy snows had caved in the roof and walls, and to the end it tried to hold up its battered head and steeple, as though trying to show what it was intended to be and had served its purpose bravely, and you had watched its death-throes to near its end. Even the setting in those aspens looked familiar to me. After all the years with no heat within, heavy loads of snow on the roof for months at a time, year after year, had caused its collapse. The original aspens around the church had been cut for fuel probably, and a new growth had grown up to replace them.

The Davis family—my father, mother, sister Mary and I came from London, England to Denver, Colorado, in 1890. I was 13 years old, and Mary was 21. My name was Annie. Father was soon asked by Rev. Horace Sanderson, Assistant Supt. of Congregational Churches for Colorado, to go to the Red Mountain District to establish a church there and then on to Ironton. They had never had a church.

Just coming from London, England, where we had been all our lives, and having not been in the mountains and having always had a good home with many conveniences, you can imagine what it meant to go and endure such things in early 1891. It was real pioneer work, but we went. Mother was not well at her age—45. It was impossible to find a decent place to rent. Everything was full up. My sister did not go with us.

Red Mountain was a thriving mining town, full of saloons, about 20 or more, and plenty of gambling houses, but no room for a church. They didn't want a church, they refused to have one, and were not even polite about it. After father made several attempts to be friendly, they threatened him with violence if he showed up in that town again. These were the business men. Some of their women very timidly said "Yes, we went to church and Sunday School before we married and came here," but they dared not assert themselves, apparently.

Then Father went to Ironton and found people quite willing to have a church. Funds were raised and in November 1891, on Thanksgiving Day, the church was dedicated by Rev. Horace Sanderson, and soon after, four good rooms were added as a parsonage.

A prominent business man of Ironton presented the new church with a lovely *new* organ. I regret that I have forgotten his name. He was a great help and a regular attendant. Mr. and Mrs. Mowray, retired elderly people, their married daughter and her husband, Mr. and Mrs. Foley, a younger daughter, Edith, and a son and his wife, all good Christian people, were very fine helpers. Mrs. Foley and I took turns at the organ. The Silver Bell Band, from the nearby Silver Bell mine came often with their music. They were Cornish, and some of them had been lay-preachers in Cornwall, England, but I think their Bibles and their religion lay in the bottom of their trunks. It is easy to forget, when there has been no church encouragement.

The Strayer family of the Strayer Hotel, Mr. & Mrs. Kilvert of the Kilvert Hotel, Mr. Giesel and Son, general grocery, the two Sampliner Bros. (Jewish) Mens Clothing & Dry Goods and Mr. and Mrs. Bussert were all very helpful and friendly. There were some saloons, but I don't recall any opposition.

Mr. Bussert operated Mule trains between Ironton and Ouray and the mines, carrying all kinds of supplies and merchandise—mining supplies, piping, lumber, frozen sides of meat, perhaps a coffin hung on the sides of two mules, etc. Mules could travel where wagons could not go. The coffin would be empty. Schools were closed in Winter, and open all Summer and Fall. Miss Stewart of Nebraska was our teacher. A lovely young lady and was useful in the church.

It was a glorious country in Summer, and majestic in Winter. We found large fields of immense Columbines and Larkspur at timberline. They grew so tall and large, we could lose each other as we picked. From November to June, all wagons and baby buggies were on runners. It snowed nearly every night and often all day. And I really mean *snow*. Large thick flakes. We had so little snow in London.

Father built the Guston church in 1892, so that it could serve Guston and Red Mountain in good weather. The very Sunday it was dedicated, at midnight the fire broke out in Red Mountain Town and burnt the entire business section to ashes. We went to see the ruins. The men who had opposed a church so strongly, came humbly toward us. Father said, "The last time I talked with you I said that sometime God would perhaps send judgement to Red Mountain Town, as he once did some thousands of years before to wicked Sodom and Gomorrah in Bible times. I believe if you men had acted differently, God would not have sent this devastation to your town today. Here are your women and children, and yourselves who need the influence of Church and Sunday School. Red Mountain seemed destined geographically and minerally to be the logical center for business, extending out to other towns in growth and usefulness in coming years. Sodom of old is gone forever. Perhaps Red Mountain Town will never be rebuilt. Business and growth go hand in hand with a church in its midst. You can't fight God and prosper."

Father was in Creede during its fire in 1892. He was supplying the Cong. Church for a few weeks. This church was very new and not in the fire area. The Catholic church was burned to the ground. Father met the Priest, who stood up in all he possessed. His lodgings and other possessions were in the fire. It was impossible to get a room. Father invited him to share his humble room back of the church. It had a bed, table, two chairs, and a cook stove. The offer was gratefully accepted, and also the privilege of using Sunday afternoons in the church for his church services until something else could be found. They became good friends, and two years later they met again on a Denver street. He was no longer a priest, having left the Catholic Church and had embraced the Protestant faith, and had married, I think.

Father conducted Bob Ford's funeral service. The church was packed full of people. Previous to this, Ford had cared for a poor, fallen woman who died of tuberculosis. I think he paid all her expenses and Father conducted that funeral also.

Soon after this, with two churches finished and ready to serve, we moved back to Denver. There were no doctors or medicines available in Ironton only at the big mines and they were only for miners' needs. Mother was far from well and needed medical advice badly. Another man was sent to carry on the work we had begun, and of course it wasn't long before the mines all closed up, and very few people left in any of the towns.

I am over 80 years old, living near my son and his family. They watch me very closely. God has been very good to me. My health is fairly good, hearing fine, and sight fair.

Now, my dear, I want to thank you so much for your interest in that part of Colorado that holds many very precious memories for me and particularly the wonderful prominence you gave my father, and little Guston church.

Chattanooga   From the summit of Red Mountain Pass the wide-ledge road follows the contour of the mountainside, passing the Silver Ledge mine whose shafthouse, which clings to the steep slope below the highway, I have watched disintegrate slowly during the past thirty-seven years. The road continues its winding course down to the site of Chattanooga, the first level area through which Mineral Creek flows.

There is nothing left to suggest that the original settlement on the meadow was a busy junction point where freight was reloaded from teamsters' wagons to burro pack trains, if starting up over Red Mountain and on to Ouray, and from burros to wagons, if the destination was Silverton or points beyond.

Among the letters I received is one from Robert Sparks Walker, formerly of Chattanooga, Tennessee (Aug. 18, 1953):

> I was especially interested in the story of Chattanooga, the ghost town. Miss Bertie Wenning of the Chattanooga Public Library knows all about this old ghost town, since her uncle, Frank Carol named it and was its first post-master. Miss Bertie visited it a few years ago. When I was editing the *Southern Fruit Grower* in Chattanooga 40 years ago, I knew Mr. Carol personally. At that time he wore a bearded face, and by his quick actions reminded me of a big horsefly. At that date he was running a printing office and got out a small paper called "The Busy Bee."

Silverton

The approach to Silverton is dramatic, whether you come from the north across the Million Dollar Highway from Ouray and descend from the 11,018-foot summit of Red Mountain Pass or whether you drive up from the south and cross Molas Divide. From the latter the highway zigzags down to Baker Park where Silverton, looking like a compact toy village, dominates the western end of the flat, dwarfed against the high mountains which surround it. Threads of roads fan out from Silverton to twist up gulches filled with mineral deposits. Only one road bisects the Park and skirts the Animas River for miles, passing through the ghosts of Howardsville, Middleton, and Eureka and continuing to rise to Animas Forks where it branches, one portion climbing over Cinnamon Pass to Lake City and the other over Engineer Pass to Ouray.

The Park was named for Capt. Charles Baker, the first prospector who ventured into the flat valley in the middle 1860's in search of gold. He and his small party sought for colors in the bed of the Animas River, panning the stream the length of the park and beyond and finding some pay dirt. But his men, and other parties which followed, knew they were intruding on Ute Indian land and were in danger of being run off or killed, as Baker was in 1868.

Even before the Brunot treaty with the Utes was signed in 1873, impatient prospectors struggled over 12,090-foot Stony Pass from the Rio Grande valley and entered Baker Park by way of Cunningham Gulch. They looked feverishly for gold but found only silver. By 1875 the whole region was alive with men and with new camps burgeoning near the most productive claims.

A note from Theodore H. Proske of Denver (Dec. 5, 1951) includes an incident of early date which shows the hardships mastered by both men and women who fought their way into the San Juan. Wagon Wheel Gap, which is mentioned in the letter, is on the Rio Grande River north of South Fork. The only trail into Baker's Park at the time was over Stony Pass.

> Today I have read Chapter Eleven of *Stampede to Timberline* and all the old stories that were in circulation years ago came back to me as memories of the past and how well they fit into the story that you have written. . . .
> Our daughter invited a Mr. and Mrs. Harry Anderson to dinner and when I told them about your book they said that he, Harry, was born at Idaho

Springs, his father had been a miner, and like many other miners in the early days had contracted Miners' Lung Trouble. . . . Harry's wife was born at Silverton, Colo. Her grandmother had walked into Silverton from Wagon Wheel Gap in the boom days carrying her baby on her back. This baby grew up in Silverton and became the mother of Mrs. Anderson.

Howardsville, at the mouth of Cunningham Gulch, was the first county seat but lost it in 1874 to Silverton, which outstripped all other camps in the District, becoming the supply center for the area.

Many of the mines became rich producers and shipped quantities of rich ore until the silver crash of 1893. Fortunately, by that date, some gold had been found in certain properties, and further development revealed larger amounts, which were profitably mined.

When the Denver & Rio Grande Railroad extended its rails in 1882 from Durango up Animas Canyon to the growing city of Silverton, it provided the needed outlet for the shipment of ore from the District.

John Herr of Whittier, California (Sept. 27, 1961), mentions the Dives mine at Silverton, whose tram crosses above the present road between the mine, far up Arastra Gulch on King Solomon Mountain, and the mill.

> Just read *Stampede to Timberline*. I am an old man in California, but was there through most of it. In 1895 I was assayer at the Durango smelter and Dan McLean was manager. Old man Bennet came down with a rich car of ore from the Dives on which he had a lease. He said he was sick and would sell his lease for seven thousand dollars, so Mac and I borrowed the money from the bank and paid him off sight unseen.
>
> Then we went up to look it over. George Hill the foreman was there and George asked us what we wanted him to do. Well we said, go ahead and take out that rich ore in the winze. Hell, he says, old man Bennet took out all the ore and swept down the walls.
>
> We thought we were stuck, but we told George to go ahead and try to find some ore. He cross cut five feet and struck three feet of fine ore and in sixty days we shipped forty thousand dollars worth.
>
> When old man Bennet heard of this he took to his bed and never recovered. We worked the mine for years and in 1906 the slide came down, took out the bunk house and killed thirteen men. The Good Lord saved my life. I was in the cook house playing poker with the cook.

In 1947 I went to the City Hall to see N. C. Maxwell, the city clerk who told me that Silverton once had a population of 3,000. "Now there are 1,500," he said, "and they move more ore than when there were 3,000 in the camp. Some properties in town are deserted and others are run down—renters don't spend money on houses; if you own property you do.

"San Juan county contains 600-700 square miles, which includes lots of mines and valuable properties. The widest flat place in the county is right here in this valley between Silverton and Eureka.

"Silverton has also produced some prominent men; perhaps you've heard of some of them. Charles E. Greene was superintendent of schools in Denver. He was here first as superintendent. C. Bowman developed the Merchants Biscuit Company in Denver. He ran a coal yard here, and he and his father had contracts with

Silverton, 1942. Main Street.

the boarding houses. John T. Barnett was an oil attorney and superintendent of schools and ran a newspaper. And of course there were a number of colorful promoters here too in the early days."

Most of Silverton's buildings date from its prosperous mining days—the City Hall, the County Courthouse, the Imperial Hotel, the Congregational church, as well as most of the business blocks and private homes with their well-tended gardens lining the city's wide streets.

The Animas River flows along the southern edge of the town, and beyond the townsite it is a narrow canyon some distance below the level of the highway. In this canyon flanking the river were the tracks of the narrow gauge road that ran to Eureka, and beyond that to Animas Forks. At the east end of Silverton, in the canyon on the narrow strip of land between the cliffs and the river, stood several important buildings. Close to the town were a couple of large mills, and further east was a brick powerhouse; but the building that always drew my attention was a large three-story private home. Upon inquiry I learned that it was the Stoiber Mansion and that it had been the home of the Edward G. Stoibers, the original owners of the Silver Lake mine.

When a friend and I stopped to see the house, it had been empty for some time. The shrubbery surrounding it half hid the lower portion of the building, but Elliot found an open door in the basement and we ventured in. It was cold and gloomy and dank, but we felt our way up a staircase and explored the main floor.

He found another staircase and investigated the two upper stories as well. It was so dark inside that I do not recall seeing any furnishings—I believe the rooms were empty.

Shortly after this glimpse inside the mansion, Mrs. Lillian Eloise Higgins of Monte Vista, Colorado, wrote the following illuminating letter (March 15, 1953):

> Silverton is my home town. For many years my brother, Charles Gardner, was superintendent of the Silver Lake Mill, during the time the Guggenheims owned and operated it.
>
> The Silver Lake was originally owned by the Stoibers, who lived in a large three-storied brick house a little way down the canyon from the mill. It was beautifully furnished; the rooms sparkled with plate glass mirrors and crystal chandeliers; the top floor had a big ball room with a little stage at one end of it, and a billiard room. "The Mansion" as it was called, was presided over by "Captain Jack," the nickname given Mrs. Stoiber by the employees. She drove the fastest team of horses in town and could out-swear any miner on the job. On Christmas her handsome sleigh was loaded with gifts for every child in Silverton, and she personally delivered them to every house. She built a tall, three-storied house in town and erected a high board fence higher than the house beside it to prevent prying neighbors from watching her gay parties and gossiping about the amount of liquor consumed. After her husband's death, she moved to Denver where she built another brick mansion and another high fence. Her second husband perished in the Lusitania disaster, and she spent her last years in a villa in Italy.
>
> Her brother, Alfred Harrison, ran the Pride of the West mine for sometime and shipped his ore to the Silver Lake where my brother, Charles Gardner, milled it for him. "Captain Jack" hated Alfred's wife so cut him off without much inheritance, but did cancel the debts he owed her. Harrison broke her will and spent his last days chasing all over the world learning new dances, Hawaiian, Cuban, etc. He would come home, wheezing and panting, and try to teach the intricate dance steps to his lady friends. His wife outlived him a short time and spent her last days playing bridge all day and often all night and drinking liquor. Thus ended the wealthy Stoibers who "struck it rich" in the Silver Lake.
>
> The mill you have drawn is not the Silver Lake that my brother operated for many years. Your drawing looks more like the Little Aspen Mill. The Silver Lake was a beautiful mill with a raised central ramp-like roof with windows the entire length of it on both sides.
>
> My family lived in several of the Colorado mining camps long before my birth. My father, George Gardner, was one of the original owners of the Tom Boy Mine in Telluride. He sold his interest before the rich strike was made. He came to Colorado after the grasshoppers and drought in Nebraska caused him to fail in the farm implement business. He had a hay and grain business in Pueblo in the early days. After burning out there, he went to Leadville where he had hay, grain and horses. Again he burned out. Then he went to Aspen and worked in the mines to earn money to start a commission business. Two days before he was to quit the mine and go into business he was killed by a falling bucket in the shaft.
>
> My brother, Charles, a boy of eighteen, took the responsibility of the fam-

Silverton, 1942. Stoiber Mansion.

Silver Lake mine and mill above Silverton. *Photo from Marvin Gregory Collection, Ouray.*

ily consisting of my mother, sister and me, then two years old. He worked at the Molly Gibson in Aspen, the Camp Bird in Ouray, the Iowa-Tiger at Silverton. He became superintendent of the Hamlet Mill and later the Silver Lake. For many years our home stood on the bank of the Animas River, over the brow of the hill and across the river from the Shenandoah-Dives Mill. Only a few years ago before the Guggenheims had all the buildings including the Mansion, torn down, my swing still stood in the back yard.

My brother and his partners sold one of the first tungsten claims located in the San Juan during World War I. For a number of years after that he was employed as a mill wright by Stearns-Rogers Co. of Denver. He built mills for them in Arkansas, Alabama, South Dakota, Arizona as well as in Colorado, and installed machinery at the School of Mines in Golden. Then he became superintendent of the Wilfley Shops in Denver.

Charles and George Wilfley had become close friends when George and his father built a little mill across the Animas River from the Silver Lake to try out their new flotation machine which replaced the first Wilfley concentrating table. The new flotation machines were used in the Sunnyside Mill at Eureka when that mine was purchased from the Terrys and a new mill erected. Charlie and George swung a little cable-suspended bridge across the river between the Silver Lake and Wilfley Mills and enjoyed many years of friendship and companionship. The Wilfley Shops now manufacture the Wilfley pump which George perfected before his death. My brother is gone now also. He and George are buried quite near each other in Fairmount Cemetery in Denver.

Of course there are many other stories of interesting families of Silverton, the Terry family of the Sunnyside, Mrs. Amanda Cotton, one of the earliest pioneers of Silverton, etc.

The Silverton homes you have drawn are especially interesting to me. The big square one belonged to the Lamonts. Mr. Lamont was a very pious preacher and miner; his wife, a refined little lady from Boston, kept a very exclusive boarding and rooming house. For many years some of the "lady" teachers lived there. The house next door belonged to a rich saloon keeper whose wife had been one of the "painted ladies" of the red light district.

From N. C. Maxwell and others in Silverton I gained additional information about the Stoiber mansion, which was called "Waldheim." It was built in 1896 by Edward G. Stoiber and was sold in 1902 to Guggenheim Bros., later the American Smelting and Refining Co. The Stoibers left Silverton after the sale. It is believed that Stoiber got two and a quarter million dollars for the mine, mill, and mansion.

A little elaboration on the Stoibers and their big house: Mrs. Stoiber ("Captain Jack") was a fiery female. Before the mansion was built, the Stoibers lived in the house where Art Walker lives now—the house below Maxwell's, and Mrs. Stoiber couldn't get along with the neighbors on either side, so she had "spite fences" built two-stories high. When they built the mansion, she had elaborate plans to outdo the women in town.

The mansion was a little ahead of the times in modernity, with three bathrooms, conservatory, Texas pine (semi-hardwood) floors, and carpeted throughout. There were three floors and a partial basement. Incidentally, they also had electricity.

On the first floor were five office rooms and a large vault in the front part; in the back there was a living room, parlor, conservatory, butler's pantry, kitchen, porch, two small rooms, and two other rooms. In the living room there was a fireplace with a large mirror over it. There were two stairways, a large one and a smaller one for the servants.

On the second floor were bedrooms—between twenty and thirty. In the wing were the servants' quarters. On the south side, near the back of the building, was a large sitting room, off this the master bedroom—quite large, and off this bedroom a small bathroom with a *small* bathtub. There were closets in most of the bedrooms, and the second bathroom was also on this floor.

There was a spacious ballroom on the third floor, and the ballroom stage was equipped with footlights and a small dressing room on each side. Rumor has it that the Stoibers occasionally got actors from New York to entertain their guests. Off the ballroom, at the head of the stairs, was a billiard room.

Mr. Stoiber made several trips to Germany to pick up new ideas for the mining industry, and he pioneered in rock drills, steam plants, and electricity generation. The two buildings just south of the mansion were the generating plant to supply electricity for the mine, the mill, the mansion, and the barns.

In the kitchen of the mansion was a dumbwaiter extending from the basement to the second floor, and one of those bell signal boards with the different room numbers on it. There was a furnace in the basement.

Mr. Stoiber was injured in an automobile accident in Paris which resulted in his death in 1905 or 1906. Mrs. Stoiber died in Italy a few years before World War II.

Eureka

Eureka in 1941 was a true ghost town with many false-fronted stores lining its main street and frame houses tucked in among a grove of trees on the opposite side of the Animas River from the Sunnyside mining property. Big mine dumps spotted the surrounding mountain slopes below mine portals, to which led steep narrow trails hacked from the hillside. But the building that dominated the town was the many-storied Sunnyside Mill. At its foot stood additional buildings and sheds, as well as the mine office and a nearby schoolhouse.

The mine was located in 1873, and the Sunnyside Extension the following year. The first mill was above timberline, close to the mine on the shore of Lake Emma; the second was at Midway, between the mine and Eureka; the third was against the mountainside in Eureka; and the fourth and largest was built just beside it. Other mines in the immediate area included the Toltec, Tom Moore, Golden Fleece, Silver Wing, Mastodon, and Lion Tunnel; but the Sunnyside and the Sunnyside Extension were the big producers.

The camp was surveyed and laid out by the Eureka Town Company in the mid-1870's, and by 1878 it contained two stores, two meatmarkets, one restaurant, a post office, and a saloon.

When Otto Mears planned his Silverton Northern railroad, he constructed a branch as far as Eureka in 1890, and then extended it four miles farther to Animas Forks. He also laid a spur up Cunningham Gulch as far as the road to Stony Pass. These two provided shipping facilities for many mines in the area.

The section between Eureka and Animas Forks was the hardest to build and maintain because of the narrow canyon with its steep mountain slopes and the several known slide areas that brought tons of snow avalanching to the valley floor and pushing up the opposite side of the gulch each winter. To protect his roadbed, Mears designed special snowsheds to be constructed at each slide area. He built

Eureka, 1904. Third Sunnyside mill. *Courtesy Josephine H. Peirce.*

Snowshed on Silverton Northern Railroad between Eureka and Animas Forks, 1904. *Courtesy Josephine H. Peirce.*

the first one at a test spot near the Silver Wing mill. It was constructed of heavy timbers bolted together and reinforced by rock foundations. That winter a slide ran from across the river and demolished the shed by pushing it up from below and reducing it to splinters. The other six were never built.

The railroad, like the town of Eureka, is only a memory, for as F. C. Evans of Wilmington, Delaware, writes (July 18, 1949):

> This morning the July issue of FORD TIMES was laid on my desk and it made me feel very much at home to read the articles on the Ghost Towns of Colorado and to see your watercolor reproductions. I say this because Mrs. Evans and I returned only last Monday from a two weeks trip by motor car of about 2700 miles into the country southwest of Denver.
>
> Being a railroad enthusiast, I again this year followed out a lot of old railroad grades which took me to all of the places shown in the article, with the exception of the Tom Boy Mine above Telluride. . . . We also went through Eureka and found that it is actually being carted away by a Salt Lake Wrecking Company.

The Sunnyside, like most of the mines that were high above the valley floor, was connected by electric tram with the mill at Eureka. Theodore H. Proske recalls the Sunnyside tram (Nov. 22, 1951):

> A number of years ago, I visited the Sunnyside Mines at Eureka, Colo., as the guest of Joe Terry one of the sons of Judge Terry who located this famous mine. I went over that Aerial Tram passing through one of the worst blizzards on the way up I ever experienced, and can assure you that to prospect in those mountains in the early days was sure some chore.

Three colorful letters and a number of old photographs sent me by Josephine H. (Mrs. Frank D.) Peirce of Leicester, Massachusetts, describe Eureka as it was in the early days:

> Dec. 12, 1949
>
> Going up to the American Antiquarian Society last week for some research, the librarian gave me your book, *Stampede to Timberline* to read. I brought it home and have been having a wonderful time, as it brought back so many pleasant memories. . . .
>
> My father was connected with Rasmus Hanson and helped him in the work on the San Juan Queen group of claims up Picayune Gulch. I was there when Mr. Hanson died, and played a piano accompaniment for the singers at his funeral.
>
> Mr. Terry I remember well too. He was a grand person. . . . I remember his family—Bill, who lived in Eureka, and Mrs. Strong, his daughter, whose husband was interested in coal in the northern part of the state. There was another son, but I don't seem to remember him as well.
>
> Mike O'Toole's Hotel had a large sign on the front with the "O" painted green. He was a policeman in South Boston before he got "itching feet." A nephew "Young Mike" came in each summer, and the mines were a great place for college boys for summer jobs. They came in from many states and earned good money wheeling "mud". . . .
>
> I have hundreds of pictures. . . . I also have one of my father's passes

on the Silverton Northern for the year 1911. There were buckskin passes. I remember my father had one. Buckskin was used for everything—postcards are one item—I think I have some of them.

Jan. 1, 1950

I have been having a lot of fun going over the old pictures. . . . As near as I can remember the snowshed was about a mile from Eureka, and I do not remember more of them. The last time I was there, there was no sign of even one.

I am trying to find a print of the Gold Prince Mill after it was finished. I remember watching the construction, and as you can see, it was just as you enter Animas Forks. The Gold Prince used to have some great dances in the boarding house. We went up on moonlight nights in the wagon with four horses, shown in front of the store. At the dance there would be miners in clean dungarees, sheep herders complete with chaps and spurs, and all the girls that were "in" visiting. About ten men to each gal. Music was a fiddle and a melodeon. Square dances were most popular, but waltz, two-step and polka were usual. Refreshments were usually ice cream and cake. . . .

The Esmerelda was owned by some English and Scotch people and I remember some of the owners coming over one Summer and staying at the hotel in Eureka. I had to guide them as no one else was available, and they had absolutely no sense of direction, nor could they remember which way to turn after many visits, so it was sort of a steady job.

Mr. Joyce was editor and publisher of the *Silverton Standard,* and I used to go to the dances with his son, Jack Joyce, Jr. His daughter eloped the last year I was there and I have a picture taken at a party they both attended, then left to go "out."

We used to have "steak-frys" on moonlight nights, and would carry huge watermelons up the steepest trails. Electricity was available in all the cabins, even up the trails in the main gulches, for the mills furnished it at absurdly low rates.

Today the thought of getting into one of the tram ore buckets gives me goose pimples, but in the old days I loved it, and came back from the Sunnyside mine that way many times. Even Mrs. Terry would do it to save time. And many another way too. I would stand on the upper edge hanging on a rope, and go down 100 feet without as many qualms as I had taking the elevator at the Chateau Frontenac to the river level.

Mrs. Terry and I were planning to go to Europe the year she died. Pneumonia takes people quickly at the altitude of Eureka, especially elderly people, and I have never been to Europe to this day. All mining people who made money had to see Paris, and the two girls in big hats in my story in *The Telegram,* taken at the Sunnyside mine, were just back. The pretty one lived permanently in Durango and had come back to Silverton to see her aunt; the other lived in Kansas, but they went to the same school outside. Many of the boys in Silverton went to Kemper Military Academy at Booneville, Mo.

A radio commentator recently mentioned Dr. Abram Van Meter. His brother, Ike, his wife and youngster went fishing with me at Lake San Cristobal. . . . On that fishing trip, a Mrs. Handy (whose husband was working for the Gold Prince) and I went alone to meet the Van Meters and we stayed over-

Eureka, 1904. Main Street. General store (center) was also post office; assay office on right. *Courtesy Josephine H. Peirce.*

Eureka, 1942. Main Street. General store. (Compare 1904 photo.)

night with some people Mrs. Handy knew who were "Holy Jumpers." They had a service after dinner that night and tried to convert us. That was somepin! However the food and beds were good, and we paid regular hotel rates. They should have been.

March 18, 1950

I found the glass negatives, one cracked all the way across, but they are clear enough for you to see the whole town of Eureka. The pictures were taken about 1904, by my father, but it looks just as I remember it in 1910, with a few exceptions—the Animas Forks branch of the railroad had not been built, and Judge Terry's "new house" and a few other buildings. If you will match up the two red crosses, you will have the whole town.

The two goodlooking houses belonged to the Terry sons. The log house with the barn where the crack shows was ours, and we had two stalls in the barn, one for my horse "Billy" and my father's "Kid." The mill was larger than here, but you have pix of it that were taken later.

Sincerely yours,
Josephine Peirce

Eureka, 1904. In background is switchback road up to Sunnyside mine. Road at right leads to Animas Forks. *Courtesy Josephine H. Peirce.*

Mrs. Mike Olivieri of Brighton, Colorado, sent a letter, originally written to her father, J. S. Wojcik, in 1955, by Marion A. Speer. Both men are rockhounds, and since the ore described came from Eureka, she thought I might be interested in the story:

The Story of This Rhodonite

This Rhodonite is from the old Sunnyside gold mine of Eureka, Colorado. I call it "Sunnyside Rhodonite." It does not have dark spots common to most Rhodonites. However, it is all gold bearing, running in gold from $25 to $450 per ton. It, also, carries silver, lead and zinc. Free Gold can be seen in the dark spots of the zinc. It is very hard and could not be milled in the stamp mills. A stamp weighing a ton would drop down on this and it would not crumble or shatter. More often it would be knocked out through the screen, causing a shut down for repairs and the like. It does take a good polish and is fine for cabinet specimens.

In 1914 when ball mills were coming into general use for fine grinding, the Sunnyside installed one of these and used this Rhodonite for fine grinding instead of steel balls. After the sucecss of using this in ball mills, a new 500 ton mill was built and fully 500 men were kept busy in getting ore out to keep

it going day and night. So far as I know, this was the only mine in the world where gold bearing Rhodonite was found. The mine has been worked out and closed down for nearly 30 years. All surface buildings have long since been torn down. Some mining has been done from time to time but nothing as of yore when I knew it and worked there.

This Rhodonite comes from what has since been known as "The Fourth of July Vein or Stope." In June of each year a certain amount of the net proceeds was set aside for development. Will Terry in charge of this work was following a new lead, when on July 4th, 1896, he found this rich deposit that made all members of the Terry family millionaires. It is from this vein this Rhodonite comes. The Sunnyside was owned by the Terry family. It was discovered by George Howard and a Mr. McNutt in 1880. It was sold or traded to Judge John Terry, as the story goes, for a few drinks of whiskey. When I knew the Terry family, sons Joe and Will and daughter, Helen, directed all work and development. The mother passed away when the children were very young. They never had an Engineering staff. The mine produced many millions and paid its own way from grass roots. The specimens I gathered from mine blastings. We used to pick it out of the ore buckets as they came down from the mine over the tramway and throw it over a cliff to the canyon below. It is this hide-away place that the specimens I got are from. The Postmaster of Silverton, Colorado, and his Jeep made it possible for me to get all there is to be had. It is no longer worth the effort to find more. This is your last chance to have some of this gold bearing Rhodonite. I worked there at several different times from 1903 to 1916. I represented the American Smelting and Refining Company. The Smelting Company bought the product of the Sunnyside for processing at their Durango plant. The Terry mansion was moved out of the mountains in sections to Farmington, New Mexico, only last year. The little mountain town of Eureka, where from 300 to 500 people lived, is all gone. The jail was still standing this July 1st and 2nd when I gathered this Rhodonite. I helped several times to drag the town marshall to this jail to sober up. Saloon men would get him drunk so that they could do as they pleased.

Foundations of the little cabin where Mrs. Speer and I lived is all that remains to remind us of our home in the rugged mountains of Colorado. I hope you will like what I am sending you. . . . My sweetheart passed away on April 29th, and this was a lonesome trip for me with that vacant seat by my side. I am,

Truly your friend,
Marion A. Speer

Miss Helen Thompson of Long Beach, California, writes of her family and of tragedies connected with Eureka (Sept. 2, 1965):

It was with great interest that we read excerpts from your chapter on "Eureka." . . . I thought you might like to know some of the early history of the Sunnyside Mine, which was such an intricate part of our lives. Judge Terry was supposed to be my father's best friend—the reason the "sale" was not completed was it required my mother's signature.

My father (Louis Thompson) owned the Sunnyside mine, gave Eureka its name; Lake Emma was named after his sister Emma who was married to Milt Engleman of Canon City, Colorado, who with my father owned a drygoods store there. My father finally sold his interest in the store so he could devote more time to his mining interests. However he had a severe attack of Grippe from which he never recovered. Consequently, he was unable to go over to the mine for that year. I was born on the 23rd of July and three months later he died on October 21st. Terry immediately stepped in and took over the mine.

My father's personal attorney, Gus Macon, of Canon, a mining expert, died also and my mother was stranded as it were. When I was about nine years of age we had a trial at Ouray at which time Terry owed my mother $450,000 for ore taken out of the mine. In spite of all the evidence against him Terry's own judge decided the case. Our attorney McLain of Canon left immediately for the East and did not return to Canon for over six months, too late to get the case into the Supreme Court. I went with my mother time and time again to Mrs. McLain's home trying to find out where he had gone and for his address, but she pretended not to know. He seemed to have plenty of money when he returned and a few years later built a lovely home. Part of the house-warming was a taffy pull in the basement and I was there as a guest of his daughter.

When my brother Herb was quite young, he went over to Silverton wanting to work in the mine, but Terry wouldn't let him near the place. Mrs. Terry was very nice to him; he stayed in their home. Terry and the two sons carried guns against each other as much as anyone else.

Through a Denver friend who had worked in the mine and who knew how one of the boys was high-grading ore from the richest vein, Herb met Attorney Morley of Denver and he had the case all ready to go to court when suddenly he became the governor of Colorado. Herb learned he later went to the penitentiary where he died.

A cousin in Denver was looking up the genealogy of the family in the library not long ago and all she found about our father was that he was one of nine children of Andrew Thompson—Louis Cass Thompson, deceased. Did not say where or when.

My father was born in Jacksonville, Illinois, and was graduated from the University of Illinois. He was a Knight Templar and an Odd Fellow. For a time he was City Treasurer of Canon. My mother was born in Independence, Missouri—she met my father while visiting her Aunt Nancy in 1872 and they were married February 4, 1879. There were five children all born in Canon. Hal died when he was two years old, when Lou was two weeks. Herb was next, then Frances and then me.

We have a picture of the McNutts on donkeys, showing how one got to the mine. The women rode side-saddle and the children were in baskets on either side of the donkey. Herb always rode on the outside as Lou didn't like looking down so far.

The Terrys had raised an adopted daughter who was married to Henry Sartor of Canon. They had a small ranch in East Canon. I spent considerable time there with one of their daughters when growing up, and Mr. Sartor was always telling me what a fine man my father was and how proud I could be of him. Both of the girls are now widows and live not far from each other on Colorado Blvd. in Denver.

It seems there was so much sadness for all of those connected with the mine. Ellen Terry Strong's two-year-old daughter fell from one of the balconies into the rotunda of the Brown Palace Hotel in Denver and was killed—they were living there at the time. Mr. Strong ran off with the wife of their best friend. She was later married to a Mr. Robinson and they lived in Denver. We heard one of the boys was killed in a train wreck. While hiking on a narrow trail in Twin Lakes, McLain's daughter stepped on a rock, which gave way and threw her in such a way that the stone fell on her chest. She died before they could get her back to camp.

Our mother died in 1940, Lou in 1951 and Herb in 1952. Frances is helpless with arthritis and I nurse and keep house. We live in a home which was left us by a very dear friend as long as we live, then it goes to her great granddaughter. After three years of carrot juice, exercise (Yoga) and manipulations which we go through she is so much better. . . .

I do hope I haven't bored you too much. Your Eureka brought back so many memories. Greed is such a monster and brings so much misery to everyone. There was plenty for all, as my father had planned there should be.

George J. Puth (deceased 1968) of Appleton, Wisconsin, reminisces in four short letters about his years in the Silverton area, especially those spent in Eureka and in Animas Forks where the Gold Prince mill stood. His widow, Marie E. Puth, wrote me in 1974 that all his photographs of his mining days were sent to Fort Lewis College, Durango, where they can be seen. She concludes: "Hoping your efforts to bring back those colorful and dangerous days will be successful." Mr. Puth wrote:

April 7, 1964

Just received as a gift the book it seems you are the author of.

I happen to be one of that breed of animals known as a hard rock miner and put in several years in the vicinity of Silverton, Telluride, and Durango. All this happened some time ago. I worked in both the Sunnyside mill and the mine. Also at Animas Forks, in Silverton, and spent several years in Colorado. I happened to be a Camera bug at the same time and have quite a collection of real pictures of that part of the country. . . .

I personally knew the Terry family: John Terry, Sr. and the family, William, Joe, and Ellen Terry Strong.

April 22, 1964

It was way back when in 1908 I first came to Colorado. The first job I received was when Joe Terry told me come up in the morning and several other unnecessary words. I little knew that he wanted me to appear at the mine and I got lost at the mill, the first opening there in several years. The next time I got a glimpse of him was across the wheel of a turbine as I finished turning off the water supply at the mill. Thru a piece of blind luck I got there first, and he was more than elated as I had closed it down the first time without a smashup. From then on I was a buddy.

I was well acquainted with Mrs. Strong and William and Joe Terry, the two brothers and their sister, who at the time owned the Sunnyside Mine. Of course some of the things are sort of dim but one thing brings on another. Many of the gang used to meet the train from Silverton every evening. The great sport was meeting the train. I was with them until the fire at the tram house which put them out of business, and as is usual the gang scattered like a flock of ducks.

May 2, 1964

My last memory of Animas Forks is the keeping awake for a whole day and when the day was finished and we got to the hotel we were informed that the mine was down and we were free men. It was different then, now there was no giving a miner a notice that he was no longer working but here is your check and you are thru. Social security was unheard of in those days. I had not slept for the last three nights and would never have worked the last two had I known that the third was the last.

I have a photo of the Gold Prince mine and the town of Animas Forks. For your information the Gold Prince up to that time had never turned a wheel. The picture shows the mill and a couple houses that would make a good sized Dog House. The Bridging of the R.R. between the Forks and Eureka was damaged the first winter after completion. The slides reversed the order of their appearance, which was unusual and filled the canyon with snow from the wrong way and the Tunnel was ripe for picking when the big slide came down. The lower end of the cover was about a mile above Eureka. I also have one of the Sunnyside Mill and Flotation Plant, except that this was not a flotation plant but a Huff Electrostatic Plant. This made the Separation by electrolosis instead of flotation and was dustier than a Tonopah Glory Hole. Right across the town from Eureka and up on the hill above was the Crown Tunnel which was our Glory Hole.

The substantial little building which stood right alongside the track in Eureka was the jail, Jug, or other names that might have been fastened to it. I also have a photo of the building at Howardsville which stood partway up the hill from the track. Halfway between Howardsville and Eureka was the Kittimac Mine. I have a nice photo of this mill before it was damaged by a slide. It was not operating.

I have always promised myself that I would take a farewell trip thru Colorado and give it a decent burial. Silverton, elected the only member of the Bar-Tenders Union to the State Legislature. (Jack Slattery from the Hob Saloon, next to the Hotel.) They used to let a miner win a bundle, and then re-mined it with a Black Jack. The Driver of the stage with a straw hat thru the winter and a Fur Cap in the summer. Four horses to pull the Stage because of the light air. His weird and woeful stories made many a tenderfoot buy his return ticket and take the morning train out of town. 24″ of snow July Fourth. These were true in Silverton.

Jan. 15, 1965

I have finally gotten to the stage where I had the courage to tackle the job of going thru the photos I had tucked away for a long time. These are the vintage of 1909 and 1910 that is the time I was out in Colo. and the only time I ever was in the Silverton area. . . .

I have the town of Eureka showing the whole town, Jail and Sunnyside Mill. Also the Sunnyside Midway. Crown Tunnel across from the town. The old smelter. . . . The Kittimac Mine and Mill and the Tram Line. I have the Ross Smelter in Silverton. The Old Hundred Mill. The Silver Lake Mill. And the Boarding House & the Sub Station at the power house. . . . And before I forget it The Curb across from the Railroad where they were smashed with the Slides. And the Bridge over the Animas River in the fall after the river had tunneled thru the snow and Ice. The Eureka Hotel. And the line of Thirst centers along the way to the Jail. Many other places along the river that will make a fine book. And all of it will be true.

When I interviewed N. C. Maxwell in 1947, I also asked him about Eureka, for he had worked for six years at the Sunnyside.

"There were a number of men connected with the history of the mine," he began, "but two stand out—Rasmus Hanson who secured a lease on the Sunnyside Extension in 1886 and developed it, and John N. Terry.

"Judge Terry owned the Sunnyside and most of the people in Eureka were employed by him. In fact, ninety percent of the Sunnyside men and their families lived there. After his death in 1910, the Estate ran the property for the heirs, two sons and a daughter. In 1917 the heirs sold the majority of shares of stock to a new company for half a million dollars. In 1925 when I went up there to work, the U.S. Smelting Co. was running the mine.

"Judge Terry was a hard worker, always trying to improve the output from the mine. In 1906 he was associated with a property between Silverton and Eureka in which he represented a Mr. Harrison of London. When the latter came over in winter to inspect the property, after being driven up in a sled from Silverton, he asked at the mine office for Judge Terry. He was told that trouble had developed in the water system and that the judge was out in the middle of the creek trying to remedy it, which just goes to show that the judge was a man who did not spare himself.

"In 1912 there were 150 men on the payroll, milling 120 tons of ore a day— one man to a ton. By 1930 there were less than 400 men on the payroll, handling 1,100 tons a day—three tons to a man. The mine produced more zinc than any other metal, but it was also an important lead and gold producer. It's all over now, because of improved machinery that has simplified milling. Three men can run a 200-300-ton mill. In the old days, when we used crushing stamps, one man was needed for every 10 to 20 stamps.

"Eureka twenty years ago had 300 people. It was one of the best of the little settlements in the District. Before prohibition, there were two saloons and several boarding houses. The big white building on the right of Main Street was the Club House where card parties, dances, prize fights, and badger fights were held. The schoolhouse was a two-teacher school. In 1931 to 1932 most of the population moved away. By 1935 only a few people were left. All the houses could be rented now, but the caretaker doesn't want to bother with renters.

"They took the railroad out in 1939. The company then built the Galloping Goose, called 'Casey Jones.' They took the engine and the transmission from a Cadillac they owned and put it in the Goose. The Goose is in the little park in Silverton now."

CHAPTER XVII

Telluride, the Uncompahgres, and La Plata Mountains

One of the most scenic drives in southwestern Colorado is from Ridgway to Telluride. The country varies from farmland, backed by a sawtooth mountain range, to a narrow red rock canyon and finally to a little city surrounded on three sides by mountains whose steep slopes have been creased by avalanches.

The rolling meadows between Ridgway and the top of the Dallas Divide are dwarfed by the rugged, snowy peaks of the Uncompahgres toward which the road heads, with Mt. Sneffels towering above the rest of the range.

From the summit of the Divide it is a long descent to Placerville and the San Miguel River. Mary K. Mott, whose letters appear later in this chapter, wrote:

> I don't remember Placerville except as a junction. The coaches changed horses there in early days. It was the outlet to various mesas and the Norwood country. Farther up the river was Saw Pit. It was more of a lumber camp and a lime kiln, as I remember.

A fire in 1919 wiped out much of Placerville, and of the buildings that were left few exist today. The town was a supply point for those who lived along the river, as well as for the entire Paradox Valley during the mining era. Other small communities farther upstream are Fall Creek, Saw Pit, whose boom lasted while the Champion Bell and Commercial mines were active, and Vanadium, whose big mill has disappeared in recent years. A steep climb to a higher bench of the valley reveals a ring of mountains at whose foot Telluride nestles.

Late in 1874, prospectors who had heard of placers on the San Miguel River crossed from Baker Park into the Marshall Basin looking for lodes. By 1875 they

Telluride and Pandora

Top: Placerville, 1941.
Middle: Saw Pit, 1950. Foundation of mill and machinery.
Bottom: Vanadium, 1919. *Photo by Robert Sterling, Boulder.*

had located several, including the Pandora, Mendota, and Smuggler. J. B. Ingram discovered the Sheridan and Union claims.

By 1876 a small camp called Columbia, whose population grew to 100, was started by prospectors near the end of the valley. The year before, however, a smaller settlement named Pandora had been started two miles nearer to the mountains and closer to the mines. As the properties were developed, trails were hacked to them up the rocky sides of the rugged peaks, over which burro trains packed the ore across the range and down to Ouray.

Columbia, renamed Telluride in 1881 when ore was being shipped from a dozen properties, grew throughout the eighties, but its pace was slow due to lack of cheap shipping facilities. Yet the town was so optimistic that it built a county courthouse of brick, a large school building, a hotel, churches, and business blocks, as well as private homes.

After the Denver & Rio Grande Southern's branch to Telluride was completed in 1890, development was rapid, and mineral output increased. The silver ban of 1893 paralyzed the camp, but in the late 1890's the discovery of gold sparked a new boom. During this period long trams were built connecting the mines (most of which were above timberline) with their mills down in the valley. From 1898 on the greatest output of the District came from the Liberty Bell, the Smuggler-Union, and the Tom Boy.

Telluride, 1971. Tom Boy mine. *Photo by Malcolm Childers.*

The entrance into Telluride is spectacular, for although the valley deadends a short distance beyond the city, it is Ingram Falls, 3,600 feet above Telluride, that holds one's attention. Dropping straight down from the lip of a hanging valley it seems to spill into the huge tailings pond of the Smuggler-Union mill at Pandora. To the right of the cataract is a dim pack trail cut into the face of the granite mountain in a series of switchbacks. It connects the Black Bear mine, which is situated in the Ingram Basin above the falls, with Pandora and Telluride.

A ski area that is being developed in an adjoining valley has changed Telluride. Its old buildings are being restored, and a whole new section west of town is proliferating with condominiums. New shops, art galleries, and gourmet restaurants cater to increasing crowds of vacationers and new residents. Quite by chance, I arrived in town on a summer morning in 1974 just in time to watch four hang gliders soar from the top of Ajax Mountain and land on the main street in front of the Sheridan Hotel. But it is Telluride's past that is presented by the writers of several letters that I received, and they tell their own stories.

Mary K. Mott of Long Beach, California, whose memories of Lake City are included in an earlier chapter, reminisces about both Telluride and Pandora (Feb. 24, 1951):

Side Saddle in the San Juan

Telluride in the early days, long before that hybrid, "The Galloping Goose" made its appearance, used the galloping *horse* as a mode of transportation.

Every mining town had a number of well equipped livery stables. They occupied the Main St. along side of the business houses. Instead of an automobile darting out while passing, one would listen for pounding hoofs and a voice saying "Whoa, steady now," thus being forewarned. Horseback riding was great sport for the ladies of my day. Believe me there were many fine riders.

Those were the days of the sidesaddle and the long sweeping riding habits, the skirts all but touching the ground. Some of the ladies were fortunate enough to have their own horses, but the majority instead of spending their husband's hard earned cash on *beauty parlors* and the like, hired their horses from the livery stable.

In all the mining towns there were horseback clubs, I think there were at least 14 ladies who belonged to the Telluride one. It was so exciting to gather together one of those beautiful summer or fall days. Our horses seemed to enter into the spirit too as we galloped away on a day's jaunt.

Some of the ladies were beautiful riders and I remember especially Mrs. Tom Elliott with her black velvet riding habit and nifty little hat to match, Mrs. "Jimmie" Johnstone with her lovely black horse, Mrs. Frank Brown at San Miguel, Mrs. James Coleman and others. The most of us weren't much to look at but we certainly enjoyed ourselves, and went places.

There was a story told about Mr. John Ingram (for whom Ingram Falls was named). It seems that quite a party went to Silverton on horseback. After arriving they went to the hotel, registered and settled down for the evening. One of the roving bachelors had invited the men folk into his room and for reasons of his own, no doubt, locked the door. One of the ladies decided it was about time for friend husband to appear so she started down the hall. Hearing considerable hilarity coming from one of the rooms, she paused, and

Trail to Black Bear mine, 1962.

knocked. All grew quiet: then the genial voice of John Ingram boomed out: "Come on in over the transit." This story never failed to bring a good laugh.

In the 80's and 90's most of the miners were Cornish. Those were the colorful days of Telluride. It is a great pity those days have not been recorded, for such an era will never be again. The Smuggler was run and owned by an English Co.

Pandora, at the foot of Bridal Falls, became quite a settlement some 1½ miles from Telluride. It was here where many of the mill men who worked at the Stamp mill and trams had little homes of their own. The office building was quite large and later, spacious rooms were added to accommodate the office men with living quarters.

The trail to the Smuggler mine started here and thousands of tons of merchandise were carried by mule and burro pack train. The miners were continually going and coming on horseback. There was also a trail on the east side—One must remember that several hundred miners were employed at the Tom Boy, Sheridan and Smuggler.

The Cornish men were beautiful singers, and how they did love it. Of course many of them had a weakness for the saloon and they squandered their money. Some of them would stay up at the mine for several months, come to town and go back broke. Their lives were tragic, in that they in their recklessness did not take care of themselves and many died with dreaded Pneumonia. Their funeral services were most unique.

After the slow and solemn march from the chapel, the larger body of them would finally reach the Congregational Church. It would be filled and crowds reaching out to the street—the choir would start some selected old hymn and instantly it was swallowed up with mass singing by the miners. They were religiously inclined, knew many of our good old hymns and they certainly could sing.

On Christmas some of their Quartettes would go to the houses of their friends and sing Christmas carols. The Main St. needed no mechanical music for from every saloon, and there were many, came the "While Shepherds watched their flocks by night." I must not tarry on this subject, it is too full of memories and there is so much that could be written. I always had a warm spot in my heart for them and considered them my friends. Why not? Did they not send to England and have some relative knit me a most wonderful warm petticoat so I could keep warm when I used to go with the pastor once a month to play the little organ. When we with some of the young folks who liked to ride horseback arrived, as usual gambling was in progress. Instantly tables were pushed against the wall and services began. They were most attentive and respectful. How they did enjoy singing later on. We could hardly get away. I do not like to think of Telluride without them.

Back to the sidesaddle and long riding habits. In my childhood and girlhood days at Lake City, after I graduated from the burro-back to the Indian pony, then to a living horse now and then, I am sure no Buick or Packard ever gave anyone the thrill that was mine.

How we enjoyed a trip up Henson Creek. I can still feel the urge of our horses as they neared my beloved Capitol City where there was a small trickling stream that came down the rocky mountain side, lush grass and watercress grew down the sides. A rusty tin was chained at the base. Here is where we rested our horses, drank of that ice cold water and threw rocks into the deep canyon where the dashing stream flowed. We listened for the echo and enjoyed the quiet stillness as we viewed that beautiful landscape. I am thinking of now—no time for echoes, too much haste and noise.

In Telluride I think Mrs. Bulkely Wells was about the first to introduce Cross Saddle for ladies. It was frowned upon at first, but it was not long until side-saddles were obsolete. Mine with its Brussell carpet upholstery edged with leather will always remain one of my *jewels.* . . .

As far as Pandora society was concerned it was more or less a part of Telluride. Mr. & Mrs. Bulkely Wells did some entertaining and those *next to the throne* did likewise. There were many dancing parties given in Telluride when the elite swept the ballroom floor with their long skirts and numerous petticoats, the last one having a dust ruffle. At the fancy dress balls or masquerades some of the bolder ones deigned to appear with skirts halfway to the knees (perish the thought). The Wells' entertained friends from Denver and Colorado Springs quite often and *their crowd* were horseback riders, thus going to the many towns. Mr. and Mrs. Charles Chase were at the Liberty Bell when Mr. Chase was manager. They entertained quite extensively. The Liberty Bell was halfway from Telluride to Pandora. It was really the "gay nineties" in every sense of the word.

This "Galloping Goose" affair intrigued me. There was a picture of it in *News Week,* where by the way I saw yours. After the R.R. was abandoned I have been told an old Studebaker auto was converted into a vehicle to ride the rails to Durango. Someone painted a goose on the front. It is *now* a sort of passenger car and is doing good service. A friend wrote that the first one was too funny for words.

Galloping Goose, Rio Grande Southern Railroad. *Courtesy Mary K. Mott.*

There are not too many living who have experienced the Barlow & Sanderson Stagecoach days and have seen transportation evolve from burro back to a possible rocket to the moon. When I think of dear old Lake City with an Airport, Ouray and Telluride without horses, and autos and jeeps creeping around the "hills of home," I know it is about time for me to sing "Swing Low Sweet Chariot, Comin' for to Carry me Home."

Am sorry about my writing. I cannot read it myself, so you will understand my incoherent sentences. Oh for good eyes. WHY did I not appreciate them. My doctor says "Yes, too bad but how fortunate to have what you have *above them.*" I sometimes wonder!

John B. Marshall of Alhambra, California, also mentions Pandora:

A few families lived in Pandora while I was there, there was a schoolhouse in which I have been, it was something like the one near Sawpit or Placerville at the present time. The people who lived in Pandora did give Christmas parties at the schoolhouse. . . .

We left Telluride in 1924. . . . I lived there in the hey day of mining, as far as the three big mines were concerned. For two summers I worked on the Survey crew—Black Bear—Triangulation work in Ingram Basin. This work was done by the Colorado Superior Mining Co.

One day on just a hiking trip, we stopped on the summit of the central peak, in the Ingram Basin: just back of the falls. There down in the Red Mt. Valley we saw the train moving up toward Red Mt. Kind of excited my curiosity, not knowing a great deal about the activity of that basic area.

In 1946 I interviewed Mr. Alfred Friedheim of Denver, and among his many recollections were those of Telluride in the early 1890's:

"I went to Telluride in 1891," he began. "My father had a store there. At that time the Town Marshall was Jim Clark. He had been with Quantrill's Raiders of the South—he was law-abiding but very brutal. He knew he had enemies that would get him someday, so he kept a Winchester rifle in each of four stores and carried two guns in his pants. In the summer he'd practice shooting at Lone Tree Cemetery, knocking letters out of the signs tacked on the cemetery fence. He was night Marshall and one evening when he was walking his regular route, a man hiding between buildings killed him.

"Many of the Cornish miners traded in my father's store. They'd keep away from it if they hadn't paid their bills. Jim Clark was collector for us. He'd walk around and tap a man on the shoulder while he was gambling. The man would give in and pay up. Clark often came to our store, and whenever he needed a hat he took one but never paid. He figured he'd earned it.

"The Telluride mines were very high up on the mountains. Many miners would get pneumonia and die. Their bodies would be brought down and, if possible, sent back to where they'd formerly lived. The population when I was in Telluride was 3,500, with more men up at the mines. There were twenty-one saloons with gambling halls and eight houses of prostitution.

"The first church was the Christian church. On Sunday nights Rev. Bradley would give a talk, not from the Bible but from articles in the newspapers. The high class gentlemen-gamblers would all go to church and donate from five to twenty dollars. Once Rev. Bradley announced that his subject would be 'Gambling.'

Everyone was there that time and he got more gold in the collection than any other Sunday.

"The trails up to the mines were narrow and steep. On any such trail you were pretty sure to meet 150 burros patiently plodding along with their heavy loads of 200 pounds each. They must have had wire nerves and steel muscles to cross the range once a day, carrying coal one way and ore the other. Now the trails are lonely."

Telluride, like other camps at the turn of the century, was torn by labor troubles. Certain incidents are described by C. C. Lindley of Monroe, Louisiana (June 29, 1955):

I lived in Telluride from 1901-1915; Born in 1884, in Ohio, was just at the age when one would be greatly impressed with the mountains and the mines.

One of my early jobs was in the Tomboy mine as "Powder Monkey." This job is to take orders, then deliver the powder, fuse and caps to the faces, upraises, stopes, etc., as ordered by the various crews.

A miner informed me that it would be best to join the union my first pay-day, so I went to town for the big event. At the conclusion of a short ceremony, I was handed a lapel button, and told to wear it. The button showed clasped hands, but no words. I inquired as to the meaning of the button, and the official said it was a Socialist insignia. When I said, "I won't wear that thing. I am not a Socialist," he replied "You can't be a good union man if you are not a Socialist." So I left in a huff, and have been anti-union ever since.

I worked at the Alta, Liberty Bell and Smuggler-Union mills, then in Mexico and Honduras for the next twelve years. Have been here for thirty years, and the Lord has blessed me.

May I point out an error on page 387-388 [*Stampede to Timberline*]. The strike was called as a gesture of sympathy for the striking smelter men at Durango. The men said they were not on strike, just a vacation.

I remember some incidents in connection with the "deportation" by the Citizens Alliance of a trainload of men who were known as agitators. Scarcely any of them were directly connected with mining. H. A. Floaten operated a general store, and had been Socialist Candidate for Governor. When the bunch from the Alliance meeting came to his house, and threatened to break down the door, Mrs. Floaten came to the head of the stairs and asked "What do you want?" "Want to talk to Harry" was the answer. She declared he had gone to Denver; but she opened the door when she realized that the men were going to enter, regardless of her protests. A search of the house was made and Floaten was found hiding in a closet, behind a trunk with clothes and bedding piled on top of him. He was given a hefty kick and dragged out and told to get his clothes on. He was so scared he forgot to put on shoes, and it was not noticed until they were at the roundup place.

Two brothers, red heads, carpenters, lived in a double house near the McMann sawmill at the edge of town. They were loud-mouths, and expert at it. The roundup squad, headed by Walter Kenley, a deputy Sheriff, entered the long hall dividing the house, and knocked at a door, one of the brothers yells out, "I know who you - - - are, and the next man who touches that door will get a rifle bullet through him." Kenley replied, "You are a dead man if you do; now I am going to count three, and if you don't want the door broken down, then open it up, because I'm coming in." So he counted one, two, three and smashed in the door. Another crash was heard at that instant, as the redhead went headfirst through a window. The other brother surrendered, as did also the first one later on. They were the meekest of the meek when they were finally permitted to return to Telluride.

Charles A. Chase, superintendent at the Liberty Bell mine in 1902, the year of the big slides, sent me an account he had prepared in 1903 as part of the company's annual report. His letter (Sept. 7, 1950):

> Last week in Salt Lake City I came again upon your book of travels among the old camps of the state and read again the allegedly-factual story of the Liberty Bell snowslide of 1902. Coming to this some months ago I regretted that your informant had presumed to speak so inaccurately. At the time I had the official story retyped but failed to send it to you. I enclose it now.
>
> I sent a copy of that text and of this letter to Mr. Hafen (Colo. State Historical Society) so his files may inform more accurately future students.
>
> My best wishes to you.
>
> Yours truly,
Charles A. Chase

Here is the mine report:

THE SNOWSLIDE OF FEBRUARY, 1902

The following excellent report upon the snowslide disaster at the Liberty Bell mine was prepared for the State Mine Inspector by our superintendent, Mr. Charles A. Chase. It is included here in order that the shareholders may have exact information as to the extent and character of the disaster and also that some record may be made of the lives lost.

Mr. Harry A. Chase, assayer, had recently been advanced to the position of Assistant Superintendent, after having been in the company's employ about one and a half years. He was a most capable and promising young man and had done most excellent service. He was one of the first of the rescuing parties that went from the town to the mine after the slide.

Mr. John R. Powell, Surveyor, had been in the company's employ about two years. He had done good and faithful service for the company and had worked most courageously with the rescuing party.

Mr. Gus Von Fintel was the mechanic at the mine and was one of our oldest employees. He was a reliable, loyal man and his loss was a great one to the company.

Mr. Bishop and Mr. Crowe were both young men and recent graduates of the School of Mines of Colorado and had taken hold of actual mine work for the benefit of the experience.

F. C. Clemmer was the steward in charge of the company's boarding houses and had been with us in that capacity something less than a year.

Messrs. Rohwer and Gregory were volunteers from town who were on their way to the mine to assist in the rescue work when the slide overtook them.

At 7:30 A.M. February 28, 1902, a snowslide swept from the mountain above the Liberty Bell mine and destroyed the boarding house, part of one bunk house, crusher house, ore bins and tram house. The results from this slide were seven (7) fatalities and eight (8) men injured as follows:

KILLED:

> F. C. Clemmer, Steward, American, 32 years, married, insured.
> Gus Swanson, crusherman, Swede, 35 years, single.
> R. Bishop, miner, American, 22 years, single.
> W. A. Crowe, miner, American, 24 years, single.
> Harry Trowbridge, flunkey, American, 24 years, single.
> H. S. Summerland, brakeman, American, 35 years, married, insured.
> Gus Kraul, bucketman, Dane, 30 years, single, not insured.

INJURED:

> Henry Bauer, second cook, since died, American, 45 years, married, internal injuries, insured.
> Jack Marshall, head cook, American, 45 years, slightly bruised.
> Jacob Golden, dishwasher, Polish, 30-35 years, single, arm cut off, injured internally.
> Charles Goodale, miner, Austrian, single, slight internal injuries.
> Ferdinando Zanzuchi, miner, Austrian, single, scalp wound.
> James Conlon, flunkey, Irish, single, ribs broken.
> John Isaacson, miner, Swedish, single, skull fractured.
> Knut Talse, miner, Finn, single, slight injuries.
> Ralike Palo, miner, Finn, single, hurt in back.

At shortly after 1 o'clock P.M. a second slide ran partially over the track of the first slide and partially over other lines. This caught part of the rescue party at work on the first slide with the following casualities:

KILLED:

> Harry A. Chase, assayer, American, 26 years, single, insured.
> L. D. Stanley, carpenter, American, 35 years, married.
> Andrew Aho, miner, Finn, 35 years, single.
> Louis Lundberg, miner, Swede, 35 years, single.
> George Rohwer, business man, German, 45 years, married, insured.
> William Gregory, miner, American, 40 years, married.

INJURED:

> George Conover, crusherman, American, 35 years, single.

The remainder of the party could find no trace of the missing men and had been shown the danger of risking more life. Our men were therefore ordered from the mine. A few delayed to wait until the last man injured, Conover, should be fit to travel. Then this last party of about a dozen started for town. At the end of a mile, and about 4 P.M., while the party was somewhat scattered along the trail, a third slide ran from a new and quite unsuspected quarter, overtaking and killing:

> John R. Powell, surveyor, American, 22 years, single.
> Gus Von Fintel, head mechanic, American, 45 years, married.
> Paul Dalprez, miner, Austrian, 35 years, single.
> Dr. J. Q. Allen was bruised.

These men were all, except Dr. Allen, buried to such an extent as to render our search futile and in fact, when found some days later, were far below where we had any idea of locating them. This appalling death list was owing to the remarkable snowfall of the last four days of February. In this period from three to four feet of very light snow fell, accompanied by wind, which carried the snow high on the range. During this period of heavy snowfall there was little or no sun to compact and settle the snow.

With reference to the ground swept by the slides I would say that we have held the opinion, and it has been the universal opinion of well informed men, that our mine buildings were in an exceptionally favored spot and that the trail was safe. This company selected the site for the mine buildings in 1898. This site was the flat top of a knoll about two acres in extent. On top of this knoll were a few trees and the hill behind was covered with them to timberline. To the right of this knoll is a gulch running from the top of the range. It comes down in a generally southerly course toward the buildings and then swings to the southeast, passing a high trestle. Mild slides have been known in this gulch and have been provided for by the high trestle mentioned. This slide left its natural course down the gulch at a point where the gulch is twenty feet deep, keeping a straight course towards the buildings.

The so-called second slide was in reality two slides a few seconds apart, coming from a somewhat wider range than the first. The part which followed in the path of the first proved fatal, the other part merely crushing sheds at the tunnel mouth.

After the first slide we felt that the danger in the gulch into which the buildings had been swept was too great for immediate work there and confined our men to working over the wreckage on top where men might still be alive. I feel that in working here we did no less than men should always do at such a time.

The third slide on the trail was something that no man could foresee. It swept out of its gulch to a height of forty or fifty feet vertically above the bottom of the gulch and 100 feet above the bottom, as measured on the hillside. It took spruce trees that had reached a diameter of fifteen inches. The gulch was a snowslide path but no slide had occurred in the past three winters. Before that only small slides were known. The storm had been remarkable and its effects were remarkable, in heavy deathlist and destruction of property.

With regard to future operations and the means taken to secure safety to life and property, I would say that in the first place we will erect the new boarding house on a piece of ground untouched by slides so far and which seems eminently safe. Change of site for ore bins and tram houses seems out of the question. For its protection we are building up the side of the gulch, where the slide left the channel, a 12 ft. x 12 ft. x 80 ft. crib filled with earth and rock. The slide would strike this at an acute angle and should be forced back into the natural channel. The trail in the track of the third slide will be changed.

I would further call your attention to the fact that the slides were all down barren gulches, eroded to bed-rock where no timber has ever grown.

Charles A. Chase
Superintendent.

The financial loss to the company caused by this slide was serious, but it was as nothing compared to the sad loss of the brave fellows who perished in that dismal storm, and to the grief and sorrow that were carried into many homes. Our sympathies have already been expressed and financial aid has been extended where needed and called for, but the loss to many is irreparable. The protection made against a repetition of such a disaster seems to be all that can be done. Absolute immunity from such catastrophes in such localities can never be attained. The chances of a recurrence of such a slide are, however, in the first place, very remote; never in the memory of residents had there been a slide across our building site, and the tree growth indicated that such had not occurred within fifty years. In the second place, in the event of a recurrence, our boarding house, where the men are segregated, has been rebuilt on another site and the snowslide crib, about 100 feet long, 15 feet high, 12 feet wide, and solidly filled and banked with earth and rock, will probably divert and break the force of the avalanche.

Respectfully submitted,

Arthur Winslow,

General Manager and Consulting Engineer.

Kansas City, Mo., Feb. 3, 1903.

Mrs. Daisy A. Dalrymple of Dove Creek wrote two letters about homesteading in the 1920's and of experiences in Telluride:

Nov. 18, 1970

In 1921 I with my husband James G. Burke homesteaded northwest of Dove Creek and like many of the others here at that time, had to take our leave of absences and go to the mining camps. Telluride was where we went, he worked for the old Smuggler Co.

Nov. 23, 1970

My married daughter, Mrs. Helen U. Cook of Lincoln, Nebraska, being my oldest, was with us while homesteading and remembers how we counted the 22 open saloons in Telluride; not to mention the "cribs" in the Red Light District, the only time we ever walked down their street was with a neighbor lady with her children very early one morning in the summer. This Sept. a very good lady friend took several of we old ladies on a trip to Ophir and Telluride.

My Helen likes to remind me of our *walk* down from the Tom Boy mine; we were really "green" about mining camps, did not know that there was no place for women to stay until we reached there by stage, and *sled,* so my husband, I and Helen walked back down to Telluride. We waded snow at the top, then our wet shoes on lower dry ground made blisters on our feet. A dear old lady who had a rooming house took us in—we had less than $2.00 and our camp things were at Tom Boy and would not be down till the stage came next day. Part of that $2.00 Jim bought cheese, crackers and milk. We ate it in our room.

I will soon be 75 years old, born 1895, have hit some rough edges but also have had a lot of fun and sunshine.

Mrs. Clyde Jewell of Grand Junction recalls experiences when she and other college students from the southwestern portion of Colorado were marooned in Telluride during part of the Christmas holidays by snow conditions that prevented the trains from running on schedule (July 4, 1951):

I note in your book that you visited at the Sheridan Hotel at Telluride and were impressed by the staircase. That sounds natural. I recall getting in to Telluride at 2:00 in the morning as I went to Dolores for Christmas one year. I had to take a room on the third floor which had almost no heat because my telegram for a reservation had arrived late. After a wonderful leisurely supper in the Sheridan Dining room, I found the room still an icebox. I washed, dressed completely, put on my coat, and retired. Awful as that sounds, it was the only way to sleep in comfort. I always loved coming up from the station to the hotel during the Christmas holidays because the lights were so lovely. In the center of the street would be the lighted community tree, and above, high on the mountains beyond like a city in the clouds, would be the lights of the big mines. There were usually a lot of us who were well acquainted as students from Boulder, Fort Collins, Greeley, Colorado College, and Western who usually came and went on the same train. I also recall one New Year's Eve at the Sheridan when the inebriated gentry tried to get in at all the wrong doors and generally rioted about the halls. We barricaded the door and got a call from the management which said to give them a ring in case the stewed gentry annoyed us too much for comfort. They didn't.

Also, on my first trip I ever made alone, I was stranded in Telluride on account of the big slide above Ophir which kept trains from running. Just as I was about to have to take a job as kitchen helper at a mine boarding house on account of running out of cash in hand, they called us and said they would take the train out again. This was in the late morning. Everyone dashed to the train. I was about the only one who had had my lunch. They parked us at the scene of the slide and let us watch the clearing of the track by men with shovels. The mud oozed down continually. It was half-past nine that night before we went across at a cautious crawl. They opened the grocery store at Ophir to let the passengers buy bologna and cheese.

Ophir
and
Old Ophir

Ophir when I first saw it in 1942 was the perfect ghost town. It had many frame buildings in a remarkably good state of preservation, water hydrants flanked the main street, the schoolhouse was easily spotted, and a restaurant was identified by a large fading sign. Farther up the valley beyond the townsite, hidden in trees, was the cemetery, and higher still, slanting up the steep mountain slope to the top of Ophir Pass, was a trail, originally the only connection with Silverton. An old be-whiskered miner with his burro was the sole resident in town.

In the mid-1870's Ophir (now known as Old Ophir) was a new mining camp set in a wide valley close to the Ophir Needles and walled in by sheer mountains whose sides showed the scars from snowslides. Just when the camp was settling down, rich carbonates were discovered at Rico and the population rushed to the new diggings. The Rico boom was short, and when it burst the men hurried back to their Ophir properties. Ore that was rich enough to ship was taken out by burro train over Ophir Pass to the smelter at Silverton. At its peak in 1898, Ophir had a population of 400, electric lights, water works, and a schoolhouse.

Old Ophir, early 1900's. Main Street. *Courtesy Denver Public Library, Western History Department.*

Old Ophir, 1960. In background is trail over Ophir Pass to Silverton.

New Ophir, 1941. Loading and railroad stations.

The town lies southwest of Telluride and is only two miles from New Ophir and State Highway 145. A steep climb from Telluride Junction to the entrance to the new ski area and beyond skirts a deep valley, at the end of which stand the remains of the famous Ophir Loop, built by Otto Mears. Its high trestles are gone, but sections of the curving roadbed reveal the vision and indomitable will of its designer.

Memories of Ophir at its peak are given by H. N. Wheeler of Washington, D.C., who wrote:

> I can't tell you of the origin of Ophir but it was a busy place when I taught school there in 1898 and 99. The Suffolk and Caribean Mines were the active ones. It is 2½ miles from Ophir Loop. Ophir country was hit hard by snowslides. At one time mail came into Ophir from Silverton. A snowslide carried the mail carrier to the bottom of the gulch. Snow didn't melt until the second summer. He was found with the mail sack on his back. That sack is in the historical museum in Denver.

After the arrival of the railroad, New Ophir was laid out at the "Loop" near the Silver Bell, Butterfly, and Terrible mines, and was centered around the railroad station, a hotel, and a store. From this location two cars of concentrates were shipped daily the year round from the best of the properties. The narrow-gauge labored from the time it left New Ophir up a series of sweeping curves to gain altitude, so as to continue to Trout Lake and Lizard Head Pass.

D. F. Holaday of Midwest City, Oklahoma, reports on the appearance of Old Ophir as seen on different visits. He writes (Sept. 13, 1962):

> Incidentally, there are some company houses, five or six, built in 1952, along the "right side" of the road, as shown in your picture. However, one or two of those buildings sketched lay in the path of a snowslide, in 1960, and were pushed around or broken up. I have visited the place about six times, and

the changes are noticeable each year.

Old Ophir, with a population of two in 1950 (now zero!), is my "home town." My folks left there before I could walk, and I didn't see the place for forty years. Just for fun, I took the family there in 1953, as the two children were old enough to get a laugh out of the place; however, it was and is a very pleasant trip, leaving the Oklahoma or Texas flatlands for the Rockies. We enjoyed it, and have repeated the jaunt several times.

From Lizard Head Pass, Highway 145 follows the course of the Dolores River   *Rico*
downstream to Rico, a once prosperous but isolated silver camp which drew pros-
pectors to its mineral-filled mountains in the late seventies and has since produced great quantities of silver, lead, zinc, and a little gold well into the twentieth century. Until Otto Mears built his Rio Grande Southern up the river and joined it to the completed northern section of the road, which had reached Ophir, Rico had no facilities for getting its ore shipped outside the district other than by pack train or wagon.

The camp has had dormant periods as well as more than one boom, but its appearance today reflects the 1880's and early 1890's, when most of its substantial buildings were erected and when its population topped 5,000.

My 1949 article on ghost towns which appeared in the *Ford Times* caused two people to write me about relatives who had lived in Rico. The first was from R. L. Prescott of Seattle (Aug. 14, 1949):

> I am interested in your reference in the July *Ford Times* to Rico, as my father was, I think, the first postmaster there and Wells Fargo agent and had a general store until he left in 1883. His name was Alonzo K. Prescott.

The second, also from Seattle, was from Mrs. Carl Fischer (Aug. 24, 1949):

> I was most interested in reading about Rico and Telluride. My father and his cousin a Mr. David Swickheimer, roughed it in Colorado before Colorado was a state, in fact they helped vote it a state. They were both young men around 21 years of age and unmarried. It was at this time that they became acquainted with Col. Bill Cody in a Cheyenne, Wyoming, saloon and the three of them became fast friends. Later my father returned to Illinois, was married, and when I was five years old my parents and I moved to a 25000-acre ranch near San Antonio, Texas, a gift of David Swickheimer. Mr. Swickheimer had remained in Colorado and it was he who located the Enterprise mine which he later sold for $1,250,000 cash, as you speak of in your article on Rico. . . .
> My mother and I spent many summers visiting the Swickheimer's in that little mining town. Mr. Swickheimer by the way practically owned the entire town. He was also a great friend of H. A. W. Tabor.

The mining engineer, D. H. Campbell, who was the business manager for the mines in White Pine and Tomichi, was also connected with mining in Rico:

> March 29, 1950
>
> Speaking of Rico, you state on page 394 [*Stampede to Timberline*] that "tin gave the Atlantic Cable new life." It was zinc, not tin. There is, I think, no commercial tin in Colorado.
> I was Supt. of the Rico Mining & Milling Co., that opened the Atlantic Cable. My father and I developed what I think was the first process to com-

mercially separate non magnetic iron sulphides from zinc ore, making both merchantable. About the time we got well started, we sold out to a big consolidation headed by Mr. A. B. Roeder of New York, that never got really started in production. I left on completion of the sale and except for a trip back to demonstrate the operation of the mill was not back there until two years ago.

Mrs. Campbell and I knew Mr. David Swickheimer and his former wife well and many of the old timers of that area.

I hope the suggestions of corrections will not offend you, and the information will be of some use.

Very Truly yours,
D. H. Campbell

It was my good fortune to reach the town in September 1950, the same day that Mrs. Elizabeth Pellet (better known as Betty) returned from Denver, and I was doubly fortunate that she graciously agreed to talk to me that evening about Rico and her life there as the wife of Robert Pellet, a mining engineer and successful mine owner.

Mrs. Pellet, a Colorado State Representative, completed her fourth term in the legislature in 1954, having represented Dolores, San Miguel, and Montezuma counties. She is a big, forceful, energetic woman with a refreshing sense of humor. During her term of office she worked for equal rights for women and for children's welfare, as well as for better mining laws. For years she fought successfully to save the Rio Grande Southern railroad between Ridgway and Durango from foreclosure, for it was the only means of shipping produce of all kinds from the Rico area. After the death of her husband, she managed their mining interests. Their home, where I talked with her, was filled with treasures collected abroad on vacations and business trips.

"If I'd known you were in town," she began, when we had settled comfortably into easy chairs, "I'd have invited you to supper. My husband and I arrived here in 1919 when there were only 300 people in the camp.

"Rico is a fairly old camp, for the first silver strike was made in 1869 and the Atlantic Cable was discovered in 1870. Later, prospectors opened up other mines, organized a mining district in 1878, and began to develop the lodes they had located. The first rich strike of oxidized silver was found on Nigger Baby Hill in 1879 and triggered a stampede to the District.

"The most valuable ore was taken out in wagons drawn by teams of oxen over the old Rockwood stage road. It was so steep in places that they had to tie ropes around trees and snub the oxen down. You can still see the burns on the trees made by the ropes.

"In 1880, when the Grand View smelter was blown in and several new mines were opened, investors with capital became interested. After the Rio Grande Southern reached Rico in 1891, freighting stopped. Silver Creek runs into the Dolores River from between Nigger Baby and Newman Hills. A spur of the railroad was run up Silver Creek onto Newman Hill to the mines. Four cars or 100 tons of ore was the most a train could draw over these mountains.

"The Johnny Bull mine which had a rich chimney of gold ore was way up a gulch at 11,000 feet. The mine was located by one group of men. Then a second group jumped it. That started trouble. First one group would take possession and work it for a week and not let the others in. Then the second bunch would fight them and get possession. This went on for some time with gun battles up and down the gulch before the matter was taken into court to be settled.

"You've heard of Dave Swickheimer, haven't you? He came to Rico after the first boom started and ran a saloon. His wife, Laura, ran an eating house in the back. Like everyone else, he did a little prospecting on the side and located a claim he called the Enterprise, on Newman Hill. There were two outfits working it—one sinking a shaft and the other digging a tunnel. Swickheimer was sinking the shaft and he kept putting all their money into it. After it was all gone, Mrs. Swickheimer would go at noon to the shaft and give whiskey to the men so they'd go back to work. Finally, when completely discouraged, she put either one dollar or sixteen dollars on the Louisiana Lottery and won $4,000. With this money the men were paid and sent back into the shaft. When it too was gone, she promised to give up working the mine that Saturday night. Sunday came and the men hadn't come up. Their women were frantic—they feared a cave-in. A few miners went down the shaft to investigate and found that those on shift had struck a vein of high-grade silver and were sacking it. Soon after that Swickheimer bought out the tunnel group and controlled the entire mine.

"In 1891 the Swickheimers sold the mine to an English company for a million dollars. Before long the couple was divorced. Laura bought an interest in Tabor's Grand Opera House and also a business block in Pueblo. She remarried a much younger man—a miner. David became a banker, but he was a better saloon keeper. He said if he hadn't had an interest in an irrigation ditch in Texas he wouldn't know how he'd have gotten along.

"After the silver crash, Rico was dormant and silver prices were down. Then in 1923 or '24 metal prices jumped and between 1925 and '30 Rico was as gay and wild a camp as it had been in the nineties. With this boom, outside capital became interested in Rico properties. The Falcon ran a force of 500 men; the St. Louis, whose big dump you see as you drive into town from the north, became a subsidiary of National Lead. The Pell-Eyre, Pellet's company, was a subsidiary of Anaconda Copper. In 1939-40 the Argentine Mining Company took over the Pell-Eyre, St. Louis, and Anaconda, and for ten years they boomed. The man who ran the properties netted them $10,000,000. The mill ran on three shifts again. During the Second World War Rico was second in the production of lead, zinc, and talc; and after that it was first in lead until Empire Zinc started up again. In 1950, 250 Navajo Indians were brought to the camp to work in the mines. There are about 125 here now. Rico has a four-teacher school. Now one teacher is assigned to teach Navajos English. There's only one mine, the Rico-Argentine, that is still operating.

"Rico was the county seat until 1946, when the voters moved it to Dove Creek. That's very unhandy for Rico, for we have to drive eighty miles through Montezuma County to get there. There's not even a courthouse at Dove Creek. Instead they use a grocery store. Next week a murder trial comes up and they'll have to move elsewhere for the hearings, for there won't be room enough for the crowd.

"Have you ever been to the Rico cemetery? You should see the gravestone of an eight-year old child with the inscription 'Too Wise to Sin.'

"We have lots of snow here each winter—the average fall is three feet on the level. Mr. Pellet's father had mining property in Rico and he grubstaked his uncle and his brother-in-law. The snow was terrific the year the two men reached here from the east. One of them, the cockswain of a winning crew in the Philadelphia Regatta, brought his wife with him. While he was shoveling snow one day, making a path to the outhouse, she called to ask him what he was doing. 'If I have to do much of this,' he replied, 'I'll build a two-story backhouse.' The first avalanche on Nigger Baby Hill in forty years ran last winter. Four steam engines and a rotary plow were caught in it. It also took out a house and left the people in bed—the trees made a canopy over the bed. A boy was killed that same afternoon in another slide. Then a third ran below town, all in one day. Rico's citizens sent rescue groups and took coffee to the rescuers at all the slides.

"Lucius Beebe was in this part of Colorado while he was writing *Mixed Train Daily*. He wrote me that he'd have the train stop at Rico if the school children could be dismissed to go to the station to see the train. He had an old coach all fitted up with silver mountings as his private car, with chandeliers and velvet drapes. He said he'd shake hands with each child. I told him, 'Hell, they'd rather see the Porter. They've never seen a nigger.' "

Parrott City
and
La Plata City

In 1942 I set out to find Parrott City and La Plata City, two mining camps lying in the southwest part of the state. From U.S. Highway 160 west of Durango I watched for the road that led north into La Plata Canyon. On the way up the canyon I stopped at a gem shop to ask questions. The shop was run by William C. Little, a miner, and his wife Olga, a woman who had been running burro pack trains to almost every mine in the La Plata District ever since 1909. By the time I met them they had a ranch in the canyon, and when I asked them what was left of Parrott City they told me that the last of the old buildings, some of which had stood on their land, had been torn down only a short time before.

As I was leaving, Mr. Little told me to keep an eye out for a depot or stage station on the road below La Plata City. I found it and sketched it. As I went on, the roadbed became progressively rougher and steeper until it reached a broad meadow through which the La Plata River runs, bringing down snow water from the lofty La Plata peaks in the distance. A row of frame and log buildings on the west side of the road and a few others half hidden by trees near the foot of the mountain slopes comprised La Plata City, whose population had once reached 500. Three miles beyond the meadow was the Gold King property, with its big abandoned mill and two-storied boarding house opposite it across the stream, yet connected with it by a swaying suspension bridge.

In 1972 I again drove up to La Plata City. The stage station was gone, and on the wide meadow stood only one of the original houses that had once been the heart of the townsite. This I had been prepared for by a letter received in 1950 and by my thirty-year absence.

> We visited the Gold King Mine north of Hesperus, Colo. in July 1950. In reference to your sketch on page 394 the Boarding House and foot bridge have already disappeared. All that remains are the gate at the end of the bridge, part of the support near it, and the tiny building in the right foreground. La Plata, however, remains approximately the same as pictured on page 394.

Twenty-two years later the "city" too was all but gone.

When John W. Sanders, of the Exploration Department of International Pe-

Rico, 1953. St. Louis Club House; Rohde Rooming House.

La Plata City, 1942.

troleum, Ltd. (Bogota, Colombia, South America), wanted information about the history of La Plata City in southwestern Colorado, he contacted the Colorado Historical Society. The letter that one of the librarians wrote in reply to his request is a thorough amassing of details from all the material at her disposal. However, it shows how sometimes archival information available to librarians needs to be doublechecked—preferably by first-hand investigation, an option obviously not open to most of the staff at a historical society.

Parrott City and La Plata City were not two names for the same place. Parrott City, established in 1873, was the older of the two, but only by a few years. La Plata City, to the north, small as it was, outlasted Parrott, as I discovered when I visited it. As the librarian's letter shows, most of her information concerns the early years of Parrott City when Moss was in charge of mining activities. It was chiefly a placer camp, whereas La Plata City, higher up the canyon, was closer to the lode mines which were discovered and developed in the later 1870's.

In *Stampede to Timberline* (2nd edition, 1974, p. 546) I quote Myron H. Broomell of Durango, who wrote me that "the site marked 'Parrott' on the U.S.G.S. topographic map (La Plata Quadrangle) is the site of what is known as May Day and the actual Parrott City was a mile or so south and a little west of it." This statement places Parrott City just where I found it.

Although I read the available material in the libraries of the state about the mining communities of Colorado, I did not write about them until, in addition, I had visited and inspected the places, had dug into the files of local newspapers where they still existed, and had talked when possible with the old-timers who were still living there. In many cases, therefore, as in this one, my statements are more authentic than those derived solely from books and journals. I also have in my files two printed Colorado maps (one 1888, the other 1897) that clearly show the "cities" of Parrott and La Plata as being four to five miles apart.

The details of the Historical Society letter (Dec. 11, 1953) are so interesting, however, that I quote it in full:

> In answer to your letter, Dec. 3, 1953, we do have information about La Plata City. The *Colorado Magazine* for November 1941 on page 230 states, "La Plata, 21 pop., in La Plata County, first known as Parrott City now bears the Spanish name meaning 'silver.' In 1882 it had a population of 100 and the postoffice was established in August of that year."
>
> Because La Plata City was first known as Parrott City most of the references we have are about Parrott City. The *Colorado Magazine,* July, 1942, page 145 says, "Parrott City now La Plata; in the summer of 1873, Captain John Moss, sponsored by the San Francisco banking house of Parrott & Company, led a party of prospectors from California into the San Juan country. After prospecting the region with satisfactory results, Moss executed a private treaty with Ignacio, Chief of the Southern Ute Indians, for the right to mine and farm thirty-six square-miles of country, with the center at a point where Parrott City now stands. For this privilege the Indians received one hundred ponies and a quantity of blankets. The settlement which they founded was named in honor of Tabucio Parrott of Parrott & Company. The name was later changed to La Plata."
>
> Mr. Barkazow Barnacle in one of a series of articles about "Characters of the Early Days" wrote one entitled "Captain John Moss, Adventurer," which

appeared in *The Trail* for March, 1920. On pages five and six of this article Mr. Barnacle says, "He [Moss] does not appear to have come to the front in Colorado until the early seventies, when for about four years he was engaged in mining on the La Plata River in the vicinity of Parrott City, which city he founded, probably in the year 1873. While living in the San Juan country Moss was elected to the Colorado Legislature, as the member from La Plata County. This occurred in the fall of 1876. . . .

"Moss did not make a brilliant record as a legislator, possibly because of his lack of interest, but surely because of his lack of attendance on the meetings . . . he never showed up to claim his seat until the session was well-nigh concluded. He was known there only as the great absentee. When he came to Denver he drew his pay and after one day's sojourn took an outgoing train for his old home in San Francisco. He was in his seat in the House of Representatives for only a few minutes, and if he ever returned to the State I never learned of the fact."

The W.P.A. Workers during 1933 and 1934 collected interviews from old settlers of the state for the State Historical Society of Colorado. These interviews have since been bound according to counties. The La Plata County volume has several references to Parrott City. A William Valliant is quoted as saying, "We went up the La Plata to Parrott City. . . . After setting up the mill we sawed enough lumber to house the mill, then built the town, it took four months to build Parrott City. . . ."

According to John Meston of Durango, Colorado, in *La Plata County,* by 1883, "Parrott City was past its glory; there were only a postoffice and two or three buildings occupied."

In 1876 a Charles Naeglin of Durango, Colorado, was in Parrott City and had this to say about it in *La Plata County,* "There were about 25 people in La Plata County, nearly all of them at Parrott City, then the county seat. Parrott was a placer camp. A California outfit operating there. John Moss was in charge. When we arrived there was a continuous celebration in progress; whiskey was carried around in a water bucket and everyone invited to help himself much the same as water is carried to gangs on a job by the water-boy. The whiskey was freighted in from Ft. Garland or Ft. Wingate and must have cost plenty."

John Moss was a very colorful figure and we have quite a lot of material about him. In April of 1921 the *Denver Post* ran a series of articles about Moss. According to these articles Moss may have had some connection with a swindle which was trying to make people to believe there were diamonds in the San Juan. The other men who helped Moss found the city were Richard Giles, John Merritt, Thomas McElmel, John McIntire, John Thompson, John Madden, Henry Lee and John Robinson.

The *Denver Post* on April 17, 1921, in the magazine section has this to say about Parrott City and John Moss, "With the first wagonload of furniture came the largest barrel of whisky obtainable. This was set up in the office belonging to Moss, tapped, a faucet set in, and tin dippers hung on the wall near it. Every visitor was invited to take a dipper, put it under the faucet, and fill it to the limit of its capacity. There were no restrictions in the number of times a day a visitor might call; nor on the size of the potation.

"In 1874 W. H. Jackson of Denver . . . accompanied by Ernest Ingersoll, came to the San Juan for the purpose of photographing its grand and marvelous scenery. Of course they were steered to Capt. John Moss at Parrott City as the fitting person to entertain guests bent on such a "boom" venture. Captain Moss put them up at his house, introduced them to all the miners and ranchers and to the barrel.

" 'I'll be glad to show you around,' said Moss, 'and I'll show you something else no white man has ever laid eyes upon except myself.' "

"It developed that he meant the ruins of the cliff and cave dwellers, known only to Moss and Indians—for he had been there."

On April 24, 1921 in the magazine section of the *Denver Post* Josiah M. Ward continued his series about Moss, "Other parts of the San Juan were being developed, but nowhere was prosperity more noticeable than at Parrott City. In June, 1874, Annerian Root located the first lode mine, named the Comstock, and commenced placer mining on his own account. He obtained several nuggets of gold, one weighing over an ounce. This discovery gave new impetus to the Parrott Ditch Company. The ditch company had nearly completed its task when winter set in and the miners were compelled to abandon their enterprise until another season. The original California party scattered to more genial climes in the lower altitudes, but John Merritt and Richard Giles wintered on their ranches on the Mancos River. Moss and many others remained with the barrel of whisky.

"In 1876 Captain Moss, with Major Cooper, came to Denver, while the first state constitutional convention was in session to urge the division of La Plata, the upper part to be known as San Juan County. Their mission was successful. Parrott City was still the county seat and remained so until Durango snatched the honor away—but not while Captain Moss was in the San Juan."

At the time John Moss was elected to be a state representative, according to the *Denver Post* article, things were not going well in Parrott City. "The La Plata and its tributaries were becoming exhausted of its gold. Work on the ditch was suspended, as the need for water grew less and less. An expert sent out from San Francisco made discouraging reports to the home office. . . . Moss could not retrieve himself in the San Juan. Parrott and Co. withdrew their credit."

The highest population we have located for Parrott City was 280. This number was in Crofutt's *Guide of Colorado,* published in 1881.

We have located no record as to just when the name of the town changed from Parrott City to La Plata. In 1881 the county seat moved to Durango. . . . La Plata was known as Cima at one time, also, according to our files.

We hope this information will be of some use to you.

Sincerely yours,
Dorothy Stuart, Asst. Librarian

Additional information is provided by Mrs. Clyde Jewell of Grand Junction (July 4, 1951):

About Parrott City in La Plata County. I notice the mention of Johnny Moss. As I heard about him from a man who knew him back in the days when he was the local representative of the Parrott Company, he was very picturesque. According to him Moss had the perfect recipe for getting on well with the Indians and pretty much so with the miners. To them all, to quote this man, H. F. Morgan, the Indians thought Moss was a more important man than the president because when they were hungry, he would feed them and otherwise help them all out. The Parrotts, Mr. Morgan said, thought this cost too much money and removed Moss. But his successors couldn't get the work out of the men, and Indians were a problem. Johnny Moss, Mr. Morgan said, thought good will was worth all it cost in beef, even if the Parrotts didn't.

Baker's Bridge and Animas City

Baker's Bridge was a short-lived settlement on the eastern bank of the Animas River and was founded in 1861 by the Charles Baker expedition. Baker, who had already done some prospecting in the Animas Valley in 1859 and 1860, returned the next year with a party composed of men, women, and children and began building a settlement of log cabins. In spite of their efforts, they found little gold, and their presence was not welcomed by the Ute Indians on whose reservation they were trespassing. After a long, cold winter the party disbanded, leaving many of their houses unfinished.

By the terms of the Brunot Treaty of 1873, the Utes relinquished their rights to the San Juan Mining District, clearing the way for the establishment of Animas City in 1876. This new settlement lost most of its residents in 1880 when the Denver & Rio Grande R.R. extended its line westward from the San Luis Valley and laid out a new camp called Durango, two miles south of their little "city."

Mrs. May Wallace Ross of Lomita, California, who sent me so much information about Logtown and Bowentown in connection with the history of Summitville, wrote three letters about Animas City's rather complicated beginnings and its later relation to Durango. Since she grew up near the first Animas City and eventually became deeply interested in preserving the history of the immediate area, her statements clarify many misunderstood facts.

Nov. 19, 1950

I read your book, *Stampede to Timberline,* last summer. There were a couple of points on which I felt prompted to write you. . . .

I feel personally grateful to you for your fairness in pointing out that the Indians had rights to the San Juan Mountains which stemmed from a treaty made in 1868. I grew up in the area ceded by the Indians in 1873, and always felt keenly the unwillingness of the people to acknowledge that the Indians had any genuine rights. Most of my generation will never be willing to see that the white men invaded land which the Indians had held for centuries, and that the white men failed to respect the treaties they made with the Indians.

The other matter is in regard to a statement on page 398 of your book: "Animas City, just north of Durango, is much older for it was laid out by Baker in 1861." Your informant was confused in regard to the name *Animas City,* first by the Baker party in 1861, and later at a site 15 miles farther down the river in 1876. The plat for the town called Animas City, 2 miles north of the site chosen five years later for the railroad town, Durango, was filed in the fall of 1876. There was no placer mining there. The Las Animas Perdidas at that point is full of silt and so sluggish that the Indians called it El Rio Arenoso. The priests called it El Rio de Las Animas Perdidas. The later comers pronounced the one word they used as though it were spelled *Anna Muss.*

The first Animas City, founded by the Baker party in 1861, was about 15 miles farther upriver, at a point where the river emerges from a box canyon. There is an old log bridge there which was thrown across the river by the Baker party, by means of felling trees so that they would fall from one rock bank to the one on the opposite side. There is a placard near the bridge which gives the names of some of the members of the Baker party. The site of the village was near the bridge, on the east side of the river.

There is now nothing to indicate the positions of the log cabins built in 1861. There are, however, a couple of people still living who were born in one of those cabins, and lived there until the family built another house. Louis Girardin of Pagosa Springs is the son of one of the members of the Baker party. His mother, daughter of Colonel J. M. Chivington, was the wife of Thomas Pollock, the trader who supplied the Baker party. After Mr. Pollock died, she married William Girardin. In 1879 she came to the site of the old Animas City, and began to homestead it. He has a sister, Ruth, who was also a classmate of mine, who lives in Grand Junction, Colorado.

There are surviving descendants of another member of the Baker party who helped in the construction of the placer village, and came back in 1879 to live about four miles south of the site of the village. Mrs. Mary Hathaway of Bandon, Oregon, Box 401 and Mrs. Jane Collier of 175 S. Elliott St., Coquille, Oregon, are daughters of Charles Idle, whose name appears on the placard near the bridge. . . .

The placer village was at the end of the valley, just before the highland which gradually leads to the Silverton level is reached. Near the highway on the west side of the river, just before the road begins to climb to the highlands, there is a sign which reads Baker's Bridge. There is a cross road which leads to the county road on the east side of the river. Along that road just after you cross the river, lies the original site of Animas City. . . .

I grew up within walking distance of the first Animas City. My father, J. W. Wallace, was the first sheriff of the large La Plata County formed in February, 1874, to provide a government for the mining area which the Indians had signed away the fall before. After statehood came, he preempted a ranch about two miles below the placer village site. When I was a child, the log buildings were still there near the log bridge.

Frank Hall, who was secretary of the Territory of Colorado during more than half of its existence, in his *History of the State of Colorado,* Vol. II, pp. 192 et seq., gives the story of the founding of the placer village called Animas City in 1861, as told by one of the backers of the expedition. His account differs a little from Louis Girardin's memory of what his mother told him. No two people ever place quite the same emphasis on given details after an event is over.

In connection with Hall, his labors in getting the histories of individual counties for his Vols. III and IV of his history, might arouse your sympathy. He didn't climb, as you did, but he had to travel by buckboard and team instead of by automobile. The last volume was published in 1895, the first three in 1889. I was nine years old when his Vol. IV was published. The general opinion seemed to be that the histories of La Plata County and San Juan County were essentially correct. If you have time to examine them, you

Animas River and site of Baker's Bridge. *Courtesy Mary K. Mott.*

a. Highway to Silverton
b. R.R. to Silverton
c. Hermosa Cliffs
d. Animas River

e. Ranch preempted by J. W. Wallace before Durango was founded
f. Site of Baker party's Animas City
g. County road to Durango on east side of river

will see that your informant at Howardsville was in error in many particulars. Howardsville and Silverton were never county seats of the same county. Howardsville was the county seat of the enormous La Plata County formed in February, 1874. My father had his office there at the foot of Cunningham Gulch. Silverton was platted in the fall of 1874, and in the spring of 1875 the county clerk took his records down to Silverton, without authority of law. The other offices remained at Howardsville until San Juan County was cut off from La Plata County in the spring of 1876. Then Silverton was made the county seat of San Juan County. I think, also, that "Gabby" Forsythe gave you the impression that he was the first postmaster at Howardsville. He was not. The first postmaster was named Nichols. He was in office when the first mail came addressed to Howardsville the first of July, 1874.

None of these matters are important. The fact that, after 65 years of struggle, the second Animas City allowed itself to be annexed to Durango, has removed the name from maps and post office lists. It would be difficult now for a stranger to find out where either Animas City lay. For that reason I thought I should offer the information for your files.

The matter of the information wrongly furnished by Forsythe is of no importance. The making of moving pictures in the San Juans, with a complete disregard for accuracy in dealing with history, makes it a hopeless task to keep the record straight. In *Ticket to Tomahawk,* filmed largely in Silverton, the traveling salesman said, in 1876, "Durango is just a hole in the rocks." The fact is that no town was planned for the site where Durango stands, until 1880. They also talked, in 1876, about the *Silverton Northern* railroad. I didn't stay after that.

Dec. 1, 1950

Since I wrote you about Animas City I have learned that John Turner is still alive in Durango. I believe he is the oldest resident in the area, since he came, as a small child, with his father, in 1876. I think the father, "Jack" Turner, was the only one who was present at the founding of the first Animas City in 1861 and also at the establishment of the second town of that name in 1876, a couple of miles north of the point where Durango was laid out in 1881.

John Turner is in the real estate business in Durango, with an office on Main Street, somewhat north of the old business section of the town. He is nearer to 80 years than to 75 years, but he still goes to the office, I am told. He might know more about the first Animas City, because of the fact that he lost one leg during childhood, and could not get around very well, hence probably stayed at home and listened to what was said there. The great difficulty with early history is that those who have listened were too much interested in something else, until it was too late.

I am very much interested in the completion of your work and if I can be of any help in any way, I shall be glad to do so. . . . The ones of my generation are dying every now and then. None of us are young enough to go into the high places. All that we can do to get the record straight is to be of any help we can to younger persons like yourself. It is not worth while to worry about errors in history of former times. 20th Century Fox has discovered the San Juan Mountains and will continue to present garbled facts in competition with the most careful research you could carry out. I believe the films called *Colorado Territory, Sand,* and *Ticket to Tomahawk,* filmed respectively near Aztec, N.M.; Hermosa Park; and Silverton, were so garbled that it would be better for the San Juan area if they had never been issued. The worst of it is that people remember what they saw on the screen while they forget the contents of the most carefully prepared book.

March 1, 1951

In regard to the placer village called Animas City, near Baker's Bridge, there is a description written in 1874, when the buildings were 13 years old. It is found in the 8th Annual Report of the U.S. Geological and Geographical Survey of the Territories (for the year 1874). You will find it in the report by Franklin Rhoda, Topographer for the Hayden Expedition, entitled *Topography of the San Juan Mountains,* at Page 486.

Mr. Rhoda says that Animas City was located on a beautiful, level patch of ground with yellow pines scattered over it. He says there was one street with rows of log cabins on either side for several hundred yards, some houses nearly finished, some half done. The sites of others were marked by 2 or 3 tiers of logs laid one above another. The elevation he gave as 6850 feet above sea level there.

When he came upon Animas City Mr. Rhoda was with the part of the Hayden Expedition which went for the first time to Mesa Verde to investigate reports of an ancient civilization there. The party made its way down the river from Silverton to Baker's Bridge in two days and camped the second night in one of the log houses at the deserted Animas City.

I think from his description you can make your sketch. If it is important to know how long that street was, that can be easily determined, but the way lies through cactus that covers the ground under the trees. If you walk south on the tract which fronts on the county road after you cross the present Baker's Bridge, you will come to a wire fence which divides the old Girardin homestead from the James Duffield preemption, which has been known since 1900 as "The Carter Place." The cabins of old Animas City strung along a line which crosses the boundary into the Duffield preemption. Four of the cabins were south of the fence.

Louis Girardin stated to me that when he was a boy growing up, there were four cabins still standing on the Girardin homestead and four on the Duffield tract. (Louis was born about 1885.) I talked with Harold Carter who grew up on the Duffield tract, moving there in 1900. He said there were only chimneys and foundations left where the old cabins had stood when he first came there.

According to the most authentic report of the founding of Animas City, the one given in Vol. 2 of Hall's *History of the State of Colorado,* beginning at Page 192, which was in the form of an interview given by one of the backers of the expedition, who also was present at the building of Animas City, the party was at that site only from May 1 to July 4th, 1861. Probably the logs from the unfinished buildings had been taken away before, say 1890, when Louis Girardin would begin to remember, and only the eight finished ones were left.

This account of the time spent there tallies with that of Charles Idle, who was present at the building, a man whom I knew well. The party were trespassing on Ute Indian land and the Indians tolerated the trespass until the party began to build a permanent settlement, then forced them to leave with the village unfinished.

The old log Baker's Bridge was in use when I was growing up. Since Louis Girardin states that he was born in the largest of the cabins, the one known as the courthouse, and lived there for a time as a child, I would assume that one was near the bridge, simply because roads were muddy at best, and the tendency would be to live as near the county road as possible.

The pines on the site of the old settlement are a second growth; the trees Mr. Rhoda found have long since been used as lumber. The ones he saw were hundreds of years old.

With all good wishes,
May Wallace Ross

I envy Mrs. Ross, because she has had the time and the persistence and the enthusiasm to ferret out data and set the record straight. In my travels and research I have always been limited by time and could never dig into any one area as thoroughly as a native could. Therefore, I am glad to conclude this collection of letters with one so filled with the desire to preserve authentic material about fast-disappearing landmarks and with personal recollections of Colorado's early mining days.

Index